Sliding Mode Control
Theory and Applications

Systems and Control Book Series
Edited by John O'Reilly and Eric Rogers

Published

- Vol. 1 *Advances in Intelligent Control*, edited by C.J. Harris
- Vol. 2 *Intelligent Control in Biomechanics*, edited by D.A. Linkens
- Vol. 3 *Advances in Flight Control*, edited by M.B. Tischler
- Vol. 4 *Multiple Model Approaches to Modelling and Control*, edited by R. Murray-Smith and T.A. Johansen
- Vol. 5 *A Unified Algebraic Approach to Linear Control Design*, by R.E. Skelton, T. Iwasaki and K. Grigoriadis
- Vol. 6 *Generalized Predictive Control in Bioengineering*, by M. Mahfouf and D.A. Linkens
- Vol. 7 *Sliding Mode Control: theory and applications*, by C. Edwards and S.K. Spurgeon

Forthcoming

Neural Network Control of Robotic Manipulators and Nonlinear Systems, by F.L. Lewis

Sliding Mode Control in Electro-mechanical Systems, by V.I. Utkin, J. Guldner and J. Shi

From Process Data to PID Controller Design, L. Wang and W.R. Cluett

Sliding Mode Control: Theory and Applications

Christopher Edwards

and

Sarah K. Spurgeon

TAYLOR & FRANCIS 1798 - 1998

UK Taylor & Francis Ltd, 1 Gunpowder Square, London EC4A 3DE
USA Taylor & Francis Inc., 1900 Frost Road, Suite 101, Bristol, PA 19007

British Library Cataloguing-in-Publication Data
A catalogue record for this book is available from the British Library.

ISBN 0-7484-0601-8

Library of Congress Cataloging-Publication-Data are available

We dedicate this book to Chris's parents and Sarah's children.

Contents

Series Introduction

Control systems has a long distinguished tradition stretching back to nineteenth-century dynamics and stability theory. Its establishment as a major engineering discipline in the 1950s arose, essentially, from Second World War driven work on frequency response methods by, amongst others, Nyquist, Bode and Wiener. The intervening 40 years has seen quite unparalleled developments in the underlying theory with applications ranging from the ubiquitous PID controller widely encountered in the process industries through to high-performance/fidelity controllers typical of aerospace applications. This development has been increasingly underpinned by the rapid developments in the, essentially enabling, technology of computing software and hardware.

This view of mathematically model-based systems and control as a mature discipline masks relatively new and rapid developments in the general area of robust control. Here intense research effort is being directed to the development of high-performance controllers which (at least) are robust to specified classes of plant uncertainty. One measure of this effort is the fact that, after a relatively short period of work, 'near world' tests of classes of robust controllers have been undertaken in the aerospace industry. Again, this work is supported by computing hardware and software developments, such as the toolboxes available within numerous commercially marketed controller design/simulation packages.

Recently, there has been increasing interest in the use of so-called intelligent control techniques such as fuzzy logic and neural networks. Basically, these rely on learning (in a prescribed manner) the input-output behaviour of the plant to be controlled. Already, it is clear that there is little to be gained by applying these techniques to cases where mature mathematical model-based approaches yield high-performance control. Instead, their role is (in general terms) almost certainly going to lie in areas where the processes encountered are ill-defined, complex, nonlinear, time-varying and stochastic. A detailed evaluation of their (relative) potential awaits the appearance of a rigorous supporting base (underlying theory and implementation architectures for example) the essential elements of which are beginning to appear in learned journals and conferences.

Elements of control and systems theory/engineering are increasingly finding use outside traditional numerical processing environments. One such general area in which there is increasing interest is intelligent command and control systems which are central, for example, to innovative manufacturing and management of advanced transportation systems. Another is discrete event systems which mix numeric and logic decision making.

It was in response to these exciting new developments that the present book series of System and Control was conceived. It publishes high-quality research texts and reference works in the diverse areas which systems and control now includes. In addition to basic theory, experimental and/or application studies are welcome, as are expository texts where theory, verification and applications come together to provide a unifying coverage of a particular topic or topics.

E. Rogers
J. O'Reilly

Preface

In the formulation of any control problem there will typically be discrepancies between the actual plant and the mathematical model developed for controller design. This mismatch may be due to unmodelled dynamics, variation in system parameters or the approximation of complex plant behaviour by a straightforward model. The engineer must ensure that the resulting controller has the ability to produce the required performance levels in practice despite such plant/model mismatches. This has led to an intense interest in the development of so-called *robust* control methods which seek to solve this problem. One particular approach to robust controller design is the so-called *sliding mode* control methodology.

Sliding mode control is a particular type of *Variable structure control.* Variable structure control systems (VSCS) are characterised by a suite of feedback control laws and a decision rule. The decision rule, termed the switching function, has as its input some measure of the current system behaviour and produces as an output the particular feedback controller which should be used at that instant in time. The result is a variable structure system, which may be regarded as a combination of subsystems where each subsystem has a fixed control structure and is valid for specified regions of system behaviour. One of the advantages of introducing this additional complexity into the system is the ability to combine useful properties of each of the composite structures of the system. Furthermore, the system may be designed to possess new properties not present in any of the composite structures alone. Utilisation of these natural ideas began in the Soviet Union in the late 1950s.

In sliding mode control, VSCS are designed to drive and then constrain the system state to lie within a neighbourhood of the switching function. There are two main advantages to this approach. Firstly, the dynamic behaviour of the system may be tailored by the particular choice of switching function. Secondly, the closed-loop response becomes totally insensitive to a particular class of uncertainty. The latter invariance property clearly makes the methodology an appropriate candidate for robust control. In addition, the ability to specify performance directly makes sliding mode control attractive from the design perspective.

The sliding mode design approach consists of two components. The first involves the design of a switching function so that the sliding motion satisfies design specifications. The second is concerned with the selection of a control law which will make the switching function attractive to the system state. Note that this control law is not necessarily discontinuous.

This text provides the reader with a thorough grounding in the sliding mode control area and as such is appropriate for the graduate with a basic knowledge of classical control theory and some knowledge of state-space methods. From this basis, more advanced theoretical results are developed. Resulting design procedures are emphasised using MATLAB mfiles.[1]

[1] MATLAB is a registered trademark of Mathworks, Inc.

Fully worked design examples are an additional tutorial feature. Industrial case studies, which present the results of sliding mode controller implementations, are used to illustrate the successful practical application of the theory.

The book is structured as follows. Chapter 1 introduces the concept of sliding mode control and illustrates the attendant features of robustness and performance specification using a straightforward example and graphical exposition. The necessary background for more formal development is presented in Chapter 2 to enhance readability.

Chapter 3 formulates the general multivariable sliding mode control problem. The interpretation of the sliding surface design problem as a straightforward linear state feedback problem for a subsystem is emphasised. Possible control strategies to enforce sliding motion are identified and the problem of smoothing undesirable discontinuous signals addressed. Controller design issues are considered in Chapter 4. A number of methods for sliding surface design are described. The so-called unit vector control structure is exploited as this lends itself to the development of simple numeric design algorithms. Frameworks for the solution of both regulation and tracking problems are presented. A design study from the flight control area is used to illustrate the application of the proposed design methods.

As developed thus far, the controllers require full state information. In practice, it may be impossible or impractical to measure all of the system states. The next three chapters address this problem. Chapter 5 investigates the possibilities of obtaining sliding mode schemes where the control law and decision rule are a function of the measured outputs. The associated limitations relating to both nominal system structure and attendant design freedoms are demonstrated. A design algorithm is presented and illustrated using numerical examples. Chapters 6 and 7 consider the possibilities of both state reconstruction and control using sliding mode techniques. Again design frameworks are emphasised.

The book concludes with industrial studies describing both sliding mode controller design and implementation. Chapter 8 considers two automotive studies. The automotive environment is harsh; components are subject to large temperature variations and shock, and parameters can be expected to vary significantly during the lifetime of the vehicle. Robustness is thus a key requirement. Further, there is the need to produce high levels of performance and safety from inexpensive components. Sliding mode design will be seen to fulfil these requirements.

Chapter 9 considers sliding mode control of a gas-fired furnace. The furnace dynamics are highly nonlinear and any control strategy is required to yield robust performance over a large operating range. Tracking requirements are essential; when firing ceramics, for example, it is essential that the furnace temperature follows a prespecified profile exactly if the end product is to be of good quality. Efficiency of combustion is also a priority; large magnitude interacting control signals result in significant gas wastage. It will be established that sliding mode control can achieve the desired specifications.

In terms of the interdependence of the individual chapters, Chapters 1 and 2 are not formally connected to the rest of the text. Chapter 1 only seeks to motivate some of the ideas that are expressed rigorously in Chapter 3. Chapter 2 is meant to indicate the topics in systems theory that are required in subsequent chapters.

Readers who have not encountered the state-space and Lyapunov theory presented there are encouraged to refer to more specialist texts in these particular areas which develop the ideas in a more measured and expert way. Chapter 3 is the cornerstone of the book; in particular Section 3.6 very much sets the tone of the control laws which are discussed in Chapters 4, 5 and 7. Chapter 4 is quite strongly linked with Chapter 3. Although its main thrust is to describe different hyperplane design strategies, it does introduce in a tutorial way the integral action and model-following approaches for reference tracking, which subsequently appear in Chapters 5 and 7. Although Chapter 5 (by necessity) considers and develops different control strategies because of the removal of the state availability assumption, parallels are still drawn with the results of Chapter 3. Chapter 6, which is primarily concerned with the design of observers, does utilise some of the ideas and results from Chapter 5 because a similar canonical form is used as the fundamental design framework. As a result, some of the proofs are more sketched but suitable cross-referencing has been made to the original development. (This is true of the run of chapters from 3 through to 7.) Chapter 7 essentially reworks some of the ideas of Chapter 4 (the reference tracking methodologies) and considers the state feedback controllers in conjunction with the state observers developed in Chapter 6. The first actuator case study in Chapter 8 could (almost) be tackled once the concepts from Chapter 1 have been understood. The idle speed control problem in Section 8.3 and the furnace case study in Chapter 9 rely heavily on the integral action controller exposition from Chapter 7.

Finally it is our pleasure to thank our families, friends and colleagues who have helped in the preparation of this book. Several deserve a special mention. First and foremost we would like to thank Aamer Bhatti for his assistance in the preparation of Chapter 8. We are especially grateful to him for allowing us to use, at the time of writing, unpublished work pertaining to the control of engine speed. We also wish to thank Russell Jones of Lucas Varity for his help with the automotive actuator problem, which also appears in Chapter 8. The furnace application described in Chapter 9 would not have taken place without the help and technical assistance of Haydn Porch, Sean Goodhart, Ruth Davies and Patrick Holmes. We thank them for their cooperation and patience during the trials initially at the Midlands Research Centre at Solihull and subsequently at the Gas Research Centre in Loughborough.

In terms of the preparation of the document we would like to thank Doug Pratt for drawing many of the figures which appear throughout the text. We are also very grateful to the brave souls who helped proofread and provide valuable feedback: we would like to thank Robert Cortez who suffered the earliest draft and also our graduate students Ashu Akoachere and Guido Hermman. We are particularly grateful to Xinghuo Yu for his thorough reading of the first draft and for providing many pertinent and insightful comments which we have tried to incorporate. Last but not least, we would like to thank Xiao-Yun Lu for helping us read and correct the final proofs. Of course, despite our best efforts, we are sure that many typographical errors still remain; for these we take full responsibility.

Chris Edwards
Sarah Spurgeon

March 1998

Chapter 1

An Introduction to Sliding Mode Control

1.1 INTRODUCTION

Variable structure control systems (VSCS) evolved from the pioneering work in Russia of Emel'yanov and Barbashin in the early 1960s. The ideas did not appear outside of Russia until the mid 1970s when a book by Itkis (1976) and a survey paper by Utkin (1977) were published in English. VSCS concepts have subsequently been utilised in the design of robust regulators, model-reference systems, adaptive schemes, tracking systems, state observers and fault detection schemes. The ideas have successfully been applied to problems as diverse as automatic flight control, control of electric motors, chemical processes, helicopter stability augmentation systems, space systems and robots. This chapter seeks to motivate and introduce the concepts which will be considered more formally later in the book.

Variable structure control systems, as the name suggests, are a class of systems whereby the 'control law' is deliberately changed during the control process according to some defined rules which depend on the state of the system. For the purpose of illustration consider the double integrator given by

$$\ddot{y}(t) = u(t) \tag{1.1}$$

Initially consider the effect of using the feedback control law

$$u(t) = -ky(t) \tag{1.2}$$

where k is a strictly positive scalar. One way of analysing the resulting closed-loop motion is by means of a phase portrait, which essentially is a plot of velocity against position. Substituting for the control action in equation (1.1), and multiplying the resulting equation throughout by $\dot{y}$, yields

$$\dot{y}\ddot{y} = -k\dot{y}y \tag{1.3}$$

Integrating this expression gives the following relationship between velocity and position

$$\dot{y}^2 + ky^2 = c \tag{1.4}$$

where c represents a constant of integration resulting from the initial conditions and is strictly positive. Importantly, time does not appear explicitly in the expression in (1.4). In the special case when $k = 1$, equation (1.4) represents a circle with centre at the origin and radius $\sqrt{c}$. More generally, a plot of $\dot{y}$ against y is an ellipse which depends on the initial conditions as shown in Figure 1.1. In terms of regulation – i.e. steering arbitrary initial conditions to the origin – a control law of the form given in (1.2) is not appropriate since, as shown in Figure 1.1, the y and $\dot{y}$ variables do not move towards the origin.[1]

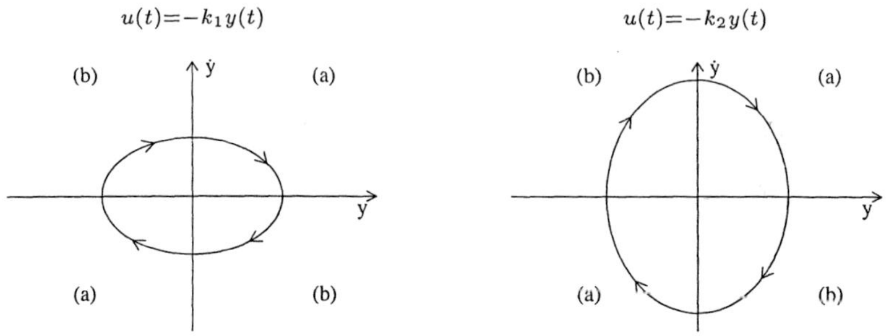

Figure 1.1: Phase portraits of simple harmonic motion

Consider instead the control law

$$u(t) = \begin{cases} -k_1 y(t) & \text{if } y\dot{y} < 0 \\ -k_2 y(t) & \text{otherwise} \end{cases} \tag{1.5}$$

where $0 < k_1 < 1 < k_2$. The phase plane $(y, \dot{y})$ is partitioned by the switching rule into four quadrants separated by the axes as shown in Figure 1.1. The control law $u = -k_2 y$ will be in effect in the quadrants of the phase plane labelled (a). In this region, the distance from the origin of the points in the phase portrait decreases along the system trajectory. Likewise, in region (b) when the control law $u = -k_1 y$ is in operation, the distance from the origin of the points in the phase portrait also decreases. This system clearly fits the description of a VSCS given earlier. The phase portrait for the closed-loop system under the variable structure control law u is obtained by splicing together the appropriate regions from the two phase portraits in Figure 1.1. In this way the phase portrait must spiral in towards the origin and an asymptotically stable motion results, as shown in Figure 1.2. This can be verified more formally by considering the function

$$V(y, \dot{y}) = y^2 + \dot{y}^2 \tag{1.6}$$

which, from Pythagoras' theorem, represents the square of the distance from the point $(y, \dot{y})$ to the origin in the phase plane and may be viewed as representing the energy of the system. The time derivative of $V(y, \dot{y})$ along the closed-loop trajectories is given by

$$\begin{aligned} \dot{V} &= 2\dot{y}y + 2\ddot{y}\dot{y} \\ &= 2\dot{y}(y + u) = \begin{cases} 2y\dot{y}(1 - k_1) & \text{if } y\dot{y} < 0 \\ 2y\dot{y}(1 - k_2) & \text{if } y\dot{y} > 0 \end{cases} \end{aligned}$$

[1]Since y and $\dot{y}$ remain bounded for all time the closed-loop is stable; it is not, however, asymptotically stable – see Section 2.2 for details.

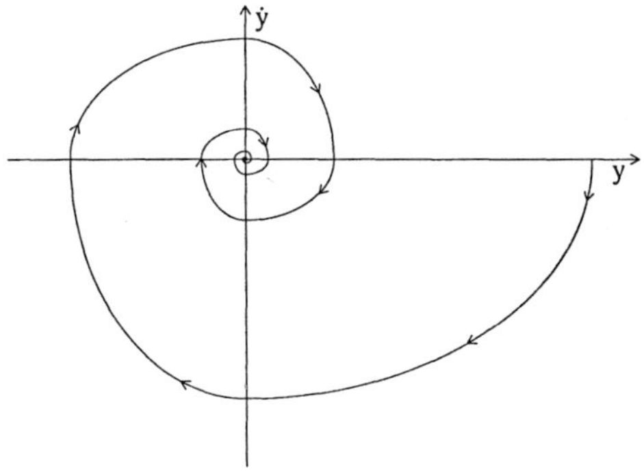

Figure 1.2: Phase portrait of the system under VSCS

This is always negative by the way the gains are constructed. Thus the distance from the origin is always decreasing, which agrees with the intuitive observation made earlier. By introducing a rule for switching between two control structures, which independently do not provide stability, a stable closed-loop system has been obtained.

A more significant example results from using the variable structure law given by

$$u(t) = \begin{cases} -1 & \text{if } s(y, \dot{y}) > 0 \\ 1 & \text{if } s(y, \dot{y}) < 0 \end{cases} \tag{1.7}$$

where the *switching function* is defined by

$$s(y, \dot{y}) = my + \dot{y} \tag{1.8}$$

where m is a positive design scalar. The reason for the use of the term 'switching function' is clear, since the function given in equation (1.8) is used to decide which control structure is in use at any point $(y, \dot{y})$ in the phase plane. The expression in equation (1.7) is usually written more concisely as

$$u(t) = -\text{sgn}\,(s(t)) \tag{1.9}$$

where sgn$(\cdot)$ is the *signum*, or more colloquially, the sign function. The signum function exhibits the property that

$$s\,\text{sgn}(s) = |s| \tag{1.10}$$

This simple result will be exploited often in the analysis which follows.

The expression given in (1.7) is used to control the double integrator. For large values of $\dot{y}$ the phase portrait, obtained from joining the parabolic components of the constituent laws, is shown in Figure 1.3. The dotted line in the figure represents the set of points for which $s(y, \dot{y}) = 0$; in this case a straight line through the origin of gradient $-m$. However, for values of $\dot{y}$ satisfying the inequality $m|\dot{y}| < 1$ then

$$s\dot{s} = s\,(m\dot{y} + \ddot{y}) = s\,(m\dot{y} - \text{sgn}(s)) < |s|\,(m|\dot{y}| - 1) < 0$$

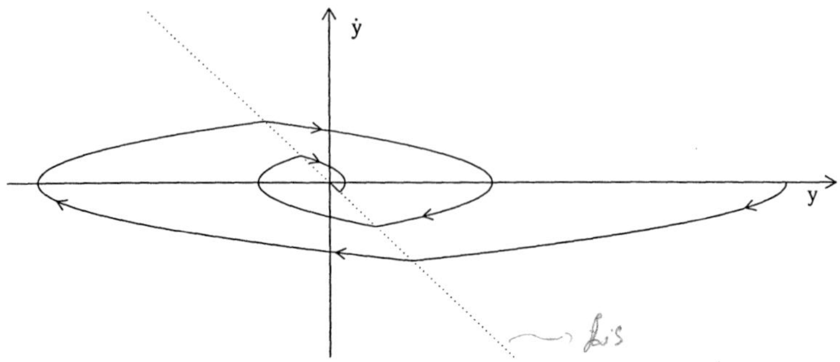

Figure 1.3: Phase portrait of the system for large $\dot{y}$

or equivalently

$$\lim_{s \to 0^+} \dot{s} < 0 \qquad \text{and} \qquad \lim_{s \to 0^-} \dot{s} > 0 \tag{1.11}$$

Consequently, when $m|\dot{y}| < 1$ the system trajectories on either side of the line

$$\mathcal{L}_s = \{(y, \dot{y}) \ : \ s(y, \dot{y}) = 0\} \tag{1.12}$$

point towards the line. This is demonstrated in Figure 1.4, which shows different phase portraits intercepting the same point on the line $\mathcal{L}$ from different initial conditions.

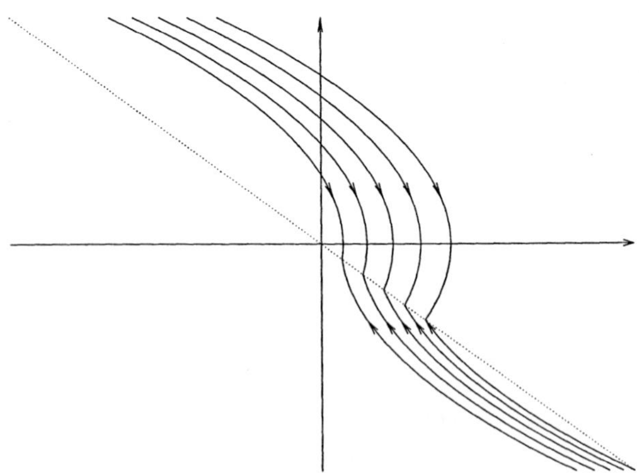

Figure 1.4: Phase portrait of the system under VSC near the origin

Intuitively, high frequency switching between the two different control structures will take place as the system trajectories repeatedly cross the line $\mathcal{L}_s$. This high frequency motion is described as *chattering*. If infinite frequency switching were possible, the motion would be trapped or constrained to remain on the line $\mathcal{L}_s$. The motion when confined to the line $\mathcal{L}_s$ satisfies the differential equation obtained from rearranging $s(y, \dot{y}) = 0$, namely

$$\dot{y}(t) = -my(t) \tag{1.13}$$

This represents a first-order decay and the trajectories will 'slide' along the line $\mathcal{L}_s$ to the origin (Figure 1.5).

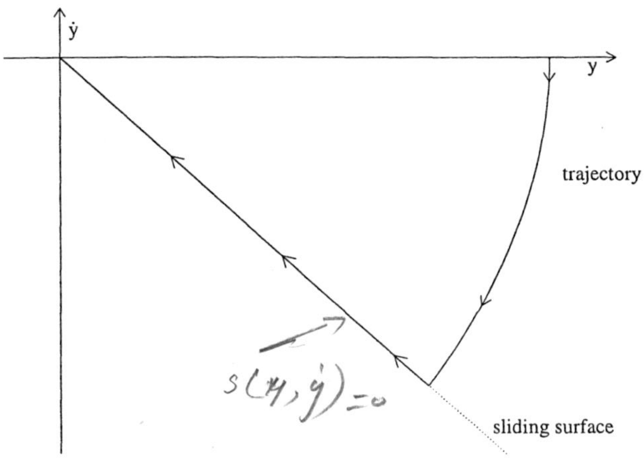

Figure 1.5: Phase portrait of a sliding motion

Such dynamical behaviour is described as an *ideal sliding mode* or an *ideal sliding motion* and the line $\mathcal{L}_s$ is termed the *sliding surface*. During sliding motion, the system behaves as a reduced-order system which is apparently independent of the control. The control action, rather than prescribing the dynamic performance, ensures instead that the conditions given in (1.11) are satisfied; this guarantees that $s(y, \dot{y}) = 0$. The conditions in (1.11) are usually written more conveniently as

$$s\dot{s} < 0 \tag{1.14}$$

which is referred to as the *reachability condition*. Thus, in terms of VSCS design, the choice of the switching function, represented in this situation by the parameter m, governs the performance response; whilst the control law itself is designed to guarantee that the reachability condition (1.14) is satisfied. In this case, as argued earlier, the reachability condition is only satisfied in a domain of the phase plane

$$\Omega = \{(y, \dot{y}) \ : \ m|\dot{y}| < 1\}$$

It should be noted that the control action required to bring about such a motion is discontinuous, and on the sliding surface is not even defined. The discontinuous requirement will be discussed in later chapters.

1.2 PROPERTIES OF THE SLIDING MOTION

This section explores in more detail the properties of the ideal sliding motion and the control action necessary to maintain such a motion. Consider once again the double integrator in equation (1.1) and the control law given in (1.7). The phase portrait in Figure 1.6 is from a simulation of the closed-loop behaviour when $m = 1$, and the initial conditions are given by $y = 1$ and $\dot{y} = 0$. The two-stage nature of the dynamics is readily observed: the initial (parabolic) motion towards the sliding surface, followed by a motion along the line $\dot{y} = -y$ towards the origin.

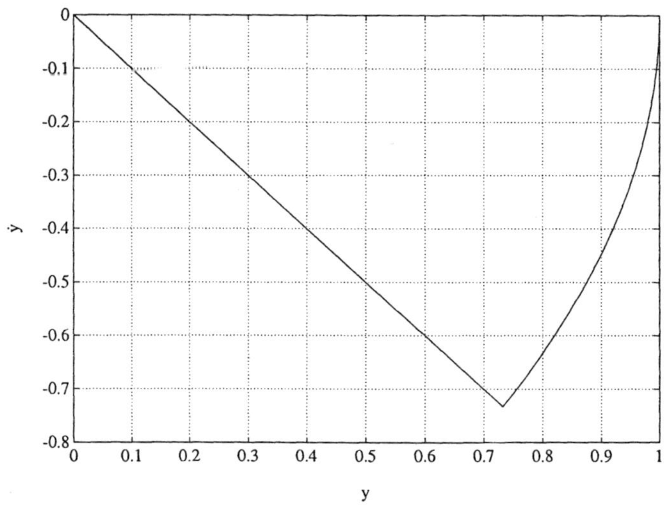

Figure 1.6: Phase portrait of a sliding motion

The control action associated with this simulation is given in Figure 1.7. It can be seen that sliding takes place after 0.732 second when high frequency switching takes place.

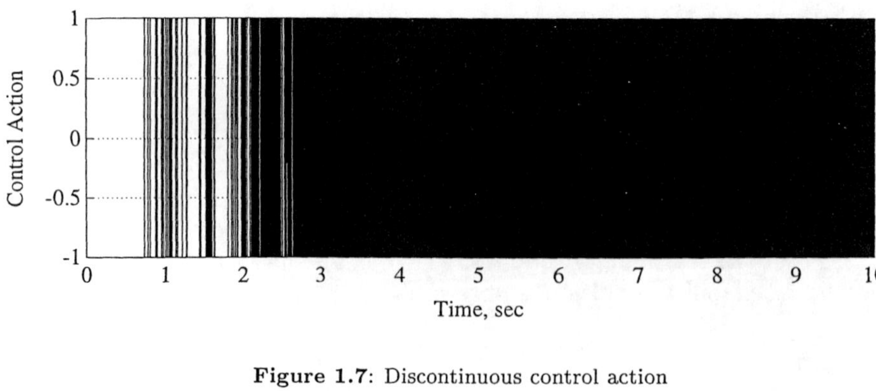

Figure 1.7: Discontinuous control action

Before considering the properties of the sliding motion, an interpretation of the control signal given in Figure 1.7 will be given in terms of its 'average' or low frequency behaviour. As argued earlier, the purpose of the control action is to ensure that the trajectories are driven towards and forced to remain on $\mathcal{L}_s$ to guarantee a sliding motion. It is natural therefore to explore the relationship between the control action and the switching function rather than between the control action and the plant output. Suppose at time t_s the switching surface is reached and an ideal sliding motion takes place. It follows that the switching function satisfies $s(t) = 0$ for all $t > t_s$, which in turn implies that $\dot{s}(t) = 0$ for all $t \geq t_s$. However, from equations (1.1) and (1.8)

$$\dot{s}(t) = m\dot{y}(t) + u(t) \tag{1.15}$$

and thus since $\dot{s}(t) = 0$ for all $t \geq t_s$, it follows from (1.15) that a control law which maintains the motion on $\mathcal{L}_s$ is

$$u(t) = -m\dot{y}(t) \qquad (t \geq t_s) \tag{1.16}$$

This control law is referred to as the *equivalent control* action. This is not the control signal which is actually applied to the plant but may be thought of as the control signal which is applied 'on average'. This can be demonstrated by passing the discontinuous control signal (in Figure 1.7) through the low pass filter

$$\tau\dot{u}_a(t) + u_a(t) = u(t) \tag{1.17}$$

to obtain the low frequency component $u_a(t)$. Figure 1.8 shows $u_a(t)$, together with the associated equivalent control when $\tau = 0.04$. It can be seen that the filtered (or averaged) control signal agrees with the equivalent control action defined in equation (1.16). Of course agreement can only take place once sliding is established – in this case after 0.732 second. Consequently, the dotted line, representing the equivalent control, is only drawn from this time onwards, as shown in Figure 1.8.

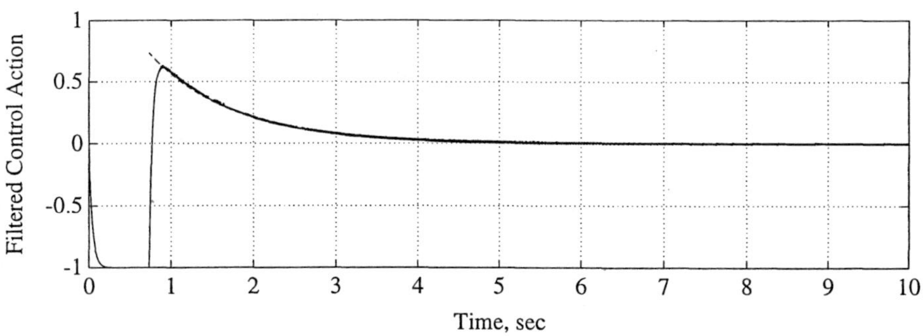

Figure 1.8: Equivalent control

The control signal applied to the plant may be thought of as comprising 'low' and 'high' frequency components so that

$$u(t) = \underbrace{u_a(t)}_{low\ frequency} + \underbrace{(u(t) - u_a(t))}_{high\ frequency}$$

For systems which may be modelled as strictly proper transfer functions,[2] for example, the high frequency component may be considered to be beyond the bandwidth; hence the control action that affects the dynamic response is $u_a(t)$, i.e. the equivalent control. A more dramatic demonstration of the strength of the notion of an equivalent control will be given later when the effects of plant/model mismatches will be examined.

Suppose the double integrator in equation (1.1) is a linear approximation of the real system on which the control law is to be implemented. Suppose that the real system is in fact a pendulum, formed from a light rod and a heavy mass, as shown in Figure 1.9.

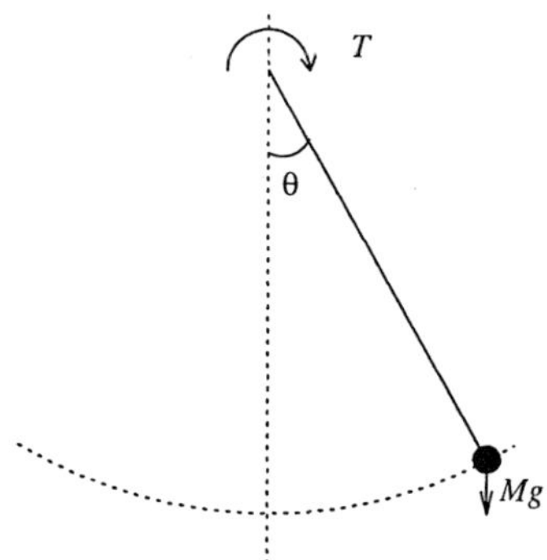

Figure 1.9: Schematic of a pendulum

The variable θ represents the angular displacement from the vertical and T represents a torque applied at the point of suspension, which will be considered to be the control input to the system. Ignoring the effects of friction, the system can be represented mathematically as

$$\ddot{\theta}(t) = -\frac{l}{g}\sin\theta(t) + \frac{1}{Ml^2}u(t) \tag{1.18}$$

where M represents the mass of the bob, l the length of the rod and g the acceleration due to gravity. By appropriate scaling, the essential dynamics of the system are captured by

$$\ddot{y}(t) = -a_1 \sin y(t) + u(t) \tag{1.19}$$

where a_1 is a positive scalar. Equation (1.19) will subsequently be referred to as the *normalised pendulum equation* or *pendulum system.* The double integrator of (1.1) may therefore be considered to be a linear approximation of the normalised pendulum equation which is obtained from ignoring the nonlinear sine term. The phase portrait of the closed-loop system obtained from using the control law (1.7) in the normalised pendulum equation (1.19), when $a_1 = 0.25$ and the initial conditions are $\dot{y} = 0$ and $y = 1$, is shown in Figure 1.10. (For comparison the dotted line represents the phase portrait of the nominal closed-loop system when $a_1 = 0$).

[2] A transfer function $G(s)$, where s is the Laplace variable, is strictly proper if the order of the numerator polynomial is strictly less than the denominator polynomial and thus $G(jw) \to 0$ as $w \to \infty$.

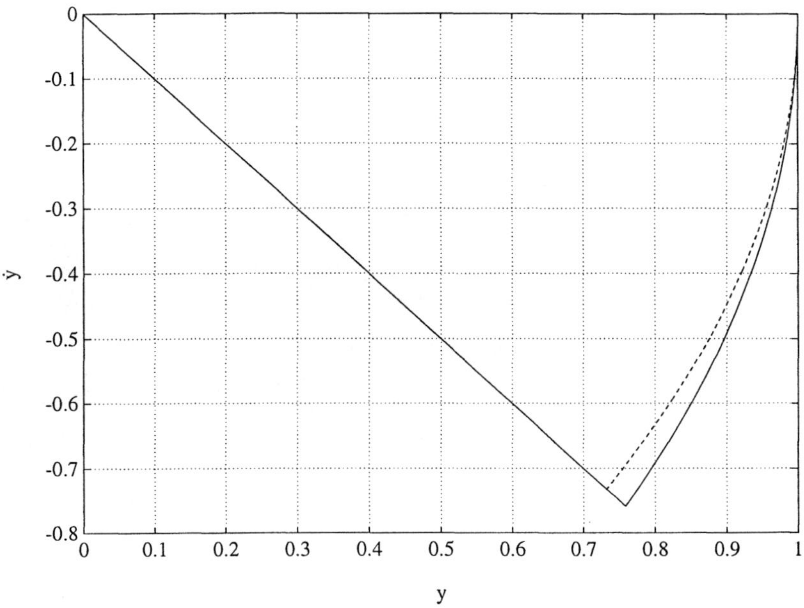

Figure 1.10: Controlled pendulum

The key result is that, in finite time, the phase portrait intercepts the sliding surface $\mathcal{L}_s$ and is forced to remain there. The significance of this is that, once ideal sliding is established, the double-integrator system and the normalised pendulum behave in an identical fashion, namely

$$\dot{y}(t) = -my(t) \tag{1.20}$$

An alternative interpretation is that the effect of the nonlinear term $a_1 \sin y(t)$, which may be construed as a disturbance or uncertainty in the nominal double-integrator system, has been completely rejected. As such, the closed-loop system is said to be *robust*, i.e. it is insensitive to mismatches between the model used for control law design and the plant on which it will be implemented.

The idea of completely rejecting a disturbance or uncertainty is decidedly different from other control approaches such as $\mathcal{H}_2$ and $\mathcal{H}_\infty$. These linear methodologies attempt to *minimise* in some sense the transfer functions relating the disturbances to the outputs of interest. Indeed, at first, it is difficult to imagine how the effect of an unknown disturbance can be *cancelled*. It is at this point that the concept of the equivalent control can be used to provide insight. Arguing as before, once an ideal sliding motion has been attained, $s(t) = 0$ and $\dot{s}(t) = 0$ for all subsequent time. From equations (1.8) and (1.19) it follows that

$$\dot{s}(t) = m\dot{y}(t) - a_1 \sin y(t) + u(t) \tag{1.21}$$

The equivalent control is obtained by equating the expression in (1.21) to zero, resulting in

$$u_{eq}(t) = -m\dot{y}(t) + a_1 \sin y(t) \tag{1.22}$$

This expression captures precisely the uncertainty or disturbance in the closed-loop system, in this case the $a_1 \sin y(t)$ term, and cancels its effect. Of course the control action applied to the plant *does not* utilise any knowledge of the uncertainty. Indeed, as shown in Figure 1.11, the control action appears no different from that used in the nominal double-integrator plant (Figure 1.7).

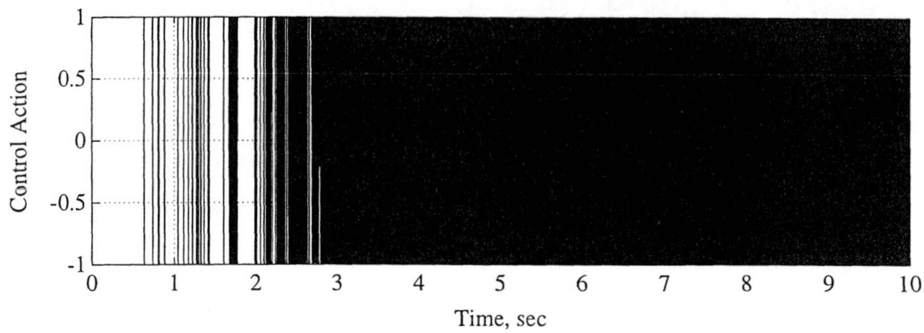

Figure 1.11: Applied control action

However, passing the control action in Figure 1.11 through the low pass filter given in (1.17) provides the average applied control signal shown in Figure 1.12. This is quite different to the average control signal obtained previously and, as expected, is visually identical to the equivalent control from (1.22), shown as a dotted line.

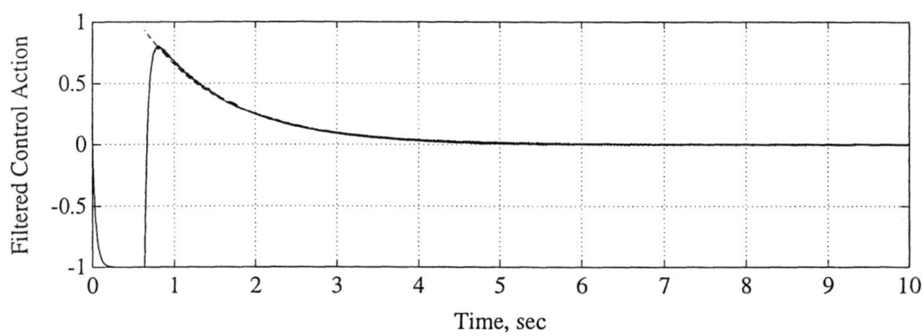

Figure 1.12: Filtered control action compared to the equivalent control

Apart from the robustness properties exhibited by the sliding motion, another benefit accruing from this situation is that the pendulum is forced to behave as a first-order system. This guarantees that no overshoot will occur when attempting to regulate the pendulum from an arbitrary initial displacement to the origin (provided sliding can be attained quickly enough). This can be seen from Figure 1.13, which displays the exponential decay characteristic typical of a first-order system. The two properties of an ideal sliding motion described previously, namely disturbance rejection and order reduction, are the key properties that have motivated the study of controllers which induce sliding motions.

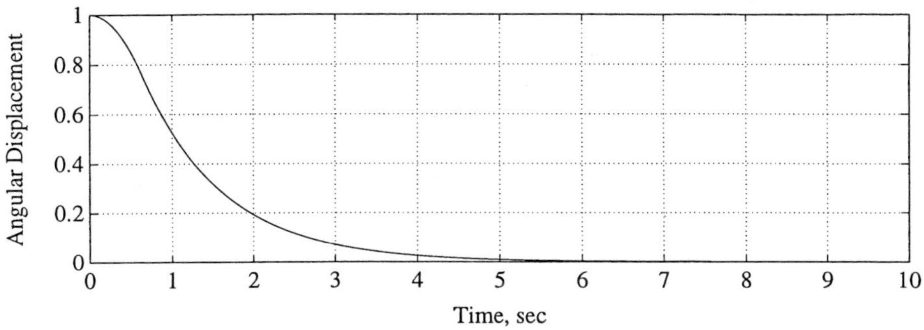

Figure 1.13: Evolution of displacement with respect to time

1.3 DIFFERENT CONTROLLER DESIGNS

Consider the pendulum example from the previous section together with the control law in (1.7) and consider the response of the system to large initial displacements. Figure 1.14 represents a plot of the switching function with initial conditions $y = 3$ and $\dot{y} = 0$.

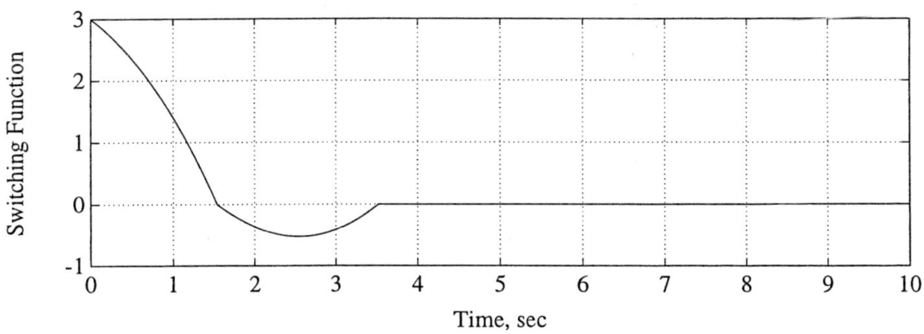

Figure 1.14: Switching function with respect to time

It can be seen that at approximately 1.5 seconds the trajectory intercepts the switching line $\mathcal{L}_s$ but insufficient control energy is available to maintain a sliding motion. The sign of the control law switches, the trajectory pierces the switching line and moves away before intercepting the line again at approximately 3.5 seconds, at which point sliding takes place. This result is not perhaps surprising since the initial conditions correspond to releasing the pendulum from near the upward vertical which will intuitively generate larger angular velocities. This also agrees with the theory presented earlier since even for the nominal double integrator, a sliding motion could only be guaranteed in the region of the phase plane for which the angular velocity is less than $1/m$. The result in this case is that the pendulum crosses the downward vertical before a sliding motion can be established and thus overshoot occurs.

Since the key properties of robustness and order reduction are only obtained once sliding is induced, from the perspective of control law design, the time taken to induce sliding should be minimised, and the region in which sliding takes place maximised. If the magnitude of the control signal is limited to the range ± 1 then the bang-bang controller given in (1.7) may well represent a legitimate control scheme. Otherwise it is instructive to consider modifying the control law given in (1.7), through the inclusion of a linear feedback component, to attempt to ensure that once the trajectories reach the sliding surface they are forced to remain there.

A candidate control structure is given by

$$u(t) = l_1 y(t) + l_2 \dot{y}(t) - \rho\, \mathrm{sgn}\,(s(t)) \tag{1.23}$$

where l_1, l_2 and ρ represent scalars yet to be designed. The intention is to choose the three parameters in (1.23) so that the inequality $s\dot{s} < 0$ is always satisfied for the normalised pendulum equation. From the definition of the switching function given in (1.8) it follows that

$$s\dot{s} = s\,(m\dot{y} + \ddot{y}) = s\,(m\dot{y} - a_1 \sin(y) + u) \tag{1.24}$$

Substituting for the control action in equation (1.24) gives

$$s\dot{s} = s\,(m\dot{y} - a_1 \sin(y) + l_1 y + l_2 \dot{y} - \rho\, \mathrm{sgn}(s)) \tag{1.25}$$

By choosing $l_1 = 0$ and $l_2 = -m$ it follows that

$$s\dot{s} = -s a_1 \sin(y) - \rho |s| < |s|(a_1 - \rho)$$

Thus finally, by choosing $\rho > a_1 + \eta$ where η is a positive design scalar, the inequality

$$s\dot{s} < -\eta |s| \tag{1.26}$$

is established. In the literature this is referred to as the *η-reachability condition.* The previous control law only satisfied the reachability condition in a region of the phase plane; the control law in (1.23) guarantees that, whenever the sliding surface is reached, an ideal sliding motion takes place. A motion such as the one obtained in Figure 1.14, with the trajectory piercing the switching surface, is now no longer possible. It should be noted that the linear component corresponds exactly to the nominal equivalent control expression given in (1.16). This is by no means accidental since, ignoring the nonlinear terms in the preceding analysis, the linear component has been chosen to ensure that $\dot{s}(t) = 0$.

Using the control law

$$u(t) = -m\dot{y}(t) - \rho\, \mathrm{sgn}(s(t)) \tag{1.27}$$

with $m = 1$ and $\rho = 1$, and the same initial conditions $y = 3$ and $\dot{y} = 0$, the switching function associated with the simulated closed-loop response is given in Figure 1.15. It reveals that no piercing of the sliding surface occurs, although it could be argued that the time taken to reach the sliding surface is a little slow. This is reflected in the angular response given in Figure 1.16. Sliding is achieved early enough so that the approach to the downward vertical is governed by the reduced-order motion (and so no overshoot can occur), but the closed-loop performance is unduly sluggish.

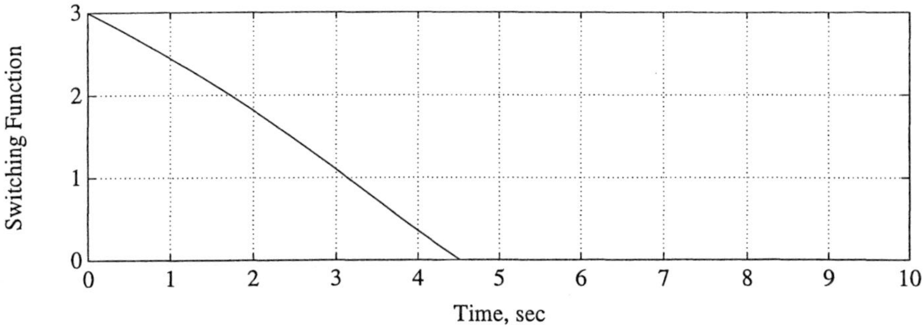

Figure 1.15: Switching function with respect to time

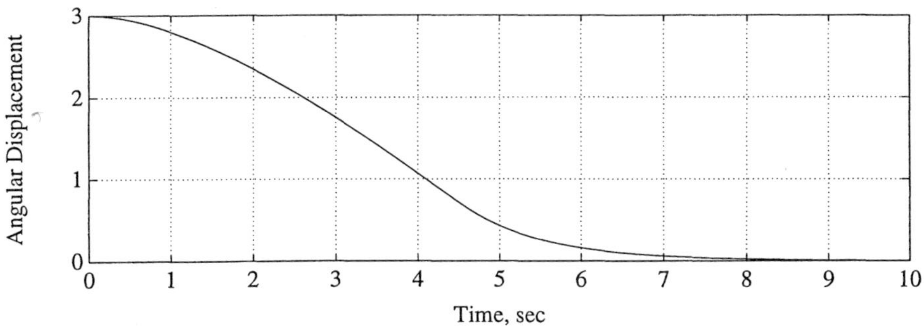

Figure 1.16: Angular displacement with respect to time

The most serious objection to the control law given earlier is perhaps that choosing $\rho = 1$ is rather conservative in the sense that it need only be greater than 0.25 for the η-reachability condition to be satisfied. A lower value of ρ would reduce the amplitude of the high frequency switching, which is advantageous from the point of view of limiting wear and tear on the actuators. Unfortunately, retaining the same linear component and reducing the scalar ρ results in even slower attainment of the sliding surface. To overcome this difficulty, a judicious modification to (1.27) is to add the term $-\Phi s$, where Φ is a positive design scalar, so that $l_1 = -\Phi m$ and $l_2 = -(m + \Phi)$ making

$$u(t) = -(m + \Phi)\,\dot{y}(t) - \Phi m\, y(t) - \rho\,\mathrm{sgn}(s(t)) \tag{1.28}$$

Arguing as before, it can be established that with this control law the inequality

$$s\dot{s} \leq -\Phi s^2 - \eta |s| \tag{1.29}$$

is satisfied. Consequently, since $\Phi s^2 \geq 0$, an η-reachability condition has been established and a sliding motion will take place. By ignoring the nonlinear term in (1.29) it follows that

$$\frac{d}{dt}|s(t)| \leq -\Phi |s(t)|$$

which implies

$$|s(t)| \leq |s(0)| e^{-\Phi t}$$

where $|s(0)|$ represents the initial distance away from the sliding surface. The parameter Φ can thus be seen to affect the rate at which the sliding surface is attained. As a result of this modification, ρ can be chosen as small as possible (to reduce the amplitude of the switching) and Φ can be chosen to determine the time taken to attain sliding.

The following simulation results are obtained from using the control law in (1.28) with $\Phi = 1$ and $\rho = 0.3$ and assuming, as before, the initial conditions $\dot{y} = 0$ and $y = 3$. Figure 1.17 represents the switching function as a function of time and demonstrates that the time to reach the sliding surface has been greatly reduced.

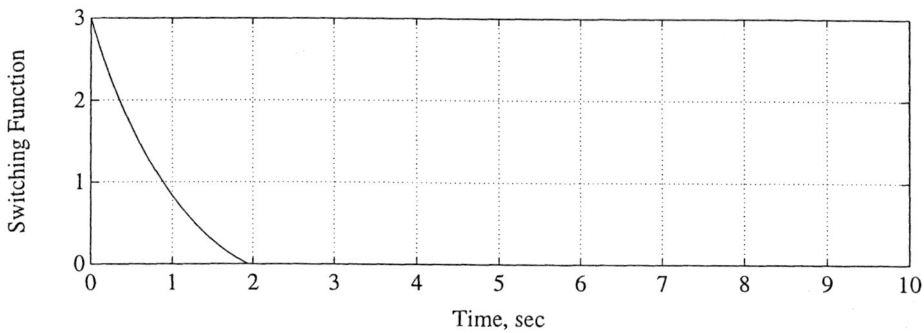

Figure 1.17: Switching function with respect to time

To obtain a faster response, the value of Φ can be increased assuming that sufficient control energy is available. As argued earlier, the reaching-time behaviour displayed in Figure 1.17 is preferable to Figure 1.15 since all the robustness and order reduction properties only occur once the surface has been attained. The corresponding angular displacement is given in Figure 1.18.

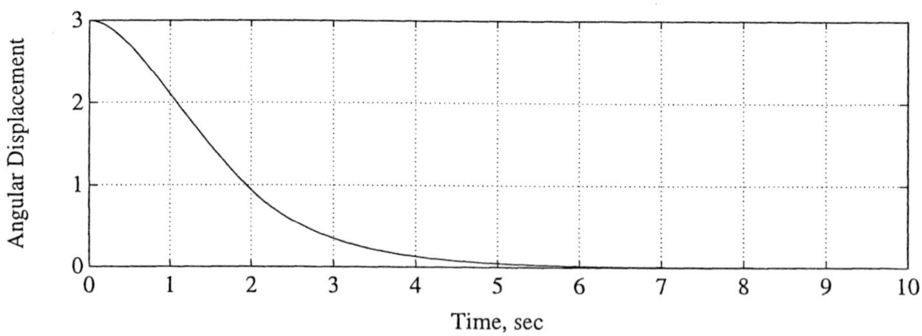

Figure 1.18: Angular displacement with respect to time

Here again the settling time is much improved when compared with the one obtained in Figure 1.16. The big advantage of the controller given in (1.28) over (1.27) is that the control signal is much less aggressive in the sense that the amplitude of the switching is now ± 0.3 as shown in Figure 1.19.

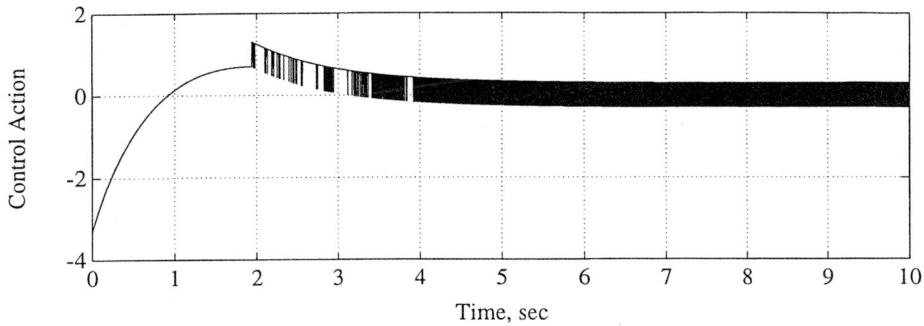

Figure 1.19: Evolution of control action with respect to time

In the nominal double-integrator case with $\rho = 0$ in equation (1.28), it follows that the closed-loop system satisfies

$$\ddot{y}(t) + (m + \Phi)\,\dot{y}(t) + \Phi m\, y(t) = 0 \tag{1.30}$$

This represents a stable motion with poles at $\{-\Phi, -m\}$. It was established earlier that the pole at $-\Phi$ corresponds to the rate at which the sliding surface is attained. The other pole, located at $-m$, corresponds to the pole of the sliding motion. It can thus be argued that the linear part of the control action establishes a sliding mode for the nominal system whilst the discontinuous component counteracts the effects of the uncertainty or nonlinearity.

1.4 PSEUDO-SLIDING WITH A SMOOTH CONTROL ACTION

In certain problems, such as control of electric motors and power converters, the control action is naturally discontinuous and sliding mode ideas can be used to obtain extremely high performance.[3] Although the control signal obtained from Figure 1.19 is preferable to the earlier designs in terms of its chattering behaviour, in many situations such a control signal would still not be considered acceptable. A natural solution is to attempt to smooth the discontinuity in the signum function to obtain an arbitrarily close but continuous approximation. One possible approximation is the sigmoid-like function

$$\nu_\delta(s) = \frac{s}{(|s| + \delta)} \tag{1.31}$$

where δ is a small positive scalar, which is shown in Figure 1.20. It can be visualised that as $\delta \to 0$, the function $\nu_\delta(\cdot)$ tends pointwise to the signum function. The variable δ can be used to trade off the requirement of maintaining ideal performance with that of ensuring a smooth control action.

Assuming the same initial conditions as in the previous section and using the control law given in (1.28) with $\nu_\delta(s)$ replacing $\mathrm{sgn}(s)$ with $\delta = 0.005$, the closed-loop response given below is obtained.

[3]For details see Utkin (1992).

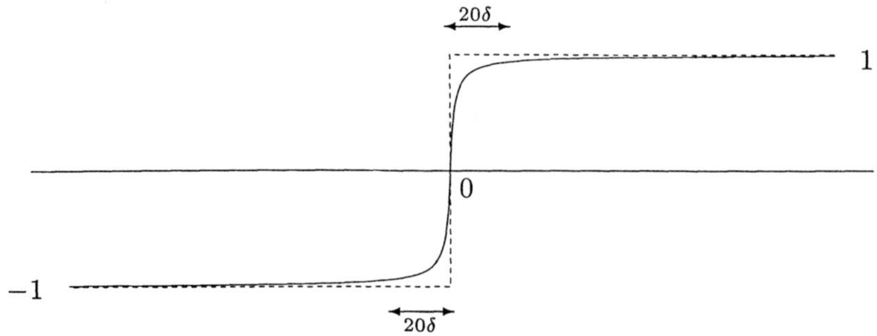

Figure 1.20: A differentiable approximation of the signum function

The angular displacement (Figure 1.21) is indistinguishable from Figure 1.18. The key effect, however, is that the control action is smooth, as shown in Figure 1.22.

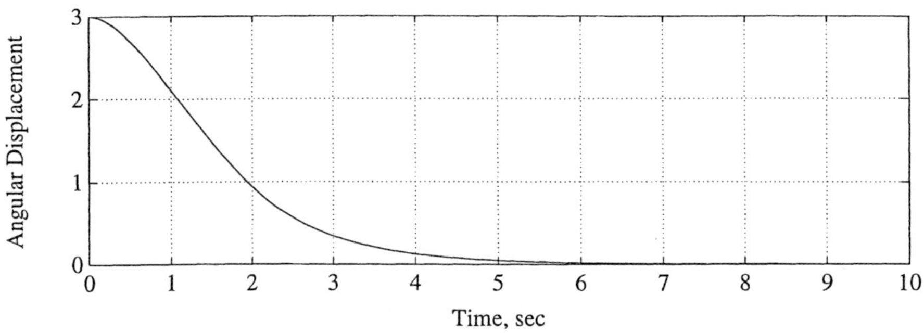

Figure 1.21: Angular displacement with respect to time

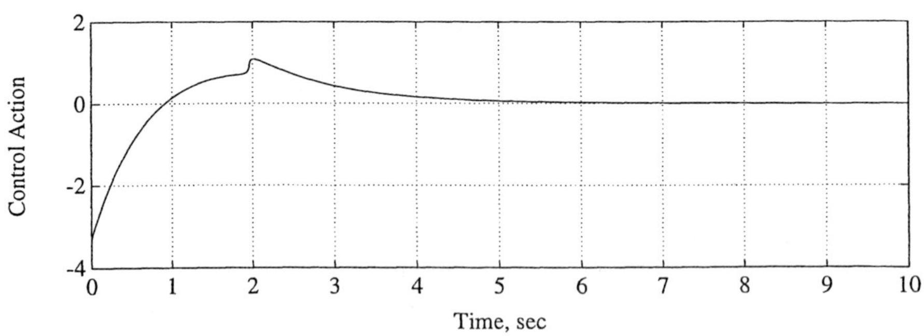

Figure 1.22: Evolution of control action with respect to time

Such continuous approximations enable 'sliding mode' controllers to be utilised in situations where high frequency chattering effects would be unacceptable. It should be stressed that ideal sliding no longer takes place: the continuous control action only drives the states to a neighbourhood of the switching surface. However, arbitrarily close approximation to ideal sliding can be obtained by making δ small. In the literature this is often referred to as *pseudo-sliding.*

1.5 A STATE-SPACE APPROACH

Phase plane analysis is an effective way of analysing the second-order systems considered so far. In future, however, multivariable systems of high order will need to be analysed and therefore a more general framework must be established. The state-space approach pioneered in the 1960s will provide an excellent means of accomplishing this objective.

For the purposes of illustration, consider once again the double integrator in (1.1). If a new vector variable

$$x \triangleq \begin{bmatrix} y \\ \dot{y} \end{bmatrix}$$

is introduced then equation (1.1) can be written in *state-space* form as

$$\dot{x}(t) = \begin{bmatrix} 0 & 1 \\ 0 & 0 \end{bmatrix} x(t) + \begin{bmatrix} 0 \\ 1 \end{bmatrix} u(t) \tag{1.32}$$

The switching function from (1.8) can also be conveniently expressed in matrix terms as

$$s(y, \dot{y}) = Sx(t) \tag{1.33}$$

where

$$S = \begin{bmatrix} m & 1 \end{bmatrix}$$

The normalised pendulum can also be written in state-space form as

$$\dot{x}(t) = \begin{bmatrix} 0 & 1 \\ 0 & 0 \end{bmatrix} x(t) + \begin{bmatrix} 0 \\ 1 \end{bmatrix} u(t) - \begin{bmatrix} 0 \\ 1 \end{bmatrix} a_1 \sin x_1(t) \tag{1.34}$$

In this representation it can be seen that the nonlinearity or uncertainty represented by $\sin(x_1)$ acts in the 'channel' of the input – i.e. in the second of the coupled pair of differential equations. It will be proved in Chapter 3 that VSCS with a sliding mode have the ability to completely reject the effect of bounded uncertainty acting in the input channels – which is referred to as *matched uncertainty.* This effect was of course noted earlier for the normalised pendulum. Uncertainty which does not act in the input channels is referred to as *unmatched uncertainty.*

In the remainder of the book, the uncertain linear time invariant system with m inputs given by

$$\dot{x}(t) = Ax(t) + Bu(t) + f(t, x, u) \tag{1.35}$$

where $A \in \mathbb{R}^{n \times n}$, $B \in \mathbb{R}^{n \times m}$ and $f(\cdot)$ is an unknown bounded function, will be studied. Switching functions of the form

$$s(x) = Sx \tag{1.36}$$

where $S \in \mathbb{R}^{m \times n}$ will be used to induce a sliding motion on the hyperplane

$$\mathcal{S} = \{x \in \mathbb{R}^n \quad : \quad s(x) = 0\} \tag{1.37}$$

Before formally exploring in a general multivariable context the properties of sliding modes, it is necessary to review some properties of linear systems and ideas of stability for nonlinear systems; these aspects are fundamental to the analysis in the remainder of the book.

1.6 NOTES AND REFERENCES

The earliest work in English on the properties of sliding modes appeared in the late 1970s in the form of two books; Itkis (1976) and Utkin (1992). References to the early work, pioneered mainly in the former USSR, are given in Utkin (1992). VSCS concepts have subsequently been utilised in the design of robust regulators, model-reference systems, adaptive schemes, tracking systems, state observers and fault detection strategies. Many of these ideas will be examined later in the book, where more extensive references will be given.

A recent survey paper which gives many references to the various application areas in which sliding mode ideas have been utilised is Hung *et al.* (1993).

The concept of equivalent control is attributed to Utkin (1977). A rigorous interpretation of the relationship between the equivalent control law and the low frequency components of a discontinuous control action maintaining a sliding motion is described in Utkin (1992). The η-reachability condition stems from the work of Slotine (1984).

The control law described in equation (1.28) is essentially the single-input simplification of the control law proposed by Ryan & Corless (1984). This will be discussed at length in Chapter 3.

All the simulations in the chapter have been performed using the Runge–Kutta integration routines in the SIMULINK libraries of MATLAB.

The example of the VSCS for the double integrator in Section 1.1 is taken from Utkin (1977). The pendulum example has been used by many authors to demonstrate various nonlinear phenomenon; see for example Khalil (1992).

Chapter 2

Multivariable Systems Theory

2.1 INTRODUCTION

This chapter provides the necessary background in system theory and control topics that is required for the sliding mode control developments which are to follow. It can be regarded as optional reading for those readers familiar with the concepts, used as a reference whilst reading the rest of the text, or considered a tutorial introduction which is read in total before addressing subsequent chapters.

2.2 STABILITY OF DYNAMICAL SYSTEMS

Consider the nonlinear system given by

$$\dot{x}(t) = f(x, t) \tag{2.1}$$

where $x \in \mathbb{R}^n$ and the function $f : \mathbb{R}^n \times \mathbb{R}_+ \to \mathbb{R}^n$ is such that a well-defined solution to the differential equation exists. Denote this solution at time t as $x(t, x_0)$ where x_0 represents the initial conditions at $t = 0$.

Suppose that $f(0, t) = 0$ for all t, i.e. the dynamical system defined in (2.1) has an equilibrium point at the origin. This imposes no undue restriction since if (2.1) has an equilibrium point at x_e then the change of coordinates $x \mapsto x + x_e$ translates the equilibrium point back to the origin.

Definition 2.1 *The origin of (2.1) is said to be stable if given any $\epsilon > 0$ there exists a $\delta > 0$ such that if $\|x_0\| < \delta$ then $\|x(t, x_0)\| < \epsilon$ for all $t > 0$.*

Less formally the definition means that by starting close enough to the equilibrium point, the solution will thereafter always remain arbitrarily close.

Definition 2.2 *The origin of (2.1) is said to be asymptotically stable if it is stable and the solution $x(t, x_0) \to 0$ as $t \to \infty$.*

The distinction here is that for stability, although the solution $x(t, x_0)$ must remain close to the origin, it does not have to return to the origin. For example, consider the simple harmonic oscillator

$$\dot{x}_1(t) = x_2(t) \tag{2.2}$$
$$\dot{x}_2(t) = -\omega^2 x_1(t) \tag{2.3}$$

where $\omega > 1$ and $x_1(t)$ and $x_2(t)$ are thought of as components of the vector $x(t)$ in equation (2.1). Arguing as in Section 1.1, it follows that the solution to (2.2) and (2.3) satisfies

$$x_2^2(t) + \omega^2 x_1^2(t) = r^2 \tag{2.4}$$

for all $t > 0$ where r is a positive constant depending on the initial conditions. This represents an ellipse (as shown in Figure 1.1) and it follows that the phase portrait lies between two concentric circles, with centres at the origin, of radius r and r/ω. Thus given any $\epsilon > 0$, if the initial conditions $x(0)$ satisfies $||x(0)|| < \epsilon/\omega$ then

$$||x(t)|| < \epsilon \qquad \text{for all } t > 0$$

and hence the simple harmonic oscillator satisfies the definition of stability. However $||x(t)||$ is bounded away from the origin by a circle of radius r and hence $||x(t)||$ cannot tend to zero. Therefore the oscillator is stable but not asymptotically stable.

Before discussing in more detail the properties of general nonlinear systems, the special case of linear systems will be examined.

2.2.1 Linear Time Invariant Systems

Consider the special case where $f(x,t)$ is a time invariant map and linear with regard to the states so that

$$\dot{x}(t) = Ax(t) \tag{2.5}$$

where $A \in \mathbb{R}^{n\times n}$. In this situation equation (2.5) yields an analytic solution

$$x(t, x_0) = e^{At}x_0$$

where x_0 represents the initial condition and the matrix function

$$e^{At} \triangleq I_n + At + \frac{1}{2}A^2t^2 + \ldots + \frac{1}{n!}A^nt^n + \ldots \tag{2.6}$$

This represents a natural extension of the scalar case. The luxury of an explicit solution to equation (2.5) enables necessary and sufficient conditions for stability to be established in terms of the eigenvalues of A. Suppose λ is a real eigenvalue of A and suppose the initial conditions x_0 represent an associated right eigenvector. Using the properties of eigenvalues and eigenvectors

$$A^n x_0 = \lambda^n x_0$$

and therefore

$$\begin{aligned} e^{At}x_0 &= (I_n + \lambda t + \frac{1}{2}\lambda^2t^2 + \ldots + \frac{1}{n!}\lambda^nt^n + \ldots)x_0 \\ &= e^{\lambda t}x_0 \end{aligned}$$

from the power series expansion of the exponential. Consequently, if $\lambda > 0$ then $x(t, x_0) \to \infty$ as $t \to \infty$ and the differential equation (2.5) is not stable because inside any arbitrarily small neighbourhood of the origin it is always possible to find an eigenvector associated with the positive eigenvalue λ for which the solution becomes arbitrarily large. If any of the eigenvalues of A are complex then the associated eigenvectors will be complex and the argument above needs to be modified since physically a complex initial condition is not tenable. Suppose that λ is a complex eigenvalue with an associated complex eigenvector z. Since all the elements in A are assumed to be real, λ^*, the complex conjugate of λ, must be an eigenvalue of A and z^* must be an associated eigenvector. Consider the initial condition $x_0 = \mathrm{Re}[z]$ or equivalently $x_0 = \frac{1}{2}(z + z^*)$; then arguing as before

$$\begin{aligned} x(t, x_0) = e^{At}x_0 &= \frac{1}{2}\left(e^{At}z + e^{At}z^*\right) \\ &= \frac{1}{2}\left(e^{\lambda t}z + e^{\lambda^* t}z^*\right) \end{aligned}$$

Writing $\lambda = \alpha + j\beta$, it follows

$$x(t, x_0) = \frac{1}{2}e^{\alpha t}\left(e^{j\beta t}z + e^{-j\beta t}z^*\right)$$

where the expression in parentheses is real and involves terms in $\cos(\beta t)$ and $\sin(\beta t)$ (essentially Euler's expression $e^{j\phi} = \cos\phi + j\sin\phi$ has been used). Hence if α is positive the solution grows arbitrarily large and the system is unstable. Therefore a necessary condition for asymptotic stability is that the eigenvalues of A have negative real parts.

Consider now the system

$$\dot{x}(t) = Ax(t) + Bu(t) \tag{2.7}$$

Essentially a forcing function or control input $u(t)$ has been introduced into the system. The general solution to this state equation may be expressed in the form

$$x(t) = Ve^{\Lambda t}W^{\mathrm{T}}x(0) + V\int_0^t e^{\Lambda(t-\tau)}W^{\mathrm{T}}Bu(\tau)\,d\tau \tag{2.8}$$

where V is a matrix whose columns consist of the linearly independent right eigenvectors corresponding to the distinct eigenvalues λ_i, $i = 1, \ldots, n$ of the matrix A, $\Lambda = \mathrm{diag}(\lambda_1, \ldots, \lambda_n)$ and the rows of the matrix W^{T} consist of the corresponding left eigenvectors, i.e. $V^{-1} = W^{\mathrm{T}}$. Here $x(0)$ represents an arbitrary initial condition. Again the time domain characteristics of the system are determined by the eigenvalues of A. In addition, the associated right eigenvectors determine the 'shape' of a given mode. As the solution depends upon a linear combination of functions of the form $v_i e^{\lambda_i t}$, appropriate eigenvector entries enable the transient $e^{\lambda_i t}$ to contribute, or not, to a particular state variable. In this way it is seen that the entire eigenstructure, and not just the eigenvalues, are effective in determining the time response of a system.

2.2.2 Quadratic Stability

The previous section considered the special case of linear systems. This enabled an analytic expression for the solution to be obtained. For general nonlinear systems this is usually impossible. An approach for studying the stability of differential equations, without the need to obtain an explicit solution, is the method of Lyapunov. Loosely speaking, if a differentiable function $V : \mathbb{R}^n \mapsto \mathbb{R}$ can be found which is positive except at an equilibrium point and whose total time derivative decreases along the system trajectories, then the equilibrium point is stable. The key point is that this approach obviates the need to solve the nonlinear differential equation when assessing its stability properties.

Unfortunately, no systematic way exists to synthesise Lyapunov functions for nonlinear systems. This section considers the special case when the scalar function $V : \mathbb{R}^n \rightarrow \mathbb{R}$ is the quadratic form

$$V(x) = x^{\mathrm{T}} P x \tag{2.9}$$

where $P \in \mathbb{R}^{n \times n}$ is some symmetric positive definite matrix. By construction the function is nonzero except at the origin. Next, form the function of time

$$V(t) = x(t)^{\mathrm{T}} P x(t) \tag{2.10}$$

where $x(t)$ represents the solution of the differential equation (2.1). Differentiating (2.10) with respect to time gives:

$$\begin{aligned} \dot{V}(t) &= \dot{x}(t)^{\mathrm{T}} P x(t) + x(t)^{\mathrm{T}} P \dot{x}(t) \\ &= 2x(t)^{\mathrm{T}} P \dot{x}(t) \\ &= 2x(t)^{\mathrm{T}} P f(x,t) \end{aligned}$$

where the second equality follows because the quantities are scalars and hence

$$\dot{x}(t)^{\mathrm{T}} P x(t) = \left(\dot{x}(t)^{\mathrm{T}} P x(t)\right)^{\mathrm{T}} = x(t)^{\mathrm{T}} P \dot{x}(t)$$

Definition 2.3 *The origin of the system (2.1) is said to be quadratically stable if there exist symmetric positive definite matrices P, $Q \in \mathbb{R}^{n \times n}$ such that the total time derivative satisfies*

$$\dot{V}(x) = 2x^{\mathrm{T}} P f(x,t) \leq -x^{\mathrm{T}} Q x$$

The inequality above implies $\|x(t)\| < \mathrm{e}^{-\alpha t}$ where $\alpha = \lambda_{min}(P^{-1}Q)$ and hence the origin is asymptotically stable. If $f(x,t) = Ax(t)$ then it is well known that A has stable eigenvalues if and only if, given any symmetric positive definite matrix Q, there exists a unique symmetric positive definite matrix P satisfying the *Lyapunov equation*

$$PA + A^{\mathrm{T}} P = -Q \tag{2.11}$$

Consequently, any stable linear system is quadratically stable. A symmetric positive definite matrix P satisfying (2.11) will be referred to as a *Lyapunov matrix* for the matrix A.

Lyapunov theory may also be used as a means of examining the *robustness* of a given linear system; suppose

$$\dot{x}(t) = Ax(t) + \xi(t,x) \tag{2.12}$$

where the matrix A is stable and $\xi(\cdot)$ is an imprecisely known function which represents uncertainty in the system. Let the pair of positive definite matrices (P, Q) satisfy the Lyapunov equation (2.11) and define

$$\mu = \frac{\lambda_{min}(Q)}{\lambda_{max}(P)} \tag{2.13}$$

and suppose that the uncertain function satisfies

$$||\xi(t, x)|| \leq \tfrac{1}{2}\mu||x(t)|| \tag{2.14}$$

then the system in (2.12) is stable. This can be established by using $V = x^{\mathrm{T}}Px$ as a Lyapunov function: the derivative along the trajectories satisfies

$$\begin{aligned} \dot{V} &= x(t)^{\mathrm{T}}PAx(t) + x(t)^{\mathrm{T}}A^{\mathrm{T}}Px(t) + 2x(t)^{\mathrm{T}}P\xi(t, x) \\ &= -x(t)^{\mathrm{T}}Qx(t) + 2x(t)^{\mathrm{T}}P\xi(t, x) \\ &\leq -x(t)^{\mathrm{T}}Qx(t) + 2||Px(t)||||\xi(t, x)|| \end{aligned} \tag{2.15}$$

where the Cauchy–Schwarz inequality (see Appendix A.2.6) has been used to obtain the last inequality. Now

$$||Px|| = \sqrt{x^{\mathrm{T}}P^2x} \leq \sqrt{\lambda_{max}(P^2)||x||^2} = \lambda_{max}(P)||x|| \tag{2.16}$$

where the Rayleigh principle has been used to obtain the middle inequality. Also directly from the Rayleigh principle

$$-x^{\mathrm{T}}Qx \leq -\lambda_{min}(Q)||x||^2 \tag{2.17}$$

Thus from the inequality in (2.15) and using (2.16) and (2.17) it follows that

$$\begin{aligned} \dot{V}(t) &\leq -\lambda_{min}(Q)||x(t)||^2 + 2\lambda_{max}(P)||x(t)||||\xi(t, x)|| \\ &= -\lambda_{max}(P)||x(t)||\left(\mu||x(t)|| - 2||\xi(t, x)||\right) \end{aligned}$$

and therefore if $\xi(\cdot)$ satisfies (2.14) the Lyapunov derivative is always negative and stability is proved.

In view of the condition (2.14), it is natural to attempt to choose Q in an effort to maximise (2.13). Patel & Toda (1980) show that the maximum is given by

$$\hat{\mu} = \frac{1}{\lambda_{max}(P)} \tag{2.18}$$

when $Q = I$. Furthermore, Patel & Toda (1980) show that

$$\hat{\mu} \leq -2 \max\left[\mathrm{Re}\ \lambda(A)\right] \tag{2.19}$$

with equality if the matrix A is *normal*, i.e. if it has n orthonormal eigenvectors.

When dealing with uncertain systems, it may not be possible to guarantee asymptotic stability. Consider the nonlinear system (2.1) and suppose it is subject to an imprecisely known exogenous signal $\xi(\cdot)$ so that

$$\dot{x}(t) = f(x, t, \xi) \tag{2.20}$$

Let $\mathcal{E} \subset \mathbb{R}^n$ be a bounded set, then the following definition can be made.

Definition 2.4 *The solution $x(\cdot)$ to the uncertain system (2.20) is said to be ultimately bounded with respect to the set $\mathcal{E}$ if*

- *on any finite interval the solution remains bounded, i.e. if $||x(t_0)|| < \delta$ then $||x(t)|| < d(\delta)$ for any $t \in [t_0,\ t_1)$*
- *in finite time the solution $x(t)$ enters the bounded set $\mathcal{E}$ and remains there for all subsequent time.*

The set $\mathcal{E}$ is usually an acceptably small neighbourhood of the origin and the concept is often termed *practical stability.*

As an example, consider the uncertain system

$$\dot{x}(t) = \begin{bmatrix} 0 & 1 \\ -4 & -1 \end{bmatrix} x(t) + \begin{bmatrix} 0 \\ 1 \end{bmatrix} \xi(t,x) \tag{2.21}$$

where the function $\xi(\cdot)$ is unknown but bounded and is assumed to satisfy

$$|\xi(t,x)| \leq 1 \qquad \text{for all } t, x \tag{2.22}$$

For the analysis which follows, A and D will be used to represent the system and uncertainty distribution matrices in equation (2.21). Let P denote the positive definite solution to the Lyapunov equation

$$PA + A^{\mathrm{T}}P = -I \tag{2.23}$$

which will exist because the system matrix A is stable. Consider the function $V(t) = x(t)^{\mathrm{T}}Px(t)$; the derivative along the system trajectories is

$$\begin{aligned} \dot{V} &= -x(PA + A^{\mathrm{T}}P)x + 2x^{\mathrm{T}}PD\xi(t,x) \\ &\leq -||x||^2 + 2|x^{\mathrm{T}}PD||\xi(t,x)| \end{aligned} \tag{2.24}$$

Now since $x^{\mathrm{T}}PD$ is a scalar

$$\begin{aligned} |x^{\mathrm{T}}PD| &= \sqrt{x^{\mathrm{T}}PDD^{\mathrm{T}}Px} \\ &\leq \sqrt{\lambda_{max}(PDD^{\mathrm{T}}P)}||x|| \end{aligned} \tag{2.25}$$

Define $\delta = 2\sqrt{\lambda_{max}(PDD^{\mathrm{T}}P)}$, then from (2.24) and (2.25) it follows that

$$\dot{V} \leq -||x||(||x|| - \delta|\xi|) \leq -||x||(||x|| - \delta) \tag{2.26}$$

Let D_δ represent the disc centred at the origin given by

$$D_\delta = \{x \ : \ ||x|| < \delta\} \tag{2.27}$$

then from equation (2.26) it follows that for $x \notin D_\delta$, $\dot{V} < 0$. Therefore there exists a $t_0 > 0$ such that the states $x(t) \in D_\delta$ for all $t > t_0$. In other words, $x(t)$ is ultimately bounded with respect to D_δ, which acts as the ellipsoid $\mathcal{E}$.

Now consider the special case when $\xi(t,x) = \sin(2t)$; this clearly satisfies (2.22). The phase portrait shown in Figure 2.1 represents a simulation of (2.21) with the initial condition $x_0 = [10 \;\; 0]$. The disc D_δ is shown as the dotted circle where in this case $\delta = 1.2748$. It can be seen that, as predicted by the theory, the states enter the disc and remain there.

Note in this case the solution tends asymptotically to the solution $x_1(t) = -\frac{1}{2}\cos(2t)$ and so ultimately the phase portrait sweeps out an ellipse with major and minor axes 1 and $\frac{1}{2}$ respectively.

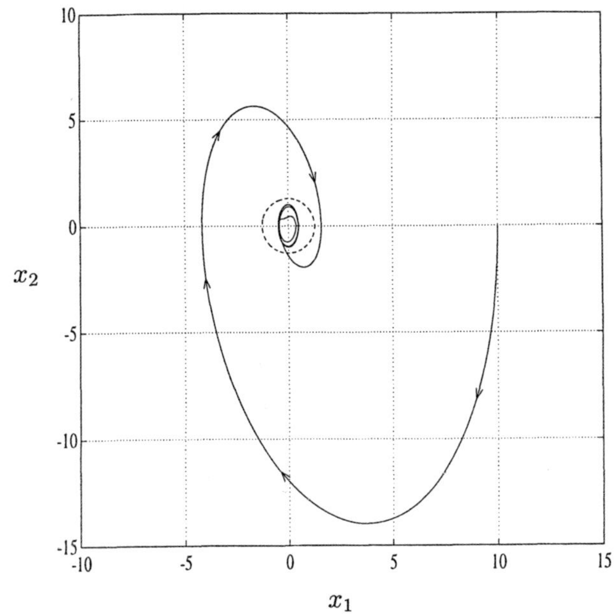

Figure 2.1: Phase portrait together with the ultimate boundedness set

2.3 LINEAR SYSTEMS THEORY

The remainder of this chapter will confine its attention to the special case of linear systems and will introduce some of the important concepts from linear systems theory.

2.3.1 Controllability and Observability

Consider the linear time invariant system

$$\dot{x}(t) = Ax(t) + Bu(t) \qquad (2.28)$$

$$y(t) = Cx(t) \qquad (2.29)$$

where $A \in \mathbb{R}^{n\times n}$, $B \in \mathbb{R}^{n\times m}$ and $C \in \mathbb{R}^{p\times n}$. The variables $u(t)$ and $y(t)$ will be referred to as the *inputs* and *outputs* respectively. The matrices A, B and C will be termed the *system*, *input distribution* and *output distribution* matrices respectively. In these terms the general linear system given in (2.5) is considered to represent a *free motion*, i.e. a motion which is independent of any control action or exogenous force. For convenience the system in (2.28) and (2.29) will be referred to as a system triple (A, B, C).

If $T \in \mathbb{R}^{n \times n}$ is nonsingular then the change of coordinates $x \mapsto Tx$ induces a new system representation with system matrix TAT^{-1}, input distribution matrix TB and output distribution matrix CT^{-1}, i.e. the triple (TAT^{-1}, TB, CT^{-1}).

Definition 2.5 *The system is said to be completely controllable if given any initial condition $x(t_0)$ there exists an input function on the finite interval $[t_0,\ t_1]$ such that $x(t_1) = 0$.*

From this definition the following theorem can be proved.

Theorem 2.1 *Given any pair (A, B) the following conditions are all equivalent:*

- *(A, B) is completely controllable*
- *the controllability matrix $[B \quad AB \quad A^2B \ \ldots \ A^{n-1}B]$ has full rank*
- *the matrix $[sI - A \quad B]$ has full rank for all $s \in \mathbb{C}$*
- *the spectrum of $(A + BF)$ can be assigned arbitrarily by choice of $F \in \mathbb{R}^{m \times n}$.*

The third condition, which is often the most convenient method of establishing controllability, is often referred to as the Popov–Belevitch–Hautus rank test or PBH test for short.

Definition 2.6 *The linear system is said to be completely observable if the output function $y(t)$ over some time interval $[t_0,\ t_1]$ uniquely determines the initial condition $x(t_0)$.*

An important duality exists between the notions of controllability and observability which can be stated as follows.

Theorem 2.2 *The pair (A, C) is completely observable if and only if the pair $(A^{\mathrm{T}}, C^{\mathrm{T}})$ is completely controllable.*

From the theorem above, the results of Theorem 2.1 can be modified to provide a list of equivalent statements for observability.

In addition to these concepts there exist two slightly weaker notions: *stabilisability* and *detectability*.

Theorem 2.3 *Given any pair (A, B) the following conditions are equivalent:*

- *(A, B) is stabilisable*
- *the matrix $[sI - A \quad B]$ has full rank for all $s \in \mathbb{C}_+$*
- *there exists an $F \in \mathbb{R}^{m \times n}$ such that the eigenvalues of $A + BF$ belong to $\mathbb{C}_-$.*

The notion of detectability can be defined as the dual of stabilisability.

If the pair (A, B) in (2.28) is not controllable then there exists a change of coordinates $x \mapsto Tx$ so that in the new coordinate system the new system and input distribution matrices have the form

$$TAT^{-1} = \begin{bmatrix} A_{11} & 0 \\ A_{21} & A_{22} \end{bmatrix} \quad TB = \begin{bmatrix} 0 \\ B_2 \end{bmatrix} \tag{2.30}$$

where the pair (A_{22}, B_2) is completely controllable. Note that because of the special structure of the canonical form in (2.30) it can readily be observed that

$$\lambda(A_{11}) \subset \lambda(A + BF)$$

for $F \in \mathbb{R}^{m \times n}$. Consequently, the pair (A, B) is stabilisable if and only if A_{11} is stable.

By duality, a similar canonical form exists for pairs (A, C) which are not observable.

2.3.2 Invariant Zeros

Consider the linear time invariant system

$$\dot{x}(t) = Ax(t) + Bu(t) \tag{2.31}$$

$$y(t) = Cx(t) \tag{2.32}$$

Assume the initial state is given by $x(0)$. Taking Laplace transforms of the system representation yields

$$\begin{bmatrix} sI - A & -B \\ C & 0 \end{bmatrix} \begin{bmatrix} X(s) \\ U(s) \end{bmatrix} = \begin{bmatrix} x(0) \\ Y(s) \end{bmatrix} \tag{2.33}$$

The polynomial system matrix

$$P(s) = \begin{bmatrix} sI - A & -B \\ C & 0 \end{bmatrix} \tag{2.34}$$

is sometimes referred to as Rosenbrock's system matrix. A necessary and sufficient condition for an input

$$u(t) = u(0)e^{zt} \tag{2.35}$$

to yield rectilinear motion in the state space of the form

$$x(t) = x(0)e^{zt} \tag{2.36}$$

such that the output of the system is identically zero for all time is that z, $x(0)$ and $u(0)$ satisfy

$$P(z) \begin{bmatrix} x(0) \\ u(0) \end{bmatrix} = 0 \tag{2.37}$$

This result defines a set of complex frequencies z which are associated with specific directions $x(0)$ and $u(0)$ in the state and input spaces for which the output of the system is zero. These elements are called *invariant zeros*. It is clear that information regarding the existence of invariant zeros comes from examining the rank of $P(z)$ in equation (2.37). For example, in the case of a square system – i.e. systems with equal numbers of inputs and outputs – in order for equation (2.37) to have a nonzero solution for $x(0)$ and $u(0)$, $\det(P(z))$ must be zero.

2.3.3 State Feedback Control

For the controllable state-space system represented by

$$\dot{x}(t) = Ax(t) + Bu(t) \tag{2.38}$$

a state feedback controller is defined by

$$u(t) = -Kx(t) \tag{2.39}$$

where $K \in \mathbb{R}^{m \times n}$. The state equation of the closed-loop system is given by

$$\dot{x}(t) = (A - BK)x(t) \tag{2.40}$$

As the system is controllable, from Theorem 2.1, the closed-loop poles can be allocated to any desired location by appropriate choice of K.

The Linear quadratic regulator (LQR) is a particular formulation of the state feedback control problem. Given the state-space system (2.38) with given initial condition $x(0)$, the input signal $u(t)$ is sought which regulates the system state to the origin by minimising the cost function

$$J = \frac{1}{2}\int_0^\infty x(t)^{\mathrm{T}} Q x(t) + u(t)^{\mathrm{T}} R u(t)\, dt \tag{2.41}$$

where Q and R are positive definite symmetric matrices which penalise the deviation of the state from the origin and the magnitude of the control signal, respectively. The optimal solution, for any initial state, is given by

$$u(t) = -Kx(t) = -R^{-1}B^{\mathrm{T}}Xx(t) \tag{2.42}$$

where X is the unique positive semidefinite solution ($X \geq 0$), of the algebraic Riccati equation

$$A^{\mathrm{T}}X + XA - XBR^{-1}B^{\mathrm{T}}X + Q = 0 \tag{2.43}$$

2.3.4 Static Output Feedback Control

It has been demonstrated in the previous subsection that it is possible to readily determine a state feedback based control strategy for a controllable linear system. In practice it may not be feasible to measure all the state variables for a given system. If only a subset of state information is available, the feedback control problem must now consider the system

$$\dot{x}(t) = Ax(t) + Bu(t) \tag{2.44}$$

$$y(t) = Cx(t) \tag{2.45}$$

where $y \in \mathbb{R}^p$ denotes an available vector of output measurements. The development of a control law

$$u(t) = -KCx(t) = -Ky(t) \tag{2.46}$$

where $K \in \mathbb{R}^{m \times p}$ is needed. It is important to note that this problem has associated with it further constraints. A sufficient condition to ensure that the eigenvalues of $A - BKC$ may be placed arbitrarily close to desired locations is that the pair (A, B) is controllable, the pair (A, C) is observable and the dimensionality requirement

$$m + p + 1 \geq n \tag{2.47}$$

is satisfied. These requirements are referred to as the *Kimura–Davison* conditions.

2.3.5 Observer-Based Control

It has been demonstrated that control solutions which depend upon the availability of the state vector are straightforward to determine. However, as argued earlier, for practical purposes, only a subset of the state vector may be available for measurement. It has been seen that if the control problem is solved upon the basis of output information only, the solution set is restricted. An alternative method to overcome problems with restricted measurements is to use a state feedback controller in conjunction with a dynamical system whose purpose is to estimate the state vector. Such a dynamical system is called an *observer*. For the observable system

$$\dot{x}(t) = Ax(t) + Bu(t) \tag{2.48}$$

$$y(t) = Cx(t) \tag{2.49}$$

the corresponding observer is defined by

$$\dot{\hat{x}}(t) = (A + LC)\hat{x}(t) + Bu(t) - Ly(t) \tag{2.50}$$

where the design matrix $L \in \mathbb{R}^{n \times p}$ is selected to ensure the eigenvalues of $A + LC$ have negative real parts. Equation (2.50) may be alternatively expressed as

$$\dot{\hat{x}}(t) = A\hat{x}(t) + Bu(t) - L(y(t) - C\hat{x}(t)) \tag{2.51}$$

and hence the observer is seen to be a model of the plant, which is driven by the mismatch between the plant and observer outputs. Define the difference between the actual state $x(t)$ and the estimated state $\hat{x}(t)$ to be

$$e(t) = x(t) - \hat{x}(t) \tag{2.52}$$

Then it is straightforward to establish that the dynamic behaviour of the error signal is given by

$$\dot{e}(t) = (A + LC)e(t) \tag{2.53}$$

As $A + LC$ is a stable matrix it follows that the error vector will converge to zero from any initial condition and thus $\hat{x}(t)$ will converge to $x(t)$. Judicious choice for the poles of $A + LC$ will ensure that the error tends to zero rapidly. It can easily be verified that when the observed state $\hat{x}$ is used to represent the state vector in the closed-loop implementation, then the characteristic equation of the resulting closed-loop system has as its roots the poles allocated via the feedback design alone together with the poles due to the observer design alone. The design of the feedback controller and the observer can thus be completed independently and a *separation principle* is said to hold.

The observer approach used thus far is designed to reconstruct all of the state variables; it is thus called a *full-order observer*. In practice, some of the state variables may be accurately measured and thus need not be estimated. An observer that estimates fewer than n state variables is called a *reduced-order observer.* Assume that $x \in \mathbb{R}^n$ and $y \in \mathbb{R}^p$ where the output vector y can be measured accurately. Since the p output variables are linear combinations of the state variables, only $n-p$ state variables need to be estimated. In this case the reduced-order observer will be an observer of order $n-p$. Without loss of generality, consider the partitioned state equations

$$\begin{bmatrix} \dot{x}_1(t) \\ \dot{y}(t) \end{bmatrix} = \begin{bmatrix} A_{11} & A_{12} \\ A_{21} & A_{22} \end{bmatrix} \begin{bmatrix} x_1(t) \\ y(t) \end{bmatrix} + \begin{bmatrix} B_1 \\ B_2 \end{bmatrix} u(t) \tag{2.54}$$

$$y(t) = \begin{bmatrix} 0 & I \end{bmatrix} \begin{bmatrix} x_1(t) \\ y(t) \end{bmatrix} \tag{2.55}$$

The state equation for the unmeasured portion of the state is thus given by

$$\dot{x}_1(t) = A_{11}x_1(t) + A_{12}y(t) + B_1u(t) \tag{2.56}$$

Rewriting the dynamic equation of the second subsystem as

$$\dot{y}(t) - A_{22}y(t) - B_2u(t) = A_{21}x_1(t) \tag{2.57}$$

an 'output equation' for the subsystem (2.56) is obtained. If in equation (2.57) the signal $\dot{y}(t) - A_{22}y(t) - B_2u(t)$ is regarded as the available output vector and $A_{12}y(t) + B_1u(t)$ represents the external input applied to the subsystem, then equation (2.50) may be used to define an observer for the system (2.56) so that

$$\dot{\hat{x}}_1(t) = (A_{11} + LA_{21})\hat{x}_1(t) + A_{12}y(t) + B_1u(t) - L(\dot{y}(t) - A_{22}y(t) - B_2u(t)) \tag{2.58}$$

The reduced-order observer defined above requires the signal $\dot{y}$ which is clearly undesirable. Define a modified state vector of the reduced-order observer by

$$\hat{\eta}(t) = \hat{x}_1 + Ly \tag{2.59}$$

then equation (2.58) yields

$$\begin{aligned} \dot{\hat{\eta}}(t) &= (A_{11} + LA_{21})\hat{x}_1(t) + (A_{12} + LA_{22})y(t) + (B_1 + LB_2)u(t) \\ &= (A_{11}+LA_{21})\eta(t) + (A_{12}+LA_{22}-(A_{11}+LA_{21})L)y(t)+(B_1+LB_2)u(t) \end{aligned} \tag{2.60}$$

The feedback controller

$$u(t) = -Kx(t) = -K_1x_1(t) - K_2y(t) \tag{2.61}$$

may now be implemented as

$$u(t) = -K_1(\hat{\eta} - Ly) - K_2y(t) \tag{2.62}$$

which depends only upon the state of the reduced-order observer (2.60) and the output of the system.

2.4 NOTES AND REFERENCES

Nonlinear systems and Lyapunov stability theory are described in Khalil (1992) and Slotine & Li (1991). Formal statements and Lyapunov arguments associated with ultimate boundedness are discussed in Ryan & Corless (1984) and Khalil (1992). Further details of quadratic stability and its relationship with other stability theory concepts, including $\mathcal{H}_\infty$ ideas, are given in Corless (1994).

Chapter 3

Sliding Mode Control

3.1 INTRODUCTION

In this chapter the intuitive concepts of Chapter 1 will be fashioned into rigorous theory for general multi-input multi-output systems. Initially the problem of regulating a finite-dimensional linear time invariant system subject to parameter uncertainty or nonlinearities will be considered. A controller will be sought to force the system states to reach, and subsequently remain on, a predefined surface within the state space. The dynamical behaviour of the system when confined to the surface is described as an *ideal sliding motion*. The advantages of obtaining such a motion are twofold: firstly there is a reduction in order; and secondly the sliding motion is insensitive to parameter variations implicit in the input channels. The latter property of invariance towards so-called *matched uncertainty* makes the methodology an attractive one for designing robust controllers for uncertain systems. The design approach comprises two components: the design of a surface in the state space so that the reduced-order sliding motion satisfies the specifications imposed on the designer; and the synthesis of a control law, discontinuous about the sliding surface, such that the trajectories of the closed-loop motion are directed towards the surface. An alternative interpretation of the latter property is that the discontinuous control action renders the sliding surface invariant and (at least locally) attractive. From this description the closed-loop dynamical behaviour obtained from using a variable structure control law comprises two distinct types of motion. The initial phase, occurring whilst the states are being driven towards the surface, and often referred to as the *reaching phase*, is in general affected by any matched disturbances present. Only when the states reach the surface and the sliding motion takes place does the system become insensitive to all matched uncertainty. Ideally the control system should be designed so that the initial (disturbance affected) reaching phase is as short as possible.

The motivation for considering differential equations with discontinuous right-hand sides is that they occur naturally in physical systems; for example, the resistance force due to dry Coulomb friction takes values of opposite sign depending on the direction of motion; for details see Utkin (1992). Also, in the case of certain electric motors and power converters, the control action is naturally discontinuous. In other systems, especially mechanical ones, the introduction of discontinuous control action will not induce an ideal sliding motion. Imperfections in the process such as delays and hysteresis will conspire to induce a high frequency motion known as *chattering*. This is characterised by the states repeatedly crossing rather than remaining on the surface. Such a motion is highly undesirable in practice and will result in unnecessary wear and tear on the actuator components. It is usual in such situations to modify the discontinuous control action so that, rather than forcing the states to lie on the sliding surface, they are forced to remain within an (arbitrarily) small boundary layer about the surface. In the literature this is often referred to as a *pseudo-sliding motion*. The total invariance properties associated with ideal sliding will be lost. However, an arbitrarily close approximation to ideal sliding can usually be obtained.

Initially a nominal linear system will be considered and a methodology will be presented for designing a hyperplane which guarantees that the reduced-order sliding motion is stable. The properties of such a sliding motion will then be examined – in particular the invariance property with regard to matched uncertainty. Sufficient conditions will be established which guarantee the existence of an ideal sliding motion. A variety of different control laws will be presented which induce and maintain sliding despite the presence of matched uncertainty. This leads to a more detailed analysis of a general multivariable problem in which the effects of both matched and unmatched uncertainty are considered. Finally, the notion of a pseudo-sliding motion will be discussed which will be motivated as the motion resulting from approximating the discontinuous control laws, required for ideal sliding, by arbitrarily close continuous ones.

3.2 PROBLEM STATEMENT

Consider the uncertain linear time invariant system with m inputs given by

$$\dot{x}(t) = Ax(t) + Bu(t) + f(t, x, u) \tag{3.1}$$

where $A \in \mathbb{R}^{n\times n}$ and $B \in \mathbb{R}^{n\times m}$ with $1 \leq m < n$. Without loss of generality it can be assumed that the input distribution matrix B has full rank. The function $f : \mathbb{R} \times \mathbb{R}^n \times \mathbb{R}^m \mapsto \mathbb{R}^n$ is assumed to be unknown but bounded by some known functions of the state. Different restrictions can be placed on this function, which represents the parameter uncertainty or nonlinearities present in the system. In general terms, the problem to be addressed is that of synthesising a control law which forces the states back to the origin from some arbitrary initial displacement, i.e. a regulation problem. In later chapters a more general (output) tracking problem will be considered. This will be achieved by using an integral action or a model-following methodology to transform the problem into one of regulation.

Let $s : \mathbb{R}^n \to \mathbb{R}^m$ be a linear function represented as

$$s(x) = Sx \tag{3.2}$$

where $S \in \mathbb{R}^{m\times n}$ is of full rank and let $\mathcal{S}$ be the hyperplane defined by

$$\mathcal{S} = \{x \in \mathbb{R}^n \quad : \quad s(x) = 0\} \tag{3.3}$$

For reasons that will become clear later, this function will be termed the *switching function*.

Definition 3.1 *Suppose there exists a finite time t_s such that the solution to (3.1) represented by $x(t)$ satisfies*

$$s(t) = 0 \qquad \text{for all } t \geq t_s$$

then an ideal sliding motion is said to be taking place for all $t > t_s$.

Remarks

- In this definition the switching function has been thought of as the function of time defined by $s(t) = Sx(t)$.
- It must be stressed that technically there is no reason for $\mathcal{S}$ to be restricted to be a hyperplane – more general (possibly time-varying) surfaces may be considered; see for example DeCarlo *et al.* (1988).

It is now possible to be more specific about the direction of this chapter. The first problem that will be addressed is that of designing a switching function so that the motion of the dynamical system when confined to the hyperplane $\mathcal{S}$ is stable. Secondly, it considers the problem of designing variable structure control laws so that in finite time the states are forced onto and subsequently remain on the hyperplane $\mathcal{S}$.

In order to meet the finite time requirement, it proves necessary to use a control action which is discontinuous about the surface $\mathcal{S}$. Potentially this presents difficulties from a mathematical viewpoint.

3.3 EXISTENCE OF SOLUTION AND EQUIVALENT CONTROL

If the control action in equation (3.1) is discontinuous, the differential equation describing the resulting closed-loop system, written for convenience as

$$\dot{x}(t) = F(t, x) \tag{3.4}$$

is such that the function $F : \mathbb{R} \times \mathbb{R}^n \mapsto \mathbb{R}^n$ is discontinuous with respect to the state vector. From a rigorous mathematical viewpoint the classical theory of differential equations is not applicable since Lipschitz conditions are usually invoked to guarantee the existence of a unique solution, i.e. it is assumed that there exists a scalar constant L such that

$$||F(t, x_1) - F(t, x_2)|| \leq L||x_1 - x_2||$$

Since any function which satisfies Lipschitz conditions is necessarily continuous, an alternative approach must be adopted. In practice, an ideal sliding motion is not attainable – imperfections such as delays, hysteresis and unmodelled dynamics will result in a chattering motion in a neighbourhood of the sliding surface. Such a system will usually fall within the scope of classical differential equation theory. The ideal sliding motion can therefore be thought of as the limiting solution obtained as the imperfections diminish. A formal discussion along these lines appears in Utkin (1992). The solution concept, proposed by Filippov (1964) for differential equations with discontinuous right-hand sides, constructs a solution which is the 'average' of the solutions obtained from approaching the point of discontinuity from different directions. If x_0 is a point of discontinuity on the surface $\mathcal{S}$ and $F_-(t, x_0)$ and $F_+(t, x_0)$ represent the limits of $F(t, x)$ as the point x_0 is approached from opposite sides of the tangent plane to $\mathcal{S}$ at x_0, then the solution is obtained from

$$\dot{x}(t) = (1 - \alpha)F_-(t, x) + \alpha F_+(t, x)$$

where the scalar $0 < \alpha < 1$ is such that the vector

$$F_a \triangleq (1 - \alpha)F_- + \alpha F_+$$

is tangential to the surface $\mathcal{S}$ (Figure 3.1).

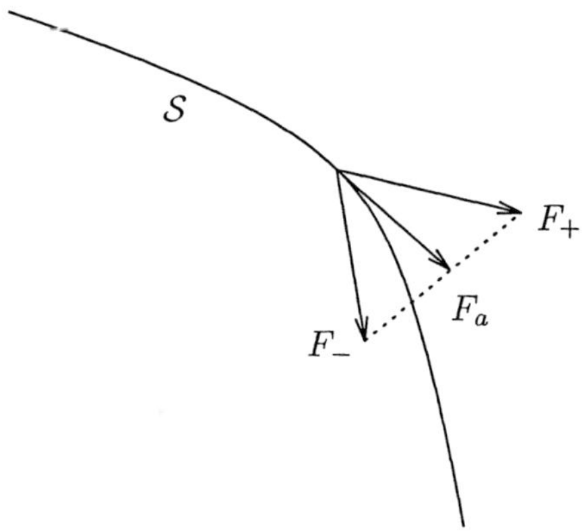

Figure 3.1: A schematic of the Filippov construction

This concept is discussed at greater length in Utkin (1992). A more recent approach using the theory of differential inclusions is taken by Ryan (1988). The most intuitively appealing approach, however, is the method of *equivalent control* proposed by Utkin (1977). Broadly speaking, the equivalent control is the control action necessary to maintain an ideal sliding motion on $\mathcal{S}$.

In describing the method of equivalent control it will initially be assumed that the uncertain function in equation (3.1) is identically zero, i.e.

$$\dot{x}(t) = Ax(t) + Bu(t) \tag{3.5}$$

Suppose at time t_s the systems states lie on the surface $\mathcal{S}$ defined in (3.3) and an ideal sliding motion takes place. Mathematically this can be expressed as $Sx(t) = 0$ and $\dot{s}(t) = S\dot{x}(t) = 0$ for all $t \geq t_s$. Substituting for $\dot{x}(t)$ from (3.5) gives

$$S\dot{x}(t) = SAx(t) + SBu(t) = 0 \qquad \text{for all } t \geq t_s \tag{3.6}$$

Suppose the matrix S is such that the square matrix SB is nonsingular. This does not present any serious difficulty since by assumption B is full rank and S is a design parameter.

Definition 3.2 *The equivalent control associated with the nominal system (3.5), written as u_{eq}, is defined to be the unique solution to the algebraic equation (3.6), namely*

$$u_{eq}(t) = -(SB)^{-1}SAx(t) \tag{3.7}$$

The necessity for SB to be nonsingular translates to a requirement that the solution to (3.6), and therefore the equivalent control, is unique.[1] The ideal sliding motion is then given by substituting the expression for the equivalent control into equation (3.5) which results in a free motion, i.e. a motion independent of the control action and given by

$$\dot{x}(t) = \left(I_n - B(SB)^{-1}S\right) Ax(t) \qquad \text{for all } t \geq t_s \text{ and } Sx(t_s) = 0 \tag{3.8}$$

Define

$$P_s \triangleq (I_n - B(SB)^{-1}S) \tag{3.9}$$

then P_s is a *projection operator* and satisfies two simple yet very important (from the viewpoint of sliding mode control) equations:

$$SP_s = 0 \qquad \text{and} \qquad P_sB = 0 \tag{3.10}$$

These can be readily established by direct evaluation. Using these tools the next section investigates the properties of the sliding motion and provides a mechanism for hyperplane design.

3.4 PROPERTIES OF THE SLIDING MOTION

Initially in this section the linear time invariant system given in (3.5) will be considered. As in the previous section it will be assumed that a controller exists which induces an ideal sliding motion on the surface $\mathcal{S}$ given in equation (3.3). The first result confirms the sliding motion is of reduced order and provides insight into its associated eigenstructure.

Proposition 3.1 *The sliding motion given in equation (3.8) is of reduced order and the eigenvectors associated with any nonzero eigenvalues of the system matrix*

$$A_{eq} = \left(I_n - B(SB)^{-1}S\right) A \tag{3.11}$$

belong to the null space of the matrix S.

Proof
From Definition 3.1 it follows that, whilst sliding, $Sx(t) = 0$ for all $t > t_s$. Since $S \in \mathbb{R}^{m \times n}$ is full rank, it follows that exactly m of the states can be expressed as a linear combination of the remaining $n - m$ states. The sliding motion therefore depends only on the dynamics of these $n - m$ states and thus a reduction in order takes place.

Further, let λ_i be a nonzero eigenvalue of A_{eq} and let v_i be the corresponding right eigenvector. Then from equation (3.10) it follows that

$$SA_{eq} = 0 \;\Rightarrow\; SA_{eq}v_i = 0 \;\Rightarrow\; \lambda_i Sv_i = 0 \;\Rightarrow\; Sv_i = 0$$

and thus the eigenvectors of the nonzero eigenvalues belong to $\mathcal{N}(S)$. ■

[1] The singular case when $\det(SB) = 0$ is considered in Utkin (1992).

From the proposition above, it is clear that the matrix A_{eq} in (3.11) can have at most $n-m$ nonzero eigenvalues. Suppose by choice of S these are distinct and that $V \in \mathbb{R}^{n\times(n-m)}$ is a matrix formed from the (right) eigenvectors associated with the $n-m$ nonzero eigenvalues. The matrix V must then satisfy the two conditions in the following proposition.

Proposition 3.2 *The matrix of right eigenvectors V satisfies*

$$SV = 0 \quad \textit{and} \quad \operatorname{rank}\begin{bmatrix} V & B \end{bmatrix} = n$$

Proof
The first result follows immediately from Proposition 3.1. To prove the second result, consider solving

$$V\eta_1 + B\eta_2 = 0 \tag{3.12}$$

where η_1 and η_2 are vectors of appropriate dimension. Multiplying (3.12) by S and using the previously established fact that $SV = 0$ implies $SB\eta_2 = 0$. This in turn implies $\eta_2 = 0$ since by assumption the square matrix SB is nonsingular. Equation (3.12) thus becomes $V\eta_1 = 0$, which implies $\eta_1 = 0$ since the eigenvectors that comprise V are linearly independent as the associated eigenvalues are distinct. Thus the only solution to (3.12) is $\eta_1 = 0$ and $\eta_2 = 0$, which implies that the columns of the matrix $[\,V \;\; B\,]$ are linearly independent. ∎

From equation (3.7) the equivalent control can be considered to be the linear state feedback component necessary to maintain the reduced-order motion. Because of the straightforward and explicit representation in equation (3.7) it would be tempting to think of using the signal

$$u(t) = Kx(t) \tag{3.13}$$

where $K = -(SB)^{-1}SA$ as a state feedback control law. In general, equation (3.13) would not in itself induce a sliding motion. However, this expression for the 'nominal' equivalent control does often form part of the overall control law. A persuasive argument against using the equivalent control law in isolation is to formally introduce some (structured) uncertainty into equation (3.5). Specifically consider the *uncertain linear system* given by

$$\dot{x}(t) = Ax(t) + Bu(t) + D\xi(t,x) \tag{3.14}$$

where the matrix $D \in \mathbb{R}^{n\times l}$ is known and the function $\xi : \mathbb{R}_+ \times \mathbb{R}^n \to \mathbb{R}^l$ is unknown. This function can be thought of as representing uncertainty in the known matrices A and B, or alternatively as an exogenous disturbance acting on the system. This is a special case of equation (3.1) where

$$f(t,x,u) = D\xi(t,x)$$

Suppose a controller exists which induces a sliding motion on the surface S despite the presence of the uncertainty or disturbance. If at time t_s the states lie on S and subsequently remain there, i.e. $\dot{s}(t) = 0$ for all $t > t_s$, then arguing as above, the control action necessary to maintain such a motion is given by

$$u_{eq}(t) = -(SB)^{-1}\left(SAx(t) + SD\xi(t,x)\right) \qquad \text{for } t \geq t_s \tag{3.15}$$

This equivalent control action is now dependent on the unknown exogenous signal and therefore cannot be realised in practice. It should therefore be stressed that the notion of the equivalent control is best thought of as a tool for the analysis of the motion obtained from constraining the system states to remain on $\mathcal{S}$.

Using the ideas introduced thus far, it is already possible to prove the key property of the reduced-order sliding motion – namely its invariance towards matched uncertainty.

Theorem 3.1 *The ideal sliding motion is totally insensitive to the uncertain function $\xi(t,x)$ in equation (3.14) if $\mathcal{R}(D) \subset \mathcal{R}(B)$.*

Proof
Substituting the equivalent control law (3.15) into the uncertain system representation (3.14), it follows that the sliding motion satisfies

$$\dot{x}(t) = P_s A x(t) + P_s D \xi(x,t) \qquad \text{for all } t \geq t_s \text{ and } Sx(t_s) = 0 \tag{3.16}$$

where P_s is the projection operator defined in (3.9). Now suppose $\mathcal{R}(D) \subset \mathcal{R}(B)$, then there exists a matrix of elementary column operations $R \in \mathbb{R}^{m \times l}$ such that $D = BR$. As a result it follows that $P_s D = P_s(BR) = (P_s B)R = 0$ by the projection property described in equation (3.10). The reduced-order motion as given in (3.16) reduces to

$$\dot{x}(t) = P_s A x(t) \qquad \text{for all } t \geq t_s \text{ and } Sx(t_s) = 0 \tag{3.17}$$

which does not depend on the exogenous signal. ∎

Definition 3.3 *Any uncertainty which can be expressed as in equation (3.14) where $\mathcal{R}(D) \subset \mathcal{R}(B)$ is described as matched uncertainty. Any uncertainty which does not lie within the range space of the input distribution matrix is described as unmatched uncertainty.*

Using this definition, Theorem 3.1 may be paraphrased as: whilst sliding, the reduced-order motion is completely insensitive to matched uncertainty. This invariance property makes VSCS a powerful tool for controlling uncertain systems and is a strong motivation for the continuing research interest in the area.

From the preceding analysis it is clear that, when matched uncertainty alone is present, it is sufficient to consider the nominal linear system representation when designing the switching function. However, whilst it can be seen from equation (3.8) that the sliding motion depends on the choice of sliding surface, the precise effect is not readily apparent. A convenient way to shed light on the problem is to first transform the system into a suitable canonical form. In this form the system is decomposed into two connected subsystems, one acting in $\mathcal{R}(B)$ and the other in $\mathcal{N}(S)$. Since by assumption rank$(B) = m$, there exists an invertible matrix of elementary row operations $T_r \in \mathbb{R}^{n \times n}$ such that

$$T_r B = \begin{bmatrix} 0 \\ B_2 \end{bmatrix} \tag{3.18}$$

where $B_2 \in \mathbb{R}^{m \times m}$ and is nonsingular. This matrix can readily be found using Gaussian elimination, which reduces the B matrix to row echelon form.[2] However,

[2]For details see for example Strang (1988).

there exists an orthogonal matrix T_r satisfying (3.18) which can be computed via 'QR' decomposition (see Appendix A.2.5). From a purely computational point of view it is more convenient and numerically stable to deal with orthogonal matrices, since inverses can be obtained by straightforward transposition. A further useful property is that orthogonal coordinate transformations preserve the Euclidean norm (see Appendix A.2.6). These computational issues will be discussed in more detail in the next chapter. By using the coordinate transformation $x \leftrightarrow T_r x$ if necessary, it can be assumed without loss of generality that the distribution matrix has the form of equation (3.18). If the states are partitioned so that

$$x = \begin{bmatrix} x_1 \\ x_2 \end{bmatrix} \tag{3.19}$$

where $x_1 \in \mathbb{R}^{n-m}$ and $x_2 \in \mathbb{R}^m$ then the nominal linear system (3.5) can be written as

$$\dot{x}_1(t) = A_{11}x_1(t) + A_{12}x_2(t) \tag{3.20}$$
$$\dot{x}_2(t) = A_{21}x_1(t) + A_{22}x_2(t) + B_2u(t) \tag{3.21}$$

This representation is referred to as *regular form.* Equation (3.20) describes the *null space dynamics* and equation (3.21) describes the *range space dynamics.* If the switching function matrix in this coordinate system is partitioned compatibly as

$$S = \begin{bmatrix} S_1 & S_2 \end{bmatrix} \tag{3.22}$$

where $S_1 \in \mathbb{R}^{m\times(n-m)}$ and $S_2 \in \mathbb{R}^{m\times m}$ then

$$\det(SB) = \det(S_2B_2) = \det(S_2)\det(B_2)$$

Therefore a necessary and sufficient condition for the matrix SB to be nonsingular is that $\det(S_2) \neq 0$ since by construction $\det(B_2) \neq 0$. By design assume this to be the case. During ideal sliding, the motion is given by

$$S_1x_1(t) + S_2x_2(t) = 0 \qquad \text{for all } t > t_s \tag{3.23}$$

and therefore formally expressing $x_2(t)$ in terms of $x_1(t)$ results in

$$x_2(t) = -Mx_1(t) \qquad \text{for all } t > t_s \tag{3.24}$$

where $M \triangleq S_2^{-1}S_1$. Substituting for $x_2(t)$ in equation (3.20) gives

$$\dot{x}_1(t) = (A_{11} - A_{12}M)x_1(t) \tag{3.25}$$

The ideal sliding motion is given therefore from the combination of equations (3.25) and (3.24). These equations give a much clearer picture than the previous description of sliding given in (3.8). In particular, two points are worth noting:

- The matrix S_2 has no direct effect on the dynamics of the sliding motion and acts only as a scaling factor for the switching function.

- In the context of designing a regulator, the matrix $A_{11}^s \triangleq A_{11} - A_{12}M$ must have stable eigenvalues.

The hyperplane design problem can therefore be considered to be one of choosing a state feedback matrix M to prescribe the required performance of the reduced-order system (A_{11}, A_{12}). The extent to which this can be achieved depends on the controllability or otherwise of (A_{11}, A_{12}). It turns out that it is possible to relate the controllability of this subsystem to the controllability of the original system.

Proposition 3.3 *The matrix pair* (A_{11}, A_{12}) *is controllable if and only if the pair* (A, B) *is controllable.*

Proof
Because of the special structure of the regular form, and using the fact that $\det(B_2) \neq 0$, it follows that

$$\begin{aligned} \operatorname{rank}\begin{bmatrix} zI-A & B \end{bmatrix} &= \operatorname{rank}\begin{bmatrix} zI-A_{11} & -A_{12} & 0 \\ -A_{21} & zI-A_{22} & B_2 \end{bmatrix} \\ &= \operatorname{rank}\begin{bmatrix} zI-A_{11} & A_{12} \end{bmatrix} + m \qquad \text{for all } z \in \mathbb{C} \end{aligned}$$

This implies

$$\operatorname{rank}\begin{bmatrix} zI-A & B \end{bmatrix} = n \;\Leftrightarrow\; \operatorname{rank}\begin{bmatrix} zI-A_{11} & A_{12} \end{bmatrix} = n-m$$

and from the Popov–Belevitch–Hautus (PBH) rank test[3] it follows that (A, B) is controllable if and only if the pair (A_{11}, A_{12}) is controllable. ∎

Using this proposition, provided the original pair (A, B) is controllable, the pair (A_{11}, A_{12}) is controllable and any robust linear state feedback method can be applied to designing M. Several approaches have been proposed, including quadratic minimisation, direct eigenvalue placement, eigenvalue placement within a region and eigenstructure assignment methods. These will be discussed in some depth in the next chapter.

Evaluating the expression for sliding motion given in (3.8) in the regular-form coordinates, it can easily be verified by direct computation that

$$A_{eq} \triangleq P_s A = \begin{bmatrix} A_{11} & A_{12} \\ -MA_{11} & -MA_{12} \end{bmatrix} \tag{3.26}$$

and furthermore

$$\begin{bmatrix} A_{11} & A_{12} \\ -MA_{11} & -MA_{12} \end{bmatrix} \equiv \begin{bmatrix} I & 0 \\ -M & I \end{bmatrix} \begin{bmatrix} A_{11}^s & A_{12} \\ 0 & 0 \end{bmatrix} \begin{bmatrix} I & 0 \\ -M & I \end{bmatrix}^{-1} \tag{3.27}$$

From the invariance properties of eigenvalues with respect to similarity transformation, it follows from equation (3.27) that $\lambda(A_{eq}) = \lambda(A_{11}^s) \cup \{0\}^m$. This is in agreement with the analysis presented earlier, which concluded that A_{eq} had at least m eigenvalues at the origin.

Thus far the notion of an ideal sliding motion has been introduced whereby at some time t_s, the linear combination of the states $Sx(t) = 0$ for all $t > t_s$. Furthermore, the control action necessary to maintain this motion can be thought of as the state

[3] The Popov–Belevitch–Hautus rank test was one of the equivalent tests for controllability stated in Theorem 2.1 in Section 2.3.1.

feedback law defined in (3.13). From this standpoint, the problem appears closely related to the so-called output zeroing problem for the (in this case fictitious) input/output system defined by (A, B, S). In fact, it will be shown that the poles of the sliding motion are given by the invariant zeros of the triple (A, B, S). At this stage, this is no more than an observation. However, when the problem of designing sliding mode controllers based upon output information only is considered, this interpretation will provide good insight into the class of systems for which the theory is applicable.

Proposition 3.4 *The poles of the sliding motion are given by the invariant zeros of the system triple* (A, B, S).

Proof
By definition, the invariant zeros are given by

$$\{z \in \mathbb{C} \; : \; P(z) \text{ loses normal rank}\}$$

where Rosenbrock's system matrix

$$P(z) = \begin{bmatrix} zI - A & B \\ -S & 0 \end{bmatrix}$$

In this case (A, B, S) has the same number of inputs as outputs and therefore $P(z)$ loses rank if and only if $\det P(z) = 0$. Substituting for the triple (A, B, S) from equations (3.20) and (3.21)

$$P(z) = \begin{bmatrix} zI - A_{11} & -A_{12} & 0 \\ -A_{21} & zI - A_{22} & B_2 \\ -S_1 & -S_2 & 0 \end{bmatrix}$$

Since by construction B_2 is nonsingular

$$\det P(z) = 0 \Leftrightarrow \det \begin{bmatrix} zI - A_{11} & -A_{12} \\ -S_1 & -S_2 \end{bmatrix} = 0$$

Because S_2 is nonsingular it can be shown by direct evaluation that for all values of $z \in \mathbb{C}$

$$\begin{bmatrix} zI - A_{11} & -A_{12} \\ -S_1 & -S_2 \end{bmatrix} \equiv \begin{bmatrix} I & A_{12}S_2^{-1} \\ 0 & I \end{bmatrix} \begin{bmatrix} zI - A_{11}^s & 0 \\ 0 & -S_2 \end{bmatrix} \begin{bmatrix} I & 0 \\ M & I \end{bmatrix}$$

where A_{11}^s and M are the matrices which appear in equations (3.24) and (3.25). Since the inner and outer matrices on the right-hand side of the above equivalence are independent of z, and both have determinant equal to unity, it follows that

$$\begin{aligned} \det \begin{bmatrix} zI - A_{11} & -A_{12} \\ -S_1 & -S_2 \end{bmatrix} &= \det \begin{bmatrix} zI - A_{11}^s & 0 \\ 0 & -S_2 \end{bmatrix} \\ &= \det(zI - A_{11}^s) \det(-S_2) \end{aligned}$$

Therefore since S_2 is nonsingular

$$\det P(z) = 0 \Leftrightarrow \det(zI - A_{11}^s) = 0$$

and thus the invariant zeros of (A, B, S) are the eigenvalues of A_{11}^s, i.e. the poles of the sliding motion.

■

Up to this point the concept of sliding modes and sliding surfaces has been introduced in a rigorous way for multi-input linear systems. The relationship between the choice of sliding surface and the resulting sliding motion has been examined and an explicit relationship has been identified which is convenient from the design viewpoint. The key notion of matched uncertainty has been introduced and the fundamental property that the sliding motion is completely insensitive to matched uncertainty has been proved. But two issues of importance, however, have not yet been addressed:

- No details have been given relating to possible controller configurations which guarantee the existence of a sliding motion.
- Unmatched uncertainties may affect a system when using a variable structure control law, and this has not been considered.

The effect of unmatched uncertainty will be explored in due course but first sufficient conditions for the existence of a sliding mode will be derived. This is vital to the design of any VSCS since the robustness properties outlined above will exist only if a sliding motion can be reached and maintained.

3.5 THE REACHABILITY PROBLEM

In an attempt to motivate this problem, it is easier to first consider a single-input problem. This is the only time in the development of the theory that a general multivariable problem is not considered at the outset. But this approach does in many ways mirror the original development of the theory.

3.5.1 The Single-Input Case

Before describing potential control structures, it is first pertinent to establish sufficient conditions which guarantee that an ideal sliding motion will take place. Intuitively, the sliding surface must be at least locally attractive, i.e. in a certain domain enclosing the surface, the trajectories of $s(t)$ must be directed towards it. This may be expressed mathematically as

$$\lim_{s\to 0^+} \dot{s} < 0 \qquad \text{and} \qquad \lim_{s\to 0^-} \dot{s} > 0 \tag{3.28}$$

in some domain $\Omega \subset \mathbb{R}^n$. In this case the sliding surface would be

$$\mathcal{D} = \mathcal{S} \cap \Omega = \{x \in \Omega \ : \ s(x) = 0\}$$

The expression given in (3.28) is often replaced by the equivalent, but more succinct criterion

$$\dot{s}s < 0 \tag{3.29}$$

The expressions in (3.28) and (3.29) are termed *reachability conditions*. An equation of this type was encountered previously in Chapter 1 for the double-integrator example

$$\ddot{y}(t) = u(t) \tag{3.30}$$

In Section 1.1, using the control law

$$u(t) = -\text{sgn}\,(s(y, \dot{y})) \tag{3.31}$$

where

$$s(y, \dot{y}) = my + \dot{y} \qquad (m > 0) \tag{3.32}$$

it was argued that the switching function satisfies the reachability condition (3.29) in the domain

$$\Omega = \{(y, \dot{y}) \ : \ m|\dot{y}| < 1\}$$

Remark

In general, if the reachability condition (3.29) is satisfied globally, i.e. $\Omega = \mathbb{R}^n$, then sincc

$$\tfrac{1}{2}\frac{d}{dt}s^2 = s\dot{s}$$

it follows that the function

$$V(s) = \tfrac{1}{2}s^2 \tag{3.33}$$

is a Lyapunov function for the state s. This observation will provide the key to extending these ideas to the multi-input situation.

Unfortunately, although equations (3.28) and (3.29) are commonly encountered in the literature, they do not guarantee the existence of an *ideal* sliding motion as given in Definition 3.1. Essentially these conditions only guarantee that the sliding surface is reached asymptotically; for example, consider once more the double-integrator in equation (3.30), this time with the linear state feedback control law

$$u(t) = -(m + \Phi)\,\dot{y}(t) - \Phi m\, y(t) \tag{3.34}$$

where Φ is a positive design scalar. It was shown in (1.30) that using the control law (3.34) results in a stable closed-loop motion with poles at $(-m, -\Phi)$. More importantly, direct computation reveals

$$\dot{s} = -\Phi s \tag{3.35}$$

which implies

$$\dot{s}s = -\Phi s^2 < 0$$

and hence the reachability condition (3.29) is satisfied. From equation (3.35) it follows that

$$s(t) = s(0)e^{-\Phi t}$$

and therefore if $s(0) \neq 0$, i.e. initially the states do not lie on the sliding surface, then $s(t) \neq 0$ for all $t > 0$, although of course $s(t) \to 0$ as $t \to \infty$. This means that the sliding surface is only reached asymptotically rather than in finite time.

A stronger condition, guaranteeing an ideal sliding motion, is the η-reachability condition given by

$$\dot{s}s \leq -\eta|s| \tag{3.36}$$

where η is a small positive constant. By rewriting equation (3.36) as

$$\tfrac{1}{2}\frac{d}{dt}s^2 \leq -\eta|s|$$

and integrating from 0 to t_s, it follows that

$$|s(t_s)| - |s(0)| \leq -\eta t_s$$

and thus the time taken to reach $s = 0$, represented by t_s, satisfies

$$t_s \leq \frac{|s(0)|}{\eta}.$$

The η-reachability condition has also been encountered in Section 1.3, where the control law

$$u(t) = -m\dot{y}(t) - \rho\,\mathrm{sgn}(s(t))$$

was used to control the double-integrator. It was shown there that

$$\dot{s}s = -\rho\,|s|$$

and hence the η-reachability condition is satisfied. Given these sufficient conditions for reachability, the next section considers the synthesis of control laws which induce ideal sliding motions.

3.5.2 Single-Input Control Structures

The control structures that are most commonly employed comprise a linear state feedback component and a nonlinear or discontinuous component. Consider an nth order single-input uncertain system represented by

$$\dot{x}(t) = (A + A_{per}(t))\,x(t) + bu(t) \tag{3.37}$$

where the nominal pair (A, b) is controllable and $A_{per}(t)$ is a time-varying uncertain matrix, which belongs to $\mathcal{R}(B)$ and represents any parameter uncertainty or nonlinearities in the system. In an appropriate coordinate system the pair (A, b) can always be written as

$$A = \begin{bmatrix} 0 & 1 & 0 & \dots & 0 \\ \vdots & 0 & 1 & & \vdots \\ \vdots & & \ddots & \ddots & 0 \\ 0 & & & 0 & 1 \\ -a_1 & -a_2 & \dots & \dots & -a_n \end{bmatrix} \qquad b = \begin{bmatrix} 0 \\ \vdots \\ \vdots \\ 0 \\ 1 \end{bmatrix} \tag{3.38}$$

which is often referred to as the *controllability canonical form*; see for example Brogan (1991). One of the advantages of this canonical form is that the characteristic equation for the matrix A is given by

$$\lambda^n + a_n\lambda^{n-1} + \dots + a_2\lambda + a_1 = 0$$

In this way the stability, or otherwise, of the matrix A can be established by considering the polynomial formed from the elements of its last row. Equation (3.38) of course has a similar structure to the regular form, hence the subsystem

(A_{11}, A_{12}) from (3.20), which plays the pivotal role in the design of the switching surface, is given by

$$A_{11} = \begin{bmatrix} 0 & 1 & 0 & \dots & 0 \\ \vdots & 0 & 1 & & \vdots \\ \vdots & & \ddots & \ddots & 0 \\ \vdots & & & 0 & 1 \\ 0 & \dots & \dots & \dots & 0 \end{bmatrix} \qquad A_{12} = \begin{bmatrix} 0 \\ \vdots \\ \vdots \\ 0 \\ 1 \end{bmatrix}$$

It can easily be verified that the pair (A_{11}, A_{12}), as predicted from Proposition 3.3, is controllable. Let

$$M = \left[\begin{array}{ccc} m_1 & \dots & m_{n-1} \end{array}\right] \tag{3.39}$$

be a row vector M of order $n-1$. Because the pair (A_{11}, A_{12}) is itself in controllability canonical form, the system matrix governing the reduced-order sliding motion $A_{11}^s = A_{11} - A_{12}M$ has the coefficients of M as its last row. Therefore the characteristic equation of A_{11}^s is given by

$$\lambda^{n-1} + m_{n-1}\lambda^{n-2} + \dots + m_2\lambda + m_1 = 0$$

Thus, provided the coefficients $m_1, \dots, m_{n-1}$ constitute a Hurwitz polynomial,[4] it follows that the eigenvalues of A_{11}^s will be stable. Consequently, the linear functional

$$s(x) = \sum_{i=1}^{n-1} m_i x_i + x_n \tag{3.40}$$

where x_i represents the ith component of the state x, is an appropriate switching function to guarantee a stable reduced-order sliding motion. Equation (3.40) is identical to the generic case given in (3.3) if

$$S = \left[\begin{array}{cc} M & 1 \end{array}\right]$$

where M is defined in equation (3.39).

It is important to note, however, that the coordinate transformation required to achieve the controllability canonical form is almost never orthogonal and thus, unless the system occurs naturally in this form, equation (3.38) should not perhaps be regarded as an appropriate regular form for the design of switching surfaces.

Since the uncertainty in the system is assumed to be matched, under the assumption that the pair (A, b) is in the form of (3.38), equation (3.37) can be expressed more conveniently as

$$\dot{x}_i(t) = x_{i+1}(t) \qquad \text{for } i = 1, \dots, n-1 \tag{3.41}$$

$$\dot{x}_n(t) = -\sum_{i=1}^{n} (a_i + \Delta_i(t))\, x_i(t) + u(t) \tag{3.42}$$

It will be assumed that, for all t, the perturbations $\Delta_i(t)$, which constitute the matched uncertainty, satisfy

$$k_i^- < \Delta_i(t) < k_i^+ \qquad \text{for } i = 1, 2 \dots, n \tag{3.43}$$

[4]A polynomial is said to be *Hurwitz* if all its roots have negative real parts.

for some fixed known scalars k_i^+ and k_i^-.

A common control structure is then

$$u(t) = u_l(t) + u_n(t) \tag{3.44}$$

where $u_l(t)$ is a state feedback law – often the nominal equivalent control – and u_n is a discontinuous or switched component. Differentiating equation (3.40) gives

$$\begin{aligned}\dot{s}(t) &= \sum_{i=1}^{n-1} m_i x_{i+1}(t) - \sum_{i=1}^{n} (a_i + \Delta_i(t))\, x_i(t) + u(t) \\ &= -a_1 x_1(t) + \sum_{i=2}^{n} (m_{i-1} - a_i) x_i(t) - \sum_{i=1}^{n} \Delta_i(t) x_i(t) + u(t)\end{aligned}$$

If the linear component of the control law is chosen so that

$$u_l(t) \stackrel{\Delta}{=} a_1 x_1(t) + \sum_{i=2}^{n} (a_i - m_{i-1}) x_i(t) \tag{3.45}$$

then substituting for (3.44) results in the expression

$$\dot{s}(t) = -\sum_{i=1}^{n} \Delta_i(t) x_i(t) + u_n(t) \tag{3.46}$$

Two different choices of $u_n(t)$ will now be explored. Firstly define a scalar function $\rho(t,x)$ such that

$$\rho(t,x) \geq |\sum_{i=1}^{n} \Delta_i(t) x_i(t)| + \eta \qquad \text{for all } t, x \tag{3.47}$$

where η is a small positive design scalar. Under the assumptions on the Δ_i's given in equations (3.43) this is always possible.[5] A potential expression for the discontinuous component is

$$u_n(t) = -\rho(t,x)\,\mathrm{sgn}(s) \tag{3.48}$$

since from (3.46) it follows that

$$s\dot{s} = -s(\sum_{i=1}^{n} \Delta_i(t) x_i(t)) - \rho(t,x)|s| \;\leq\; |s|(|\sum_{i=1}^{n} \Delta_i(t) x_i(t)| - \rho(t,x)) < -\eta|s|$$

and therefore ideal sliding in finite time is guaranteed.

The control structure in (3.48) is often referred to as a *scaled relay structure*.

An alternative choice is to let

$$u_n(t) = \sum_{i=1}^{n} k_i x_i - \eta\,\mathrm{sgn}(s) \tag{3.49}$$

[5] A conservative choice is $\rho(t,x) = \sum_{i=1}^{n} \max\{|k_i^+|, |k_i^-|\}|x_i(t)| + \eta$.

where

$$k_i = \begin{cases} k_i^- & \text{if } sx_i > 0 \\ k_i^+ & \text{if } sx_i < 0 \end{cases} \tag{3.50}$$

and again η is a small positive scalar. Using the nonlinear control component above in the expression (3.46) it can be established that

$$s\dot{s} = \sum_{i=1}^{n} s\,x_i(t)(k_i - \Delta_i(t)) - \eta|s| \leq -\eta|s|$$

since by construction

$$s\,x_i(t)(k_i - \Delta_i(t)) \leq 0 \qquad \text{for } i = 1, 2 \ldots, n$$

Once again an η-reachability condition has been established and therefore ideal sliding in finite time is guaranteed.

Remarks

- It should be noted that many different control structures appear in the literature although the ones discussed above are probably the most common. A slightly different control structure will be given in the actuator case study in Chapter 8 and others can be found in Utkin (1977).
- It should be pointed out that the linear control structure defined in equation (3.45) is in fact the 'nominal equivalent control' since it can easily be shown that

$$a_1 x_1(t) + \sum_{i=2}^{n}(a_i - m_{i-1})x_i(t) = -(Sb)^{-1}SAx(t)$$

The next subsection seeks to broaden the scope of the ideas presented above to a more general multivariable situation.

3.5.3 An Example: The Normalised Pendulum Revisited

Consider once more the pendulum arrangement in Figure 1.9. Now assume that a friction term exists which results in a new normalised pendulum equation:

$$\ddot{y}(t) = -a_1 \sin y(t) - a_2\dot{y}(t) + u(t) \tag{3.51}$$

where, as before, the control input $u(t)$ represents a torque applied at the point of suspension. Define a scalar function

$$\xi(z) \triangleq \begin{cases} \sin(z)/z & z \neq 0 \\ 1 & z = 0 \end{cases} \tag{3.52}$$

It can be verified that $\xi(\cdot)$ is bounded and, more precisely, $\xi : \mathbb{R} \mapsto [-0.2173, 1]$. Choosing y and $\dot{y}$ as states, it follows that (3.51) can be written in the controllability canonical form as

$$\begin{aligned} \dot{x}_1(t) &= x_2(t) \\ \dot{x}_2(t) &= -a_2 x_2(t) - a_1\xi(t)x_1(t) + u(t) \end{aligned} \tag{3.53}$$

In obtaining the description in (3.53) the fact that the term $\sin(y)$ can be written as $y\,\xi(y)$ has been used. This is a different uncertainty formulation than that used in Chapter 1, where essentially the whole term $\sin(y)$ constituted the uncertainty. The advantage of this approach is that in (3.53) the uncertainty bound is smaller. Choosing the switching function

$$s(t) = mx_1(t) + x_2(t)$$

it follows that the reduced-order sliding motion will be given by

$$\dot{x}_1(t) = -mx_1(t)$$

and so the positive design scalar m can be chosen to specify the rate of decay. From equation (3.45) an appropriate linear component of the control action is

$$u_l = (a_2 - m)\,x_2(t) \tag{3.54}$$

If the (nominal) values of a_1 and a_2 are taken to be 0.25 and 0.1 respectively then, using these values and knowledge of the limits on $\xi(\cdot)$, from equations (3.49) and (3.50) a suitable nonlinear component is given by

$$u_n(t) = kx_2(t) - \eta\,\mathrm{sgn}(s) \tag{3.55}$$

where

$$k = \begin{cases} -0.25 & \text{if } s\,x_2 > 0 \\ 0.50 & \text{if } s\,x_2 < 0 \end{cases}$$

and η is a small positive constant.

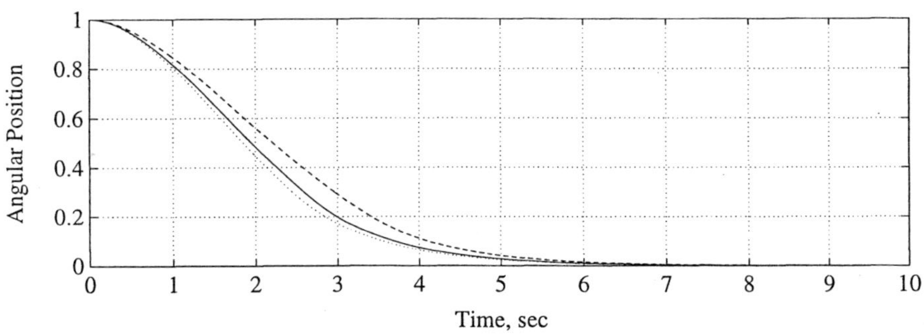

Figure 3.2: Angular position of shaft versus time

The solid line in Figure 3.2 shows a simulation of the closed-loop system obtained from using the control law in equations (3.54) and (3.55) and the nominal values of a_1 and a_2 with initial conditions $y = 1$ and $\dot{y} = 0$. The dashed and dotted lines represent the responses when the parameter a_1 is perturbed to the values 0.2 and 0.3 respectively. The corresponding switching functions are given in Figure 3.3.

3.5.4 The Multivariable Case

The earliest work to tackle the multivariable situation consisted of attempts to reduce such problems to a series of single-input subsystems. Here the natural

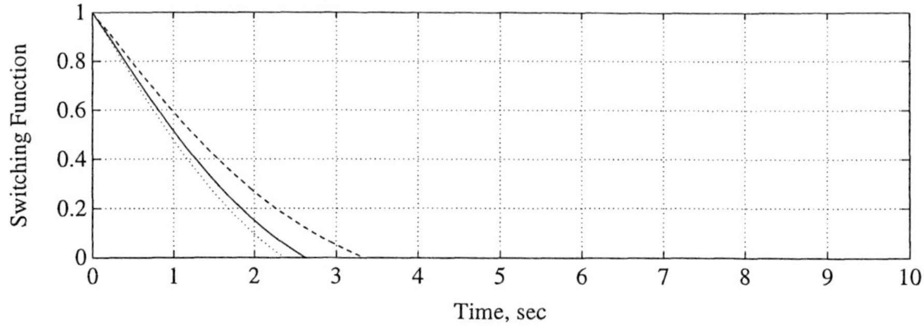

Figure 3.3: Switching function versus time

extension of the single-input problem described earlier will be considered, i.e. the control of the uncertain system

$$\dot{x}(t) = (A + A_{per}(t))\,x(t) + Bu(t) \tag{3.56}$$

where $A \in \mathbb{R}^{n\times n}$ and $B \in \mathbb{R}^{n\times m}$ and $A_{per}(t)$ represents any uncertainties or nonlinearities which, as before, are assumed to be matched so that

$$A_{per}(t) = B\Delta(t) \tag{3.57}$$

where $\Delta(t)$ is some bounded time-varying matrix. In Section 3.2 the hyperplane $\mathcal{S}$ was defined as the null space of a design matrix $S \in \mathbb{R}^{m\times n}$. An alternative and equivalent definition of the same hyperplane is

$$\mathcal{S} = \bigcap_{i=1}^{m} \mathcal{S}_i \tag{3.58}$$

where the hyperplane

$$\mathcal{S}_i = \{x \in \mathbb{R}^n \mid S_i x = 0\}$$

and $S_i \in \mathbb{R}^{1\times n}$ is the ith row of the matrix S. The *hierarchical control method* seeks to use the ith component of the control action to induce a sliding motion on the surface S_i under the assumption that the control components $u_1, \ldots, u_{i-1}$ have already induced sliding motions on $\mathcal{S}_1, \ldots, \mathcal{S}_{i-1}$. The first control component is chosen to induce a sliding motion on $\mathcal{S}_1$ for all values of $u_2, u_3, \ldots, u_m$ by considering the single-input system

$$\dot{x}(t) = \tilde{A}(t)x(t) + \begin{bmatrix} b_2 & \ldots & b_m \end{bmatrix} \begin{bmatrix} u_2(t) \\ \vdots \\ u_m(t) \end{bmatrix} + b_1 u_1(t) \tag{3.59}$$

where $\tilde{A}(t) = A + A_{per}(t)$ and b_i represents the ith column of the input distribution matrix. Once ideal sliding has been established on $\mathcal{S}_1$, the reduced-order motion is given by

$$\dot{x}(t) = \tilde{A}_{eq1}x(t) + \begin{bmatrix} b_3 & \ldots & b_m \end{bmatrix} \begin{bmatrix} u_3(t) \\ \vdots \\ u_m(t) \end{bmatrix} + b_2 u_2(t) \tag{3.60}$$

where $\tilde{A}_{eq1}$ represents the equivalent dynamics obtained from assuming that sliding takes place on $\mathcal{S}_1$. The second control component is then designed to induce sliding on $\mathcal{S}_2$ by considering equation (3.60) as a single-input equation in u_2. Since it is assumed that sliding takes place on $\mathcal{S}_1$ for any control signals $u_2, \ldots, u_m$, it follows that sliding takes place on $\mathcal{S}_1 \cap \mathcal{S}_2$. Using this inductive argument, it follows that sliding will eventually take place on $\bigcap_{i=1}^{m} \mathcal{S}_i = \mathcal{S}$.

Although this approach is conceptually appealing, in practice difficulties often occur. This method is obviously dependent on the order in which sliding takes place on the constituent surfaces $\mathcal{S}_1, \ldots, \mathcal{S}_m$. In general there is no requirement that $\mathcal{S}_1$ be the surface on which the system slides first. This explains why the method is often referred to in the literature as a *fixed-order approach.* In practice the ordering of the constituent sliding surfaces is paramount in avoiding unnecessarily large control signals.

A more recent and perhaps more practical method is the so-called *diagonalisation approach.* It was noted in Section 3.3 that in order for a unique equivalent control to exist, the matrix S must be chosen to ensure that $\det(SB) \neq 0$. In addition it has been demonstrated that the sliding motion is invariant to nonsingular scaling of the switching function. As a result, it can be assumed that, by design,

$$SB = \Lambda \tag{3.61}$$

where $\Lambda \in \mathbb{R}^{m \times m}$ is a diagonal matrix of full rank. As in the single-input case described earlier, the control signal is assumed to have the form

$$u(t) = -(SB)^{-1}SAx(t) + u_n(t) \tag{3.62}$$

where u_n represents a discontinuous control component. Premultiplying (3.56) by the switching function matrix S, and substituting for the control law (3.62) gives

$$\dot{s}(t) = \Lambda u_n(t) + \Lambda\Delta(t)x(t) \tag{3.63}$$

If the ith component of u_n is chosen to depend only on the ith component of the switching function $s_i(t)$, then equation (3.63) can be thought of as the collection of m (almost) independent equations

$$\dot{s}_i(t) = \Lambda_i\left((u_n)_i(t) + \Delta_i(t)x(t)\right) \qquad \text{for } i = 1, 2 \ldots, m \tag{3.64}$$

where $\Delta_i(t)$ is the ith row of $\Delta(t)$ and Λ_i is the ith diagonal element of Λ. Thus, for example, if

$$(u_n)_i = -\rho_i(x,t)\operatorname{sgn}(s_i)$$

where

$$\rho_i(t,x) > |\Delta_i(t)x(t)|$$

then, using the arguments given in the previous section, a sliding mode is induced on each equation given in (3.64). In summary this approach reduces the problem of multivariable controller design to m single-input problems. The advantage of the diagonalisation approach just described over the hierarchical approach is that it is not necessary to define *a priori* the order in which sliding occurs on the constituent subsurfaces. In both approaches, however, controller switching takes place at points in the state space other than those constituting $\mathcal{S}$, whenever the closed-loop trajectory crosses any of the subsurfaces $\mathcal{S}_1, \ldots, \mathcal{S}_m$. In a practical situation this will result in undue wear and tear on the actuators, which may be considered unacceptable. From this standpoint a control structure which changes only around the surface $\mathcal{S}$ is preferable. An approach will be described in the next section which fulfils this requirement and is in a sense a true multivariable extension of the single-input situation.

3.6 THE UNIT VECTOR APPROACH

A more recent and more convenient control structure for multivariable systems is the so-called *unit vector* approach. Consider the uncertain system

$$\dot{x}(t) = Ax(t) + Bu(t) + f_u(t,x) + f_m(t,x,u) \tag{3.65}$$

where the functions $f_u : \mathbb{R} \times \mathbb{R}^n \to \mathcal{R}(B))^\perp$ and $f_m : \mathbb{R} \times \mathbb{R}^n \times \mathbb{R}^m \to \mathcal{R}(B)$ are *unknown* but bounded and satisfy

$$\|f_u(t,x)\| \leq k_1\|x\| + k_2 \tag{3.66}$$

$$\|f_m(t,x,u)\| \leq k_3\|u\| + \alpha(t,x) \tag{3.67}$$

where $k_1, k_2, k_3 \geq 0$ are known constants with

$$k_3 < \sqrt{\lambda_{min}(B^{\mathrm{T}}B)} \tag{3.68}$$

and $\alpha(\cdot)$ is a known function. Here, by definition, the functions $f_u(\cdot)$ and $f_m(\cdot)$ represent the unmatched and matched uncertainty components respectively. Without loss of generality it can be assumed that equation (3.65) is already in regular form and so

$$\dot{x}_1(t) = A_{11}x_1(t) + A_{12}x_2(t) + \bar{f}_u(t,x) \tag{3.69}$$

$$\dot{x}_2(t) = A_{21}x_1(t) + A_{22}x_2(t) + B_2u(t) + \bar{f}_m(t,x,u) \tag{3.70}$$

where $\bar{f}_u$ and $\bar{f}_m$ represent projections of f_u and f_m into the coordinates of the subspaces $\mathcal{N}(S)$ and $\mathcal{R}(B)$ respectively, which is possible since $\mathcal{N}(S) \oplus \mathcal{R}(B) = \mathbb{R}^n$. It can be assumed that the transformation used to obtain the canonical form was an orthogonal one. Hence the Euclidean norm is preserved, which implies

$$\|\bar{f}_u(t,x)\| \leq k_1\|x\| + k_2 \tag{3.71}$$

$$\|\bar{f}_m(t,x,u)\| \leq k_3\|u\| + \alpha(t,x) \tag{3.72}$$

and $B^{\mathrm{T}}B = B_2^{\mathrm{T}}B_2$. Consequently

$$\sqrt{\lambda_{min}(B^{\mathrm{T}}B)} = \underline{\sigma}(B_2) \tag{3.73}$$

and furthermore

$$\underline{\sigma}(B_2) = \left(\bar{\sigma}(B_2^{-1})\right)^{-1} = \|B_2^{-1}\|^{-1} \tag{3.74}$$

where the first equality is a well-known property of singular values and the second follows from the definition of the spectral norm. Combining equations (3.73) and (3.74) it follows that inequality (3.68) can be written as

$$k_3\|B_2^{-1}\| < 1 \tag{3.75}$$

In the notation of Section 3.4 the switching function can be expressed as

$$s(t) = S_2 M x_1(t) + S_2 x_2(t) \tag{3.76}$$

where $M \in \mathbb{R}^{m \times (n-m)}$ has been chosen by some appropriate design procedure to stabilise the pair (A_{11}, A_{12}). The choice of $S_2 \in \mathbb{R}^{m \times m}$ is arbitrary but here it has been chosen so that

$$S_2 B_2 = \Lambda \tag{3.77}$$

where Λ is a nonsingular diagonal matrix which satisfies

$$k_3 \kappa(\Lambda) \|B_2^{-1}\| < 1 \tag{3.78}$$

and $\kappa(\Lambda)$ represents the condition number. This is always possible since, for example, if Λ is a scaled identity matrix then $\kappa(\Lambda) = 1$ and the above equation reduces to (3.75), which is satisfied by assumption.

Define a linear change of coordinates by

$$T_s \triangleq \begin{bmatrix} I & 0 \\ S_1 & S_2 \end{bmatrix} \tag{3.79}$$

which is nonsingular since S_2 is nonsingular. If new coordinates are defined as

$$\begin{bmatrix} x_1 \\ s \end{bmatrix} = T_s \begin{bmatrix} x_1 \\ x_2 \end{bmatrix} \tag{3.80}$$

then

$$\dot{x}_1(t) = \bar{A}_{11} x_1(t) + A_{12} S_2^{-1} s(t) + \bar{f}_u(t, x) \tag{3.81}$$

$$\dot{s}(t) = S_2 \bar{A}_{21} x_1(t) + S_2 \bar{A}_{22} S_2^{-1} s(t) + \Lambda u(t) + S_2 \bar{f}_m(t, x, u) + S_1 \bar{f}_u(t, x) \tag{3.82}$$

where $\bar{A}_{11} = A_{11} - A_{12} M$, $\bar{A}_{21} = M \bar{A}_{11} + A_{21} - A_{22} M$ and $\bar{A}_{22} = M A_{12} + A_{22}$. The proposed control law comprises two components; a linear component to stabilise the nominal linear system; and a discontinuous component. Specifically

$$u(t) = u_l(t) + u_n(t) \tag{3.83}$$

where the linear component is given by

$$u_l(t) = \Lambda^{-1} \left(-S_2 \bar{A}_{21} x_1(t) - (S_2 \bar{A}_{22} S_2^{-1} - \Phi) s(t) \right) \tag{3.84}$$

where $\Phi \in \mathbb{R}^{m \times m}$ is any stable design matrix. The nonlinear component is defined to be

$$u_n(t) = -\rho(t, x) \Lambda^{-1} \frac{P_2 s(t)}{\|P_2 s(t)\|} \qquad \text{for } s(t) \neq 0 \tag{3.85}$$

where $P_2 \in \mathbb{R}^{m \times m}$ is a symmetric positive definite matrix satisfying the Lyapunov equation

$$P_2 \Phi + \Phi^{\mathrm{T}} P_2 = -I \tag{3.86}$$

and the scalar function $\rho(t, x)$, which depends only on the magnitude of the uncertainty, is any function satisfying

$$\rho(t, x) \geq \frac{\|S_2\| \left(\|M\| (k_1 \|x(t)\| + k_2) + k_3 \|u_l(t)\| + \alpha(t, x) \right) + \gamma_2}{(1 - k_3 \kappa(\Lambda) \|B_2^{-1}\|)} \tag{3.87}$$

where $\gamma_2 > 0$ is a design parameter.

Remark

Comparing this approach with the previous section, it should be noted that in the single-input case, the unit vector component

$$\frac{P_2 s}{\|P_2 s\|} = \text{sgn}(s) \qquad (s \neq 0)$$

and hence the control structure (3.85) becomes a scaled relay structure as described in equation (3.48).

The next subsection demonstrates that the control law (3.83) induces a sliding motion on $\mathcal{S}$ in finite time despite the presence of uncertainty.

3.6.1 Existence of an Ideal Sliding Mode

Substituting equations (3.83) to (3.85) into (3.82) and simplifying, produces the system

$$\dot{x}_1(t) = \bar{A}_{11}x_1(t) + A_{12}S_2^{-1}s(t) + \bar{f}_u(t,x) \tag{3.88}$$

$$\dot{s}(t) = \Phi s(t) - \rho(t,x)\frac{P_2 s(t)}{\|P_2 s(t)\|} + S_2\bar{f}_m(t,x,u) + S_1\bar{f}_u(t,x) \tag{3.89}$$

Choosing $V(s) = s^{\mathrm{T}}P_2 s$ and differentiating, it can be verified that

$$\begin{aligned}\dot{V}(s) &\leq -\|s\|^2 - 2\rho(t,x)\|P_2 s\| + 2s^{\mathrm{T}}P_2 S_2\left(\bar{f}_m(t,x,u) + M\bar{f}_u(t,x)\right) \\ &\leq -\|s\|^2 - 2\|P_2 s\|\left(\rho(t,x) - \|S_2\|(\|\bar{f}_m\| + \|M\|\|\bar{f}_u\|)\right)\end{aligned} \tag{3.90}$$

where the fact that $s^{\mathrm{T}}P_2P_2 s = \|P_2 s\|^2$ has been used in establishing the first inequality. Since $S_2 = \Lambda B_2^{-1}$ it follows that $\|S_2\| < \|B_2^{-1}\|\|\Lambda\|$. Using this inequality and rearranging (3.87) it is straightforward to show that

$$\rho(t,x) \geq \|S_2\|\left(\|M\|(k_1\|x\| + k_2) + k_3\left(\|u_l\| + \|\Lambda^{-1}\|\rho(t)\right) + \alpha(t,x)\right) + \gamma_2 \tag{3.91}$$

Using the definition of the discontinuous control component in equation (3.85) it follows that

$$\|u_n(t)\| \leq \rho(t,x)\|\Lambda^{-1}\|$$

and therefore from the definition of $u(t)$ in (3.83) and using the triangle inequality property of norms

$$\|u(t)\| \leq \rho(t,x)\|\Lambda^{-1}\| + \|u_l(t)\| \tag{3.92}$$

Using the inequalities in (3.91) and (3.92) together with the bounds on the uncertainty given in (3.71) and (3.72) it follows that

$$\begin{aligned}\rho(t,x) &\geq \|S_2\|\left(\|M\|(k_1\|x\| + k_2) + k_3\|u(t)\| + \alpha(t,x)\right) + \gamma_2 \\ &\geq \|S_2\|\left(\|M\|\|\bar{f}_u\| + \|\bar{f}_m\|\right) + \gamma_2\end{aligned} \tag{3.93}$$

Combining inequality (3.90) and (3.93) it follows that the derivative of the Lyapunov function satisfies

$$\dot{V}(s) \leq -\|s\|^2 - 2\gamma_2\|P_2 s\| \tag{3.94}$$

This inequality will be used to show that sliding on $\mathcal{S}$ takes place in finite time. From the Rayleigh principle

$$||P_2 s||^2 = (P_2^{1/2}s)^{\mathrm{T}} P_2 (P_2^{1/2}s) \ \geq \ \lambda_{min}(P_2)||P_2^{1/2}s||^2 = \lambda_{min}(P_2)V(s) \qquad (3.95)$$

which together with (3.94) implies that

$$\dot{V} \leq -2\gamma_2\sqrt{\lambda_{min}(P_2)}\sqrt{V} \qquad (3.96)$$

Integrating equation (3.96) implies that the time taken to reach the sliding surface $\mathcal{S}$ denoted by t_s satisfies

$$t_s \leq \gamma_2^{-1}\sqrt{V(s_0)/\lambda_{min}(P_2)} \qquad (3.97)$$

where s_0 represents the initial value of $s(t)$ at $t = 0$. This is a natural multivariable analogue of the single-input case: a Lyapunov function has been found which depends on the switching function states and which satisfies (3.96), in a sense the natural extension of the η-reachability condition. The next subsection investigates the dynamical behaviour when confined to the hyperplane $\mathcal{S}$.

3.6.2 Description of the Sliding Motion

The equation representing the motion when confined to the sliding surface is obtained by substituting $s = 0$ into equation (3.88), giving

$$\dot{x}_1(t) = \bar{A}_{11}x_1(t) + \bar{f}_u(t, x) \qquad (3.98)$$

Let the matrix $P_1 \in \mathbb{R}^{(n-m)\times(n-m)}$ be the unique symmetric positive definite solution to the Lyapunov equation

$$P_1\bar{A}_{11} + \bar{A}_{11}^{\mathrm{T}}P_1 = -Q_1 \qquad (3.99)$$

where $Q_1 \in \mathbb{R}^{(n-m)\times(n-m)}$ is a symmetric positive definite matrix. This situation is identical to the one considered in Section 2.2.2. Let the function $V(x_1) = x_1^{\mathrm{T}}P_1x_1$ be a candidate quadratic Lyapunov function for (3.98). Taking the total time derivative along the system trajectories gives

$$\begin{aligned}
\dot{V}(x_1) &= -x_1^{\mathrm{T}}Q_1x_1 + 2x_1^{\mathrm{T}}P_1\bar{f}_u(t,x) \\
&\leq -x_1^{\mathrm{T}}Q_1x_1 + 2||P_1x_1||||\bar{f}_u|| \\
&\leq -\lambda_{min}(Q_1)||x_1||^2 + 2\lambda_{max}(P_1)||x_1||||\bar{f}_u|| \\
&= -||x_1||\lambda_{max}(P_1)\left(\mu||x_1|| - 2||\bar{f}_u||\right)
\end{aligned}$$

where $\mu \triangleq \lambda_{min}(Q_1)/\lambda_{max}(P_1)$. Therefore, if

$$||\bar{f}_u(t,x)|| < \tfrac{1}{2}\mu||x_1|| \qquad (3.100)$$

then $\dot{V}(x_1) < 0$ and the system (3.98) is quadratically stable. As discussed in Section 2.2.2 the optimal choice of Q_1 to maximise the value of the scalar μ is to let $Q_1 = I$, giving

$$\hat{\mu} = 1/\lambda_{max}(P_1) \leq -2\max\left[\mathrm{Re}\ \lambda(\bar{A}_{11})\right]$$

The expression (3.100) is clearly not satisfied unless $k_2 = 0$. In such a situation the best that can be guaranteed is that the states $x_1(t)$ are ultimately bounded. As discussed in Section 2.2.2, ultimate boundedness ensures that, in finite time, the solution enters a set $\mathcal{E}$ (usually an acceptably small neighbourhood of the origin), and remains there for all subsequent time.

During the sliding motion $x_2 = -M_1 x_1$ and therefore $||x|| \leq (\sqrt{1 + ||M||^2})||x_1||$. Consequently, the bound on the unmatched uncertainty (3.71) can be written as

$$||\bar{f}_u(t,x)|| < \bar{k}_1||x_1|| + k_2 \tag{3.101}$$

where $\bar{k}_1 = k_1\sqrt{1 + ||M||^2}$. Then, arguing as before and assuming $2\bar{k}_1 < \hat{\mu}$, it follows that

$$\begin{aligned} \dot{V}(x_1) &\leq -||x_1||\lambda_{max}(P_1)\left(\hat{\mu}||x_1|| - 2||\bar{f}_u||\right) \\ &\leq -||x_1||\lambda_{max}(P_1)\left(\hat{\mu}||x_1|| - 2\bar{k}_1||x_1|| - 2k_2\right) \end{aligned}$$

and consequently $\dot{V}(x_1) < 0$ if $x_1 \notin \mathcal{E}_1$ where

$$\mathcal{E}_1 = \{x_1 \in \mathrm{I\!R}^{(n-m)} \;:\; ||x_1|| < 2k_2/(\hat{\mu} - 2\bar{k}_1) + \epsilon\} \tag{3.102}$$

for some small scalar $\epsilon > 0$. Therefore the states $x_1(\cdot)$ are ultimately bounded with respect to the ellipsoid $\mathcal{E}_1$.

3.6.3 Practical Considerations

Whilst the particular coordinates resulting from the linear transformation described in (3.79) facilitate the analysis of the closed-loop behaviour and sliding motion robustness properties, it is instructive to express the linear component of the control law in terms of the regular-form coordinates. Recall from (3.84) that

$$u_l(t) = \Lambda^{-1}\left(-S_2\bar{A}_{21}x_1(t) - (S_2\bar{A}_{22}S_2^{-1} - \Phi)s(t)\right)$$

Substituting for $s(t)$ and the matrices $\bar{A}_{21}$ and $\bar{A}_{22}$, some straightforward algebra reveals

$$u_l(t) = -\Lambda^{-1}\left[\; MA_{11} + A_{21} - \Phi S_1 \;\;:\;\; MA_{12} + A_{22} - \Phi S_2 \;\right]x(t) \tag{3.103}$$

Further, substituting for Λ from equation (3.77), the control law (3.103) can be written as

$$u_l(t) = -(SB)^{-1}(SA - \Phi S)\,x(t) = -(SB)^{-1}SAx(t) + (SB)^{-1}\Phi s(t) \tag{3.104}$$

from which it is clear that the 'nominal equivalent control' appears in the formulation. Viewed from this perspective, the approach which has been described in this section is not dissimilar to the single-input approaches of Section 3.5.2. The result in equation (3.104) is not that surprising: in the absence of uncertainty, as shown in equation (3.89), the linear control component ensures that

$$\dot{s}(t) = \Phi s(t) \tag{3.105}$$

The inclusion of the nominal equivalent control guarantees that $\dot{s}(t) = 0$ and hence the superposition of the term $(SB)^{-1}\Phi s(t)$ gives (3.105).

An example will now be given to demonstrate the efficacy of the control scheme that has been described in this section.

3.6.4 Example: Control of a DC Motor

Consider the problem of controlling the angular position of the shaft in a DC motor. The differential equation governing the mechanical part of the system, modelled as a disc rotating about the shaft, is given by

$$J\ddot{\theta}(t) + b\dot{\theta}(t) = T(t) \tag{3.106}$$

where θ represents angular displacement, J represents the moment of inertia of the disc and b is a friction coefficient which opposes the direction of motion. The torque T generated from the current in the electrical windings (armature) and the field generated by the magnets in the housing is given by

$$T(t) = K_t i_a(t) \tag{3.107}$$

where $i_a(t)$ is the armature current. This in turn is assumed to satisfy Kirchhoff's voltage law

$$L\frac{d}{dt}i_a(t) + Ri_a(t) = v_a(t) - K_e\dot{\theta}(t) \tag{3.108}$$

where R and L are the armature resistance and inductance respectively and the last term represents the electromotive force. The applied armature voltage v_a is taken to be the control variable.

For illustration assume that the inductance L is not known precisely; specifically suppose that the error ratio

$$\xi_L \triangleq \frac{L_0 - L}{L} \tag{3.109}$$

satisfies $|\xi_L| < 0.1$, where L_0 is some nominal inductance value (or in other words the value of L is known to an accuracy of 90%). Similarly, suppose the inertia coefficient (or the motor gain) is not known perfectly, so that

$$\frac{K_t}{J} = \frac{K_t}{J_0} + \xi_J \tag{3.110}$$

where J_0 is some nominal value and $|\xi_J| < 0.5$. Finally, assume that the friction effects are negligible, i.e. $b = 0$. Using the variables $\theta, \dot{\theta}$ and i_a as states, the nominal system from equations (3.106) to (3.108) can be written in state-space form as

$$\dot{x}(t) = Ax(t) + Bu(t) \tag{3.111}$$

where

$$A = \begin{bmatrix} 0 & 1 & 0 \\ 0 & 0 & \frac{K_t}{J_0} \\ 0 & -\frac{K_e}{L_0} & -\frac{R}{L_0} \end{bmatrix} \quad \text{and} \quad B = \begin{bmatrix} 0 \\ 0 \\ \frac{1}{L_0} \end{bmatrix} \tag{3.112}$$

It can be seen that the chosen coordinates constitute a regular form and therefore, in the notation of Section 3.6, the matrices

$$A_{11} = \begin{bmatrix} 0 & 1 \\ 0 & 0 \end{bmatrix} \qquad A_{12} = \begin{bmatrix} 0 \\ \frac{K_t}{J_0} \end{bmatrix} \tag{3.113}$$

and $B_2 = 1/L_0$. Choosing the matrix which defines the switching function as

$$M = \begin{bmatrix} m_1 & m_2 \end{bmatrix} \triangleq \begin{bmatrix} \frac{J_0}{K_t}\omega_n^2 & 2\frac{J_0}{K_t}\zeta\omega_n \end{bmatrix} \tag{3.114}$$

the characteristic equation of $A_{11} - A_{12}M$ is the second-order quadratic equation

$$\lambda^2 + 2\zeta\omega_n\lambda + \omega_n^2 = 0 \tag{3.115}$$

Therefore the parameters ζ and ω_n represent the damping ratio and natural frequency respectively. The sliding motion governed by (3.115) can thus be assigned an appropriate dynamics by choice of ζ and ω_n. From the choice of M given in (3.114), the switching function, obtained from selecting $S_2 = 1$, is given by

$$s(t) = \begin{bmatrix} \frac{J_0}{K_t}\omega_n^2 & 2\frac{J_0}{K_t}\zeta\omega_n & 1 \end{bmatrix} x(t)$$

From equation (3.104) the linear component of the control law is

$$u_l(t) = -L_0 S A x(t) + L_0 \Phi s(t)$$

where, in this single-input situation, Φ represents a negative scalar which determines the linear rate of decay onto the switching surface. Next consider the range space dynamics; it follows from (3.111) and (3.112) that

$$\dot{x}_3(t) = -\frac{K_e}{L_0}x_2(t) - \frac{R}{L_0}x_3(t) + \frac{1}{L_0}u(t) + f_m(t,x,u) \tag{3.116}$$

where x_1, x_2 and x_3 are the components of the state-space vector,[6] and the matched uncertainty component

$$f_m(t,x,u) = -\frac{1}{L_0}\xi_L\left(K_e x_2(t) + R x_3(t) - u(t)\right) \tag{3.117}$$

Comparing the specific uncertainty in (3.117) with the general configuration in (3.67), and using the fact that from (3.109) the parameter $|\xi_L| < 0.1$, it follows that the gain associated with the input uncertainty is

$$k_3 = \frac{1}{10L_0}$$

and the bounding function

$$\alpha(t,x) = \frac{1}{10L_0}\left(K_e|x_2(t)| + R|x_3(t)|\right)$$

In this case $\sqrt{\lambda_{min}(B^{\mathrm{T}}B)} = 1/L_0$ and so k_3 satisfies the requirement of (3.68).

Next consider the reduced-order motion. It can easily be confirmed that the sliding motion is given by

$$\dot{x}_1(t) = x_2(t) \tag{3.118}$$
$$\dot{x}_2(t) = -2\zeta\omega_n x_2(t) - \omega_n^2 x_1(t) - \xi_J\left(m_1 x_1(t) + m_2 x_2(t)\right) \tag{3.119}$$

[6] This should not be confused with the notation of Section 3.6 where x_1 and x_2 represent partitions of the state-space vector not its components.

Hence in the notation of equation (3.66)

$$f_u^{\mathrm{T}}(t,x) = \left[\ 0 \quad : \quad -\xi_J(m_1 x_1(t) - m_2 x_2(t))\ \right]$$

which implies

$$\|f_u(t,x)\| \leq \tfrac{1}{2}\|M\|\,\|x(t)\| \tag{3.120}$$

since $|\xi_J| < 0.5$. Consequently, comparing this with the general description of the uncertainty class in equation (3.66), appropriate values for the coefficients are

$$k_1 = \tfrac{1}{2}\|M\| \qquad \text{and} \qquad k_2 = 0$$

Because of the special structure of the uncertainty in this particular case, provided ξ_J is such that K_t/J and K_t/J_0 have the same sign, the reduced-order sliding motion will be stable. This can easily be established from the Routh criteria; for details see for example Brogan (1991).

An appropriate scalar function to premultiply the unit vector component can now be described. Since $\Lambda = 1/L_0$ and $\kappa(\Lambda) = 1$ it follows that

$$1 - k_3\kappa(\Lambda)\|B_2^{-1}\| = 0.9$$

Hence from (3.87) the scaling function

$$\rho(t,x) = \frac{1}{9L_0}\left(|u_l(t)| + k_e|x_2(t)| + R|x_3(t)| + 5L_0\|M\|^2\|x(t)\| + 10L_0\gamma_2\right)$$

Suppose $R = 1.2$, $L_0 = 0.05$, $K_e = 0.6$, $K_t = 0.6$ and $J_0 = 0.135$. From these values, $K_t/J_0 = 4.444$ and hence, since $|\xi_J| < 0.5$, the sign of K_t/J will also be positive from (3.110). The sliding mode is thus guaranteed to be stable.

The following mfile fragment describes the MATLAB commands necessary to realise the control law presented above.

mfile: generate a unit vector controller for a DC motor

```
% An mfile to setup the nominal parameters and design a unit-vector
% controller for a dc motor

% sets up the linear model for control design
R=1.2;
LO=0.05;
Ke=0.6;
Kt=0.6;
J=0.1352;
b=0;
A=[0 1 0; 0 -b/J Kt/J; 0 -Ke/LO -R/LO];
B=[zeros(2,1); 1/LO];
C=eye(3);

% identifies the matrix sub-blocks of the regular form
A11=A(1:2,1:2);
A12=A(1:2,3);
```

```
A21=A(3,1:2);
A22=A(3,3);

% values of the natural frequency and damping ratio
wn=2;
z=0.95;

% Computes the switching function
M=[wn*wn 2*z*wn ]/A12(2,1);
S=[M 1]

% Computes the linear components of the control law
Caphi=-20;
Leq= -inv(S*B)*S*A
Lrc=Leq+inv(S*B)*Caphi*S

L=0.046;
Aper=[0 1 0; 0 -b/J Kt/J; 0 -Ke/L -R/L];
Bper=[zeros(2,1); 1/L];
Cper=eye(3);

X=[1; 0; 0];
gamma2=0.01;
```

In the simulation of Figure 3.4 the sliding motion has been assigned the properties $\zeta = 0.95$ and $w_n = 2.5$. The design parameter Φ, governing the rate of attainment of the sliding motion, has been chosen to be -20. This reflects an attempt to ensure that the control action lies in the region of $[-5,\ 1]$.

The solid line represents a simulation of the closed-loop system with initial conditions $\theta = 1, \dot{\theta} = 0$ and $i_a = 0$.

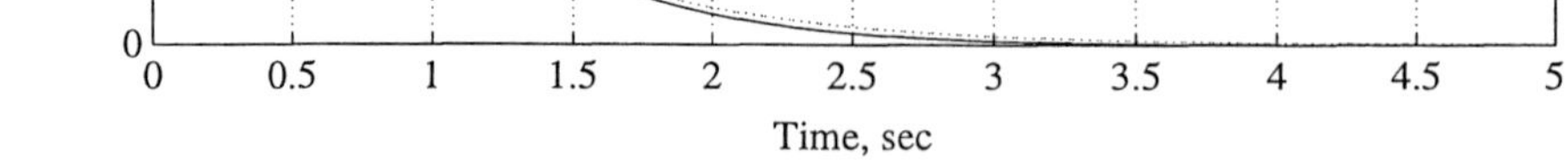

Figure 3.4: Angular position of shaft versus time

The dashed line represents the closed-loop response when the inductance is given as $L = 0.046$ and the motor gain/inertia is perturbed to $K_t/J = 6.0$. Both of these perturbations satisfy the bounds on ξ_J and ξ_L. Because the system just considered is single-input and of low order it is amenable to the sort of analysis described above. For more general systems and in particular for multi-input systems, the hyperplane design problem possesses more degrees of freedom which can be profitably exploited. A general discussion exploring these possibilities is described in Chapter 4.

3.6.5 Concluding Remarks

The guaranteed attainment of an ideal sliding mode in finite time requires that the control action is discontinuous across $\mathcal{S}$. As noted earlier, in the control of certain electric motors the control action is naturally discontinuous. However, for mechanical systems the implementation of such a control law would produce a chattering motion in a boundary of the surface $\mathcal{S}$ rather than an ideal sliding motion. Such high frequency switching might excite unmodelled dynamics and impose undue wear and tear on the actuators, so the control law would not be considered acceptable. Several modifications have been proposed to overcome this difficulty and they are described in the next section.

3.7 CONTINUOUS APPROXIMATIONS

In this section a more general definition of a sliding motion (as opposed to *ideal* sliding) will be given. In the literature this is often referred to as *pseudo-sliding*. It will obviate the need for the control law to be discontinuous and as a result presents a more practical solution from the viewpoint of implementation. The notion of ideal sliding will thus represent in a sense the limiting case or the (idealised) best performance which could be obtained.

In this section the essential difference is that, rather than requiring the states to be forced to remain on the surface $\mathcal{S}$, the states will only be required to remain arbitrarily close to the surface. It should be noted that the total invariance properties with respect to matched uncertainty will be lost, although very close approximation to the ideal sliding case can usually be achieved. In terms of designing a realisable control law for practical application, a trade-off must be made between achievable performance and the smoothing of the discontinuous control action.

Definition 3.4 *The domain $\mathcal{D}$ on the surface $\mathcal{S}$ is the domain of a sliding mode if for any given $\epsilon > 0$ there exists a $\delta > 0$ such that any motion starting within a δ-neighbourhood of $\mathcal{D}$ can leave an ϵ-neighbourhood of $\mathcal{D}$ only through an ϵ-neighbourhood of the boundary of $\mathcal{D}$.*

This may be represented pictorially as shown in Figure 3.5

Again this definition is closely associated with the notion of stability of the switching function, thought of as a subset of the states. As for the ideal sliding case, Lyapunov-like stability notions can be used to formulate sufficient conditions.

Theorem 3.2 *For the domain $\mathcal{D} \subset \mathcal{S}$ to be the domain of a sliding mode it is sufficient that in some region $\Omega \subset \mathbb{R}^n$ where $\mathcal{D} \subset \Omega$ there exists a continuously differentiable scalar function $V : \mathbb{R}_+ \times \Omega \times \mathbb{R}^m \to \mathbb{R}$ satisfying*

(*1*) *$V(t,x,s)$ is positive definite with respect to s, i.e. $V(t,x,s) > 0$ if $s \neq 0$ for all $x \in \Omega$, and on the spheres $\|s\| = r$ for all $x \in \Omega$*

- $\inf_{\|s\|=r} V(t,x,s) = h_r$ *and* $h_r > 0$ *for* $r \neq 0$
- $\sup_{\|s\|=r} V(t,x,s) = H_r > 0$

where h_r and H_r depend on r.

(*2*) *The total time derivative of $V(t,x,s)$ has a negative supremum for every $x \in \Omega$ except at points on the switching surface where the control action may be undefined and hence the derivative of $V(t,x,s)$ does not exist.*

Proof See Utkin (1977). ■

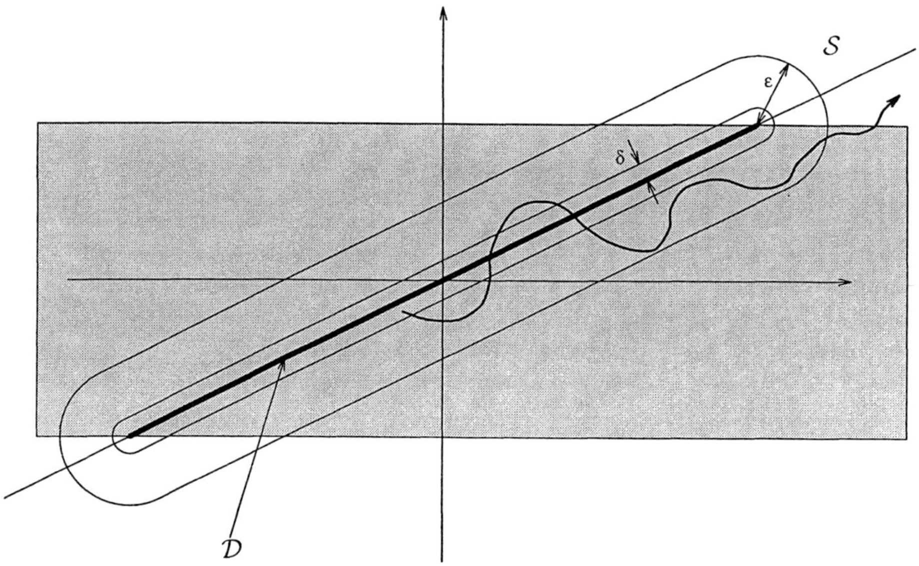

Figure 3.5: Definition of a sliding motion in pictorial terms

The simplest Lyapunov function is the quadratic form

$$V(s) = \sum_{i=1}^{m} s_i^2(x) \tag{3.121}$$

This is clearly positive definite with respect to $\mathcal{S}$ and satisfies the supremum and infimum conditions. The total time derivative is given by

$$\dot{V} = 2\sum_{i=1}^{m} s_i \dot{s}_i$$

and therefore a sufficient condition for the derivative to be strictly decreasing is that

$$s_i \dot{s}_i < 0 \qquad \text{for } i = 1, 2, \ldots, m \tag{3.122}$$

This condition is the one most often cited in the literature when demonstrating the existence of a sliding mode and it was encountered earlier in Section 3.5. It should be noted that the state feedback controller given in (3.34), namely

$$u(t) = -(m + \Phi)\,\dot{y} - \Phi m\, y(t)$$

satisfies the notion of sliding given above. In Section 3.5 it was shown that using this feedback law to control the double-integrator (3.30) resulted in the switching function, thought of as a function of time, satisfying

$$|s(t)| = |s(0)|e^{-\Phi t}$$

Sliding in the sense of Definition 3.4 now occurs because, given any $\delta > 0$, there exists a time t_s (given explicitly in this case by $[\log(|s(0)|/\delta)]/\Phi$ such that for all $t > t_s$ the switching function satisfies $|s(t)| < \delta$.

The most common approach to control law design is to 'soften' the discontinuity in the control laws. This can easily be accomplished in the case of relays and unit vector strategies. Probably the earliest and most intuitive approach is to utilise a *boundary layer* approach, whereby the discontinuous component is replaced by the continuous nonlinear approximation

$$u_n^\delta(t) = \begin{cases} -\rho(t,x)\Lambda^{-1}\frac{P_2 s(t)}{\|P_2 s(t)\|} & \text{if } \|P_2 s\| \geq \delta \\ -\delta^{-1}\rho(t,x)\Lambda^{-1}P_2 s(t) & \text{otherwise} \end{cases} \tag{3.123}$$

for some small positive scalar δ.

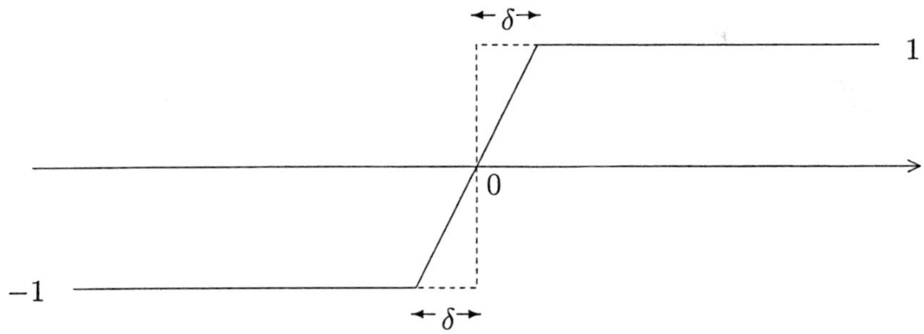

Figure 3.6: An approximation of the signum function

The size of δ defines the size of the boundary layer in which a high gain linear feedback control law operates. In the single-input case this can be visualised with the unit vector component (shown as a dotted line) being replaced by the function (shown as a solid line) in Figure 3.6. An ideal sliding motion will no longer take place and the system will *not* be totally invariant to matched uncertainty. However, the switching function state s will be uniformly ultimately bounded with respect to the ellipsoid

$$\mathcal{E}_b = \{s \in \mathbb{R}^m \;\; : \;\; s^{\mathrm{T}} P_2^2 s < \delta^2\} \tag{3.124}$$

Although no ideal sliding takes place, the states $x_1(\cdot)$ remain ultimately bounded with respect to a neighbourhood of the origin (albeit a larger one than before).

A variation on this approach is the *power law interpolation* structure

$$u_n^\delta(t) = \begin{cases} -\rho(t,x)\Lambda^{-1}\|P_2 s(t)\|^{-1} P_2 s(t) & \|P_2 s(t)\| > \delta \\ -\rho(t,x)\delta^{q-1}\Lambda^{-1}\|P_2 s(t)\|^{-q} P_2 s(t) & 0 < \|P_2 s(t)\| \leq \delta \\ 0 & P_2 s(t) = 0 \end{cases} \tag{3.125}$$

where the design scalar $q \in [0,\ 1)$ and as before δ is a small positive scalar. For the single-input case, the smoothing of the discontinuous relay function in a neighbourhood of the origin is shown in Figure 3.7 for different values of q.

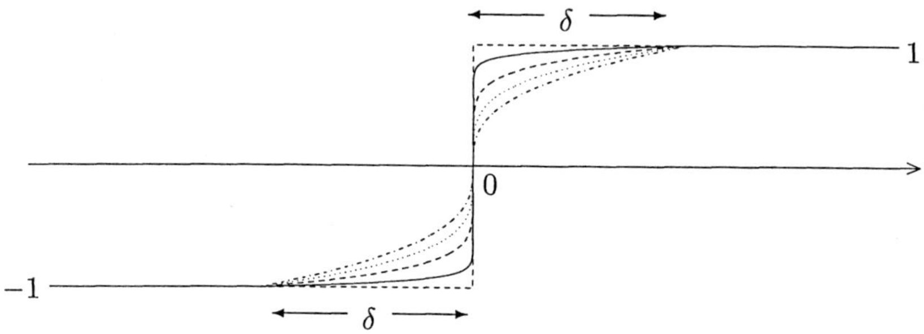

Figure 3.7: A power law approximation of the signum function

Again the dotted line represents the signum function. As $q \to 0$ the slopes of the functions in the neighbourhood of the origin become steeper and the approximations become more precise. (The solid line represents $q = 0.05$ and the chain dotted line $q = 0.5$.)

An alternative differentiable approximation is

$$u_n^\delta(t) = -\rho(t,x)\Lambda^{-1}\frac{P_2 s(t)}{\|P_2 s(t)\| + \delta} \tag{3.126}$$

where again δ is a small positive constant.

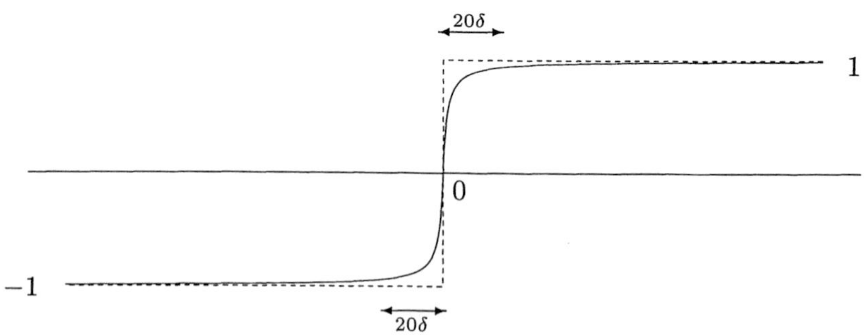

Figure 3.8: A differentiable approximation of the signum function

As shown in Figure 3.8, the scalar δ does not define the boundary layer (at $\|s\| = \delta$

the nonlinear control action is half of its maximum value), but ultimate boundedness results can be demonstrated just as for the other parallel boundary techniques. The motion resulting from using the fractional approximation (3.126) is studied in detail in Burton & Zinober (1986) where, in particular, a formal examination of its chatter elimination properties is made.

An experimental comparison of these different approaches to eliminate chattering (technically for the case of a tracking controller) is made in deJager (1992), which compares the steady-state errors that result from using the continuous approximations in the presence of a constant disturbance. It is reported that no significant difference arises between the various methods.

3.8 SUMMARY

The key ideas of VSCS with a sliding mode have been presented. The notions of a sliding surface and an ideal sliding mode have been introduced, together with sufficient conditions for their existence. The exposition has been restricted to uncertain linear systems, and only hyperplanes have been considered as potential sliding surfaces. Implicitly it has been assumed that all the states are available to the controller. This restriction is often cited as a limitation of the technique since in practice only certain 'outputs' may be directly measurable. In later chapters VSCS concepts will be examined for uncertain systems for which only outputs are available. In this situation the class of hyperplanes and controllers considered must be restricted to those requiring only the system output information, or else estimates of the unavailable internal states must be generated for use by the control law. Both these possibilities are investigated. For the time being, the assumption that all the states are available will be retained. The next chapter considers different approaches for the design of the switching surface.

3.9 NOTES AND REFERENCES

Because of the nature of this chapter, it is difficult not to have been influenced either directly or subconsciously by the work of other authors. Several excellent introductory tutorial papers exist in the literature, most notably those of Utkin (1977), DeCarlo *et al.* (1988) and Hung *et al.* (1993). Introductory material also appears in the early chapters of two essentially research-orientated collections by Zinober (1990,1994).

For those interested in the rigorous fundamental ideas concerning the existence of solution and the Filippov construction, the reader is directed to the seminal work of Utkin (1992).

The η-reachability condition and the estimation of the time to reach the sliding surface appears in the work of Slotine & Li (1991) as does the use of VSCS controllers for electric motors and power converters. This area is also discussed in more detail by Sira-Ramírez & Lischinsky-Arenas (1991) and the references contained within.

This chapter considers only systems with nominal linear parts and only hyperplanes as potential sliding surfaces. It should be stressed that in principle these ideas can be used for more general nonlinear systems and sliding surfaces. More information is given in DeCarlo *et al.* (1988) and Utkin (1992).

The original formulation of the invariance principle given in Theorem 3.1 was stated and proved by Draženović (1969). The projection operator approach is introduced by El-Ghezawi *et al.* (1983) and the use of 'QR' decomposition to separate the original system to expose the null and range space dynamics first appears in Dorling & Zinober (1986). This material also appears in the early chapters of Zinober (1990).

The use of the controllability canonical form as a platform for designing sliding surfaces and switched controllers for single-input systems with matched uncertainty is widespread throughout the early literature, for example Utkin (1977). The pitfalls of using such an approach when unmatched uncertainty is known to be present is analysed in Spurgeon (1991). A worked example relating to the design of a controller using the hierarchical approach is given in DeCarlo *et al.* (1988).

The relationship between the poles of the reduced-order sliding motion and invariant zeros is considered in Young *et al.* (1977) and by El-Ghezawi *et al.* (1983).

The 'unit vector' approach described in Section 3.6 was introduced by Ryan & Corless (1984). In many ways this work is a hybrid of VSCS concepts and the uncertain systems approach of Leitmann. Further details and relevant references are given in the original paper of Ryan & Corless (1984). The exposition in Section 3.6 does not explicitly consider the issue of existence of solution. This is treated in careful detail by Ryan & Corless (1984): the uncertain functions $f_m(\cdot)$ and $f_u(\cdot)$ are assumed to be Carathéodory functions, which guarantee the existence of solution away from the sliding surface; they also prove, using the Brouwer fixed point theorem, that $s(t) = 0$ is a feasible solution to the differential inclusion representing the dynamics on the hyperplane.

The parameter values for the DC motor example in Section 3.6.4 are taken from the case study in Bélanger (1995).

The more general definition of sliding given in Section 3.7 is discussed in detail in the work of Utkin (1992) and in the tutorial paper by DeCarlo *et al.* (1988) from which Figure 3.5 is taken. Other more complex approximations to smooth discontinuous unit vector controllers are proposed by Zhou & Fisher (1992) and Luo & Feng (1989). An analysis of the smoothed unit vector controller described in Section 3.7 is considered in Ryan & Corless (1984) and more recently by Spurgeon & Davies (1993).

Chapter 4

Sliding Mode Design Approaches

4.1 INTRODUCTION

It has been established in Chapter 3 that the switching surface design problem may be interpreted as a static state feedback selection problem for a particular subsystem. This chapter will consider the development of appropriate methods to solve this problem. The emphasis will be on numerical tractability and MATLAB mfiles will be provided to implement the design procedures. The first algorithm is based upon robust eigenstructure assignment as this offers an effective method of minimising the effects of unmatched parameter variations on the sliding mode dynamics. The second will be based upon optimal control theory. A final switching surface design procedure will be synthesised whereby a desired modal structure is specified for the ideal sliding mode dynamics. This method is particularly effective for the solution of aerospace and vibration control problems.

The exposition thus far has considered regulation problems. This chapter will review methods for incorporating a reference signal within the sliding mode design procedure. This is clearly a nontrivial problem in that it has been demonstrated that a sliding system completely rejects a class of forcing functions. It is thus imperative that the scheme is configured in such a way as to ensure the reference signal is not rejected. Two methods for incorporating a tracking requirement will be considered. The first involves a model-reference approach and the second involves the addition of integral action. The chapter will conclude with a design study concerning the development of a flight control law. This will illustrate the application of many of the techniques introduced in the chapter.

4.2 A REGULAR FORM BASED APPROACH

In Chapter 3 a particular canonical form, the so-called regular form, was shown to yield a convenient interpretation of the reduced-order sliding mode dynamics. Consider the nominal linear model of an uncertain system, given by

$$\dot{x}(t) = Ax(t) + Bu(t) \tag{4.1}$$

where $\text{rank}(B) = m$ and (A, B) is a controllable pair. Define an associated switching function

$$s(t) = Sx(t) \tag{4.2}$$

The configuration (4.1), (4.2) may be expressed in the regular form

$$\begin{aligned}\dot{z}_1(t) &= A_{11}z_1(t) + A_{12}z_2(t) && (4.3)\\ \dot{z}_2(t) &= A_{21}z_1(t) + A_{22}z_2(t) + B_2u(t) && (4.4)\end{aligned}$$

with the switching function written as

$$s(t) = S_1z_1(t) + S_2z_2(t) \tag{4.5}$$

where the change of coordinates is defined by an orthogonal matrix T_r so that

$$z(t) = T_rx(t) \tag{4.6}$$

The matrix sub-blocks in (4.3) and (4.4) can be obtained in terms of the original pair (A, B) from

$$T_rAT_r^{\mathrm{T}} = \begin{bmatrix} A_{11} & A_{12} \\ A_{21} & A_{22} \end{bmatrix} \qquad T_rB = \begin{bmatrix} 0 \\ B_2 \end{bmatrix} \tag{4.7}$$

Likewise the elements of the switching function in (4.5) satisfy

$$ST_r^{\mathrm{T}} = \begin{bmatrix} S_1 & S_2 \end{bmatrix} \tag{4.8}$$

This transformation may be effectively accomplished for any given controllable pair (A, B), where B is full rank, using the MATLAB commands described below.

mfile: changes coordinates to bring about regular form

```
% Establish the size of the input distribution matrix
[nn,mm]=size(B);

% Perform QR decomposition on the input distribution matrix
[Tr temp]=qr(B);
Tr=Tr';
Tr=[Tr(mm+1:nn,:);Tr(1:mm,:)];
clear temp

% Obtain (areg,breg); regular form description
Areg=Tr*A*Tr';
Breg=Tr*B;

% Obtain matrix sub-blocks for sliding mode controller design
A11 = Areg(1:nn-mm,1:nn-mm);
A12 = Areg(1:nn-mm,nn-mm+1:nn);
A21 = Areg(nn-mm+1:nn,1:nn-mm);
A22 = Areg(nn-mm+1:nn,nn-mm+1:nn);
B2 = Breg(nn-mm+1:nn,1:mm);
```

During sliding motion, the switching function $s(t)$ will be identically equal to zero. Equation (4.5) thus gives

$$S_1z_1(t) + S_2z_2(t) = 0 \tag{4.9}$$

In this chapter, techniques for switching surface selection are developed and it is thus reasonable to make the *a priori* assumption that S will be selected so that the matrix product SB is nonsingular. It has been seen in Chapter 3 that this assumption implies that the matrix S_2 is nonsingular. Condition (4.9), which defines the sliding mode, may therefore be re-expressed as

$$\begin{aligned} z_2(t) &= -S_2^{-1}S_1 z_1(t) \\ &= -M z_1(t) \end{aligned} \tag{4.10}$$

where $M \in \mathbb{R}^{m\times(n-m)}$ is defined to be

$$M = S_2^{-1}S_1 \tag{4.11}$$

The development of the z_2 partition in the sliding mode is thus seen to be linearly related to the z_1 partition. The sliding mode is governed by equations (4.3) and (4.10):

$$\dot{z}_1(t) = A_{11}z_1(t) + A_{12}z_2(t) \tag{4.12}$$

$$z_2(t) = -Mz_1(t) \tag{4.13}$$

This is an $(n-m)$th order system in which z_2 has the role of a linear full-state feedback control signal. Closing the loop in (4.12) with the feedback from (4.13)

$$\dot{z}_1(t) = (A_{11} - A_{12}M)z_1(t) \tag{4.14}$$

The minimum design requirement is that asymptotically stable dynamics must be ensured during sliding so that $z_1 \to 0$ as $t \to \infty$. The *existence problem* becomes that of fixing $M = S_2^{-1}S_1$ to give $(n-m)$ negative poles to the closed–loop system (4.14). Any additional degrees of freedom may be used to further tailor the performance during the sliding mode. In particular, it will be seen that making the system (4.14) maximally robust to system uncertainty which occurs within this subsystem has the effect of rendering the sliding mode dynamics minimally sensitive to the unmatched uncertainty in the system. Recall that a sliding system can be made totally insensitive to the matched uncertainty in the problem but will be effected by any unmatched uncertainty contributions. It is thus desirable to minimise the effects of any unmatched signals. The existence problem may be solved by using a modified form of any standard design procedure which prescribes a linear full-state feedback controller for a linear dynamical system. The two methods which will be considered here are based upon a robust pole assignment approach and also the minimisation of an integral cost functional with quadratic integrand. However, it is straightforward to formulate the switching surface design problem in terms of any preferred state feedback design procedure. It should be noted that, whichever scheme is chosen, fixing M does not uniquely determine S since there are m^2 degrees of freedom in the relationship

$$S_2M = S_1 \tag{4.15}$$

In the work which follows, the hyperplane matrix S will be determined from M by letting $S_2 = I_m$, giving

$$ST_r^{\mathrm{T}} = \begin{bmatrix} M & I_m \end{bmatrix} \tag{4.16}$$

This approach minimises the calculation in proceeding from M to S and so reduces the possibility of numerical errors.

4.2.1 Robust Eigenstructure Assignment

The eigenvalues of the system during sliding will consist of the set of $(n - m)$ stable eigenvalues assigned to the spectrum of $A_{11} - A_{12}M$ from equation (4.14), plus the value zero repeated m times. If the state is to remain within the null space of S, the effect of the parameter variations which do not occur in the range space of B, i.e. any unmatched parameter variations, must be minimised. This will be achieved by developing a *robust control design approach* to solve the existence problem; the aim will be to make the nonzero sliding mode eigenvalues insensitive to perturbations using robust eigenstructure assignment. This will minimise the effect of parameter variations outside the range space of B. Consider the closed-loop system equations (4.12) and (4.13) to be replaced by the perturbed representation

$$\dot{z}_1(t) = (A_{11} - A_{12}M)z_1(t) + (\Delta A_{11} - \Delta A_{12}M)z_1(t) \tag{4.17}$$

where the second term on the right-hand side of the equation represents a perturbation of the nominal system due to parameter variations. This term will usually be both time-varying and arbitrary, and the robust assignment approach will seek to minimise its effect. The nominal sliding mode system defined by equations (4.12) and (4.13) is of order $(n - m)$, with m control inputs. The control distribution matrix $A_{12} \in \mathbb{R}^{(n-m)\times m}$ must have rank l, which satisfies

$$1 \leq l \leq \min\{n - m, m\}$$

The rank l is clearly greater than or equal to 1, otherwise the controllability property of Proposition 3.3 is violated. If $l < m$ the sliding system has redundant inputs which must be removed. If $P \in \mathbb{R}^{m\times m}$ is an appropriate transformation matrix, as described in Appendix B, then

$$A_{12}P = \begin{bmatrix} A_{121} & A_{122} \end{bmatrix}$$

where A_{121} has l linearly independent columns while the matrix A_{122} is zero. Let

$$P^{\mathrm{T}}M = \begin{bmatrix} M_1 \\ M_2 \end{bmatrix} \begin{matrix} \updownarrow l \\ \updownarrow m-l \end{matrix} \tag{4.18}$$

The matrices from equation (4.17) may be re–expressed as

$$\begin{aligned} A_{11} - A_{12}M &= A_{11} - A_{121}M_1 - A_{122}M_2 \\ \Delta A_{11} - \Delta A_{12}M &= \Delta A_{11} - \Delta A_{121}M_1 - \Delta A_{122}M_2 \end{aligned}$$

where

$$\Delta A_{12}P = \begin{bmatrix} \Delta A_{121} & \Delta A_{122} \end{bmatrix} \tag{4.19}$$

The effect of perturbations can be minimised by setting the matrix M_2 equal to zero. The remaining feedback matrix partition M_1 is then determined by applying a robust control system design procedure to the matrix pair (A_{11}, A_{121}). The matrix M can then be calculated from a simple permutation of the rows of the partitioned feedback matrix using equation (4.18).

An appropriate methodology for determining M_1 is described below. It has been shown by Wonham (1967) that a full-state feedback matrix can be found to assign an arbitrary self-conjugate set of eigenvalues to a given linear system if it is controllable. Moore (1976) described the freedom available to assign eigenvectors for an arbitrary self-conjugate set of eigenvalues using state feedback. He gives both necessary and sufficient conditions for a feedback matrix to exist which satisfies

$$(A - BF)v_i = \lambda_i v_i \qquad i = 1, ..., n \tag{4.20}$$

Associate with each λ_i a matrix $G(\lambda_i) \in \mathbb{C}^{n\times(n+m)}$ such that

$$G(\lambda_i) = \left[\begin{array}{cc} \lambda_i I - A & B \end{array} \right] \tag{4.21}$$

and a compatibly partitioned matrix

$$K(\lambda_i) = \left[\begin{array}{c} N(\lambda_i) \\ M(\lambda_i) \end{array} \right] \begin{array}{c} \updownarrow n \\ \updownarrow m \end{array} \tag{4.22}$$

where the columns of $K(\lambda_i) \in \mathbb{C}^{(n+m)\times m}$ form a basis for the null space of $G(\lambda_i)$. Using this notation it is possible to present the following well-known result, which will be quoted without proof.

Theorem 4.1 *The necessary and sufficient conditions to find a real matrix F satisfying (4.20) are*

(i) $v_i \in \mathbb{C}^n$ *are linearly independent vectors*

(ii) $v_i = v_j^*$ *whenever* $\lambda_i = \lambda_j^*$

(iii) $v_i \in span(N(\lambda_i))$.

Furthermore, if such a matrix F exists and rank$(B) = m$ then F is unique.

A desirable property of any closed-loop system is that the eigenvalues are rendered insensitive to perturbations in the coefficient matrices of the system equations; such a well-conditioned solution will be termed *robust*. It has been seen in Theorem 4.1 that the solution to the problem of designing a state feedback controller using eigenvalue placement is generally underdetermined with many degrees of freedom. The objective of the design process under consideration here will be to determine a robust control system by restricting the degrees of freedom of the assignment problem. Let v_i and w_i be the right and left eigenvectors of the closed-loop system matrix corresponding to an eigenvalue λ_i. Therefore

$$(A - BF)v_i = \lambda_i v_i \tag{4.23}$$
$$w_i^{\mathrm{T}}(A - BF) = \lambda_i w_i^{\mathrm{T}} \tag{4.24}$$

If the closed-loop system matrix $(A - BF)$ has n linearly independent eigenvectors, then it is diagonalisable and it has been shown by Wilkinson (1967) that the sensitivity of an eigenvalue λ_i to perturbations in A, B and F depends upon the magnitude of the condition number, c_i, where

$$c_i = \frac{||w_i|| ||v_i||}{|w_i^{\mathrm{T}} v_i|} \geq 1 \tag{4.25}$$

For the case of multiple eigenvalues, a particular choice of eigenvectors is assumed. A bound upon the individual eigenvalue sensitivities is given by

$$\max\{c_i\} \leq \kappa(V) = ||V|| ||V^{-1}|| \tag{4.26}$$

where $\kappa(V)$ denotes the condition number of the matrix $V = [\, v_1, \ldots, v_n \,]$ of right eigenvectors. Essentially the condition number is a measure of orthogonality of the eigenvectors v_i. The closer the eigenvectors of a matrix are to being orthogonal, the smaller is the associated condition number and the greater is the robustness of the eigenvalue locations to changes in the elements of the matrix. The feedback matrix is obtained by assigning a set of linearly independent right eigenvectors corresponding to the required eigenvalues such that the matrix of eigenvectors is as well conditioned as possible. The assigned eigenvalues are then as insensitive to perturbations as possible. The robust eigenvalue assignment problem may therefore be stated as follows:

Given A, B and a set of desired eigenvalues, find a real matrix F and a nonsingular matrix V satisfying

$$(A - BF)V = V\Lambda \tag{4.27}$$

such that some measure of the conditioning, or robustness, of the eigenproblem is optimised.

Some flexibility is available in the choice of measure of the conditioning of the eigenproblem. For example, $||c||$ where $c^{\mathrm{T}} = [\, c_1, \ldots, c_n \,]$ is the vector of individual condition numbers corresponding to the matrix of eigenvectors V could be chosen. Alternatively the condition number of the matrix V defined by equation (4.26) could be used. The degrees of freedom that are available in the choice of feedback matrix F are reflected by the degrees of freedom available in the selection of the matrix of eigenvectors. For the case of a single-input system, V is uniquely determined except for scaling, and the condition numbers c_i cannot be controlled. If $m = n$ then V may always be chosen to be orthogonal, hence $c_i = 1$ for all i. However, for the general multi-input system the sensitivities of the assigned eigenvalues can be controlled only to a limited extent by appropriate choice of V. Kautsky *et al.* (1985) have developed a result which demonstrates that given A, B and a set of desired eigenvalues, minimising the conditioning of the eigenproblem also minimises a bound on the feedback gains and a bound on the transient response of the closed-loop system, for any initial condition. This work also demonstrates computation of the maximum perturbation which can be made to the closed-loop system such that stability is maintained. Such results show that if the conditioning of the assigned eigensystem is minimised, then a lower bound on the stability margin of the closed-loop system is maximised over all feedback matrices F which assign the given stable eigenvalues. The merits of a robust eigenstructure assignment approach having been discussed, a practical implementation of the theory will now be considered.

The objective of the method is to choose a set of right eigenvectors such that each is maximally orthogonal to the space spanned by the remaining vectors. This is equivalent to choosing v_i as close as possible to the normalised vector[1] q_i, where

[1] Any nonzero vector v can be normalised by scaling so that $||v|| = 1$.

q_i is defined to be orthogonal to the space

$$\mathcal{L}_i = \text{span}(v_j : j \neq i) \tag{4.28}$$

for all i. However, from Theorem 4.1, it is necessary for $v_i \in \text{span}(N(\lambda_i))$ and it may not be possible to satisfy this requirement with a vector which is also orthogonal to $\mathcal{L}_i$. It follows that the vector q_i must be projected into the subspace $N(\lambda_i)$ to yield an allowable eigenvector. The procedure is accomplished computationally using an iterative process in which each vector v_i is replaced by a vector with maximum misalignment angle to the current space $\mathcal{L}_i$ for each $i = 1, \ldots, n$ in turn. The new vector is obtained using the QR method. The decomposition of V_i is first found as follows:

$$\begin{aligned} V_i &= \begin{bmatrix} v_1 & \ldots & v_{i-1} & v_{i+1} & \ldots & v_n \end{bmatrix} && (4.29) \\ &= \begin{bmatrix} Q_i & q_i \end{bmatrix} \begin{bmatrix} R_i \\ 0^{\mathrm{T}} \end{bmatrix} && (4.30) \end{aligned}$$

where $Q_i \in \mathbb{R}^{n \times (n-1)}$, $R_i \in \mathbb{R}^{(n-1) \times n}$. Thus a $q_i \in \mathbb{R}^n$ orthogonal to $\mathcal{L}_i$ is found. This q_i, however, is not necessarily an allowable eigenvector. It is necessary to project q_i into the allowable subspace $N(\lambda_i)$ in equation (4.22). The columns of this allowable subspace are othonormalised to give $Q(\lambda_i)$. The vector q_i is then projected into $Q(\lambda_i)$ to give

$$v_i^{new} = Q(\lambda_i) \frac{Q(\lambda_i)^{\mathrm{T}} q_i}{\|Q(\lambda_i)^{\mathrm{T}} q_i\|} \tag{4.31}$$

This vector has minimum misalignment angle to q_i and follows directly from the formula (A.37) in Appendix A.2.6 using the fact that $Q(\lambda_i)^{\mathrm{T}} Q(\lambda_i) = I_m$. The scalar term $\|Q(\lambda_i)^{\mathrm{T}} q_i\|$ ensures that v_i^{new} is normalised. The procedure is continued iteratively until there is either no improvement in the condition number or a prespecified maximum number of iterations has been exceeded.

Having determined an appropriate set of eigenvectors, it is relatively straightforward to recover the state feedback matrix F which prescribes the desired modal structure to the closed-loop system (4.20). Equation (4.20) may be rewritten as

$$(\lambda_i I - A) v_i + B F v_i = 0 \tag{4.32}$$

If $G(\lambda_i)$ and $K(\lambda_i)$ are as defined by equations (4.21) and (4.22) respectively, then it follows that any arbitrary vector k_i which post-multiplies $K(\lambda_i)$ will give a resulting vector which lies in the null space of $G(\lambda_i)$. Thus

$$\begin{bmatrix} \lambda_i I - A & B \end{bmatrix} \begin{bmatrix} N(\lambda_i) \\ M(\lambda_i) \end{bmatrix} k_i = 0 \tag{4.33}$$

as $K(\lambda_i)$ is a basis for the null space of $S(\lambda_i)$. Expanding equation (4.33) yields

$$(\lambda_i I - A) N(\lambda_i) k_i + B M(\lambda_i) k_i = 0 \tag{4.34}$$

Comparison of equation (4.34) with (4.32) gives

$$\begin{aligned} v_i &= N(\lambda_i) k_i && (4.35) \\ F v_i &= M(\lambda_i) k_i && (4.36) \end{aligned}$$

The vector k_i effectively determines where in the allowable subspace v_i exists and is thus known for given v_i, as seen in equation (4.31). It follows from (4.36) that the feedback matrix may be determined from

$$F = \left[\begin{array}{cccc} M(\lambda_1)k_1 & M(\lambda_2)k_2 & \ldots & M(\lambda_n)k_n \end{array} \right] V^{-1} \tag{4.37}$$

where the k_i are those which have been used to determine the corresponding robust, allowable eigenvector v_i at the end of the iteration procedure. In this way the feedback matrix M in (4.14) and hence the switching surface are specified. MATLAB commands to establish a robust switching surface selection for a given (A, B) pair are given below.

mfile: hyperplane design using a robust pole placement method

```
% Assumes a regular form description has been computed. Requires a
% vector p which contains the desired n-m sliding mode poles

M=vplace(A11,A12,p);

%Recover the switching function matrix in the original coordinates
S2=eye(mm);        % For simplicity
S=[M eye(mm)]*Tr;
```

Remark

The MATLAB command *place* from the control system toolbox implements the robust eigenstructure assignment approach described in this section. However, in the mfile fragment above, a modification of the *place* command, termed *vplace*, has been used in preference; for details see Appendix B.

4.2.2 Quadratic Minimisation

The design of the switching hyperplane by minimising a cost functional in which the integrand is quadratic in terms of the state will now be considered. This method was proposed by Utkin & Young (1978) and can be particularly useful for model-following variable structure controller design where VSCS are required for a model-following error system. The method will enable desirable weightings to be placed upon particular elements; for example, it is obviously very important that the control surface deflections of the real system follow those of the model very closely. Consider the problem of minimising the quadratic performance index

$$J = \frac{1}{2} \int_{t_s}^{\infty} x(t)^{\mathrm{T}} Q x(t) \, dt \tag{4.38}$$

where Q is both symmetric and positive definite and t_s is the time at which the sliding motion commences. The aim is to minimise equation (4.38) subject to the system equation (4.1). It is assumed that the state of the system at time t_s, $x(t_s)$, is a known initial condition and is such that $x(t) \to 0$ as $t \to \infty$. The matrix Q from equation (4.38) is transformed and partitioned compatibly with z:

$$T_r Q T_r^{\mathrm{T}} = \begin{bmatrix} Q_{11} & Q_{12} \\ Q_{21} & Q_{22} \end{bmatrix} \tag{4.39}$$

where $Q_{21} = Q_{12}^{\mathrm{T}}$. Equation (4.38) may then be expressed in terms of the z coordinate system as

$$J = \frac{1}{2}\int_{t_s}^{\infty} z_1^{\mathrm{T}} Q_{11} z_1 + 2 z_1^{\mathrm{T}} Q_{12} z_2 + z_2^{\mathrm{T}} Q_{22} z_2 \; dt \tag{4.40}$$

In order to solve this problem it is desirable to express it in the form of the standard LQR problem met in Chapter 2, where z_1, which determines the system dynamics in the ideal sliding mode, has the role of the state and the effective control input is a function of z_2. To this end, it is necessary to eliminate the cross term $2z_1^{\mathrm{T}} Q_{12} z_2$ from equation (4.40). Noting that the last two terms in equation (4.40) may be factored to yield

$$2z_1^{\mathrm{T}} Q_{12} z_2 + z_2^{\mathrm{T}} Q_{22} z_2 = (z_2 + Q_{22}^{-1} Q_{21} z_1)^{\mathrm{T}} Q_{22} (z_2 + Q_{22}^{-1} Q_{21} z_1) - z_1^{\mathrm{T}} Q_{21}^{\mathrm{T}} Q_{22}^{-1} Q_{21} z_1$$

it is straightforward to verify that equation (4.40) may be re-expressed as

$$J = \frac{1}{2}\int_{t_s}^{\infty} z_1^{\mathrm{T}} (Q_{11} - Q_{12} Q_{22}^{-1} Q_{21}) z_1 + (z_2 + Q_{22}^{-1} Q_{21} z_1)^{\mathrm{T}} Q_{22} (z_2 + Q_{22}^{-1} Q_{21} z_1) \; dt \tag{4.41}$$

Define

$$\hat{Q} = Q_{11} - Q_{12} Q_{22}^{-1} Q_{21} \tag{4.42}$$

$$v = z_2 + Q_{22}^{-1} Q_{21} z_1 \tag{4.43}$$

Equation (4.41) may then be written as

$$J = \frac{1}{2}\int_{t_s}^{\infty} z_1^{\mathrm{T}} \hat{Q} z_1 + v^{\mathrm{T}} Q_{22} v \; dt \tag{4.44}$$

Recall the original constraint equation

$$\dot{z}_1(t) = A_{11} z_1(t) + A_{12} z_2(t) \tag{4.45}$$

Eliminating the z_2 contribution using equation (4.43), the modified constraint equation becomes

$$\dot{z}_1(t) = \hat{A} z_1(t) + A_{12} v(t) \tag{4.46}$$

where

$$\hat{A} = A_{11} - A_{12} Q_{22}^{-1} Q_{21} \tag{4.47}$$

The problem thus becomes that of minimising the functional (4.44) subject to the system (4.46) and has thus been interpreted as a standard linear quadratic optimal state regulator problem (see Section 2.3.3). The positive definiteness of Q ensures that $Q_{22} > 0$, so that Q_{22}^{-1} exists, and also that $\hat{Q} > 0$ (see Appendix A.2.7). Furthermore, the controllability of the original (A, B) pair ensures that the pair

(A_{11}, A_{12}) is controllable (Proposition 3.3) and this in turn is sufficient to ensure the controllability of $(\hat{A}, A_{12})$. It follows that a unique positive definite solution P_1 is guaranteed for the algebraic matrix Riccati equation which is associated with the problem defined by equations (4.44) and (4.46)

$$P_1\hat{A} + \hat{A}^{\mathrm{T}}P_1 - P_1A_{12}Q_{22}^{-1}A_{12}^{\mathrm{T}}P_1 + \hat{Q} = 0 \tag{4.48}$$

The optimal v minimising equation (4.44) is given by

$$v = -Q_{22}^{-1}A_{12}^{\mathrm{T}}P_1z_1 \tag{4.49}$$

This expression can be inserted into equation (4.43) to yield

$$z_2 = -Q_{22}^{-1}(A_{12}^{\mathrm{T}}P_1 + Q_{21})z_1 \tag{4.50}$$

Comparing this with (4.13) yields the expression $M = Q_{22}^{-1}(A_{12}^{\mathrm{T}}P_1 + Q_{21})$. The MATLAB mfile fragment below illustrates this design procedure.

mfile: hyperplane design using a linear quadratic cost function

```
% The regular form description and associated transformation matrix is
% assumed to have been determined already. The symmetric positive definite
% matrix Q is the state weighting matrix in the original coordinates

% Transform weighting matrix to regular form coordinate system
Qt=Tr*Q*Tr';

% Compatibly partition with regular form description
Q11 = Qt(1:nn-mm,1:nn-mm);
Q12 = Qt(1:nn-mm,nn-mm+1:nn);
Q21 = Qt(nn-mm+1:nn,1:nn-mm);
Q22 = Qt(nn-mm+1:nn,nn-mm+1:nn);

% Form reduced order system description and associated weighting matrix
Qhat=Q11-Q12*inv(Q22)*Q21;
Ahat=A11-A12*inv(Q22)*Q21;

% Solve the LQR problem
[K,P1,E]=lqr(Ahat,A12,Qhat,Q22);

% Obtain the switching function matrix in terms of the original coordinates
M=inv(Q22)*(A12'*P1+Q21);
S2=eye(mm);                              % For simplicity
S=S2*[M eye(mm)]*Tr;
```

4.3 A DIRECT EIGENSTRUCTURE ASSIGNMENT APPROACH

For the case of a scalar controlled problem, specification of the $(n-1)$ eigenvalues associated with the sliding mode will uniquely determine the matrix M of equation (4.11). There are thus no degrees of freedom available to specify the associated eigenvectors. For the multiple-input case, it has been shown in Proposition 3.3 that if the matrix pair (A, B) defining the system (4.1) is controllable, then the matrix pair (A_{11}, A_{12}) of equation (4.3) is also controllable. The direct consequence of this result is that the eigenvalues of the sliding mode system (4.14) can be placed arbitrarily in the complex plane by suitable choice of the matrix M. The remaining degrees of freedom available in the assignment problem will be used to modally shape the system response using a judicious choice of eigenvector form. For this application of the technique, the objective will be to modally shape the system response during the sliding mode. Assuming that sliding motion has commenced on the null space of S, the dynamics exhibited by the system can now be represented by

$$\dot{x}(t) = (A - BK)x(t) \tag{4.51}$$

where

$$K = (SB)^{-1}SA \tag{4.52}$$

This describes the equivalent system motion developed in Section 3.4. During sliding motion, the system state must lie within the null space of S by definition and so

$$\begin{aligned} S\dot{x} = 0 \quad &\Leftrightarrow \quad S(A - BK)x = 0 \\ &\Leftrightarrow \quad \mathcal{R}(A - BK) \subseteq \mathcal{N}(S) \end{aligned} \tag{4.53}$$

Let λ_i, $i = 1, \ldots, n$ be the eigenvalues of the equivalent system, $A - BK$, with corresponding right eigenvectors v_i, $i = 1, \ldots, n$. Then, from equation (4.53):

$$S(A - BK)v_i = \lambda_i S v_i = 0 \tag{4.54}$$

which shows that either $\lambda_i = 0$ or $v_i \in N(S)$. It has been shown in the discussion of the equivalent system motion in Section 3.4 that $(A - BK) = A_{eq}$ has precisely m zero eigenvalues. Assuming that λ_i, $i = 1, \ldots, n - m$ are the nonzero distinct eigenvalues of the equivalent system motion, the corresponding eigenvectors v_i, $i = 1, \ldots, n - m$ will determine the null space of S, since $\dim[\mathcal{N}(S)] = n - m$.

It has been shown in the previous section that the eigenvector corresponding to a given eigenvalue must lie in an allowable subspace which is determined by the plant matrix A, the input matrix B and the eigenvalue itself. If information is known about a desirable weighting of the system states for each mode, it is possible to choose a desired eigenvector specification. This will not necessarily be achievable because it may not lie within the prescribed allowable subspace. In the theoretical discussion below it will be assumed for simplicity that the desired eigenvalues are real. The modifications required to cover the complex case are outlined in Appendix B. Assume that $q \leq n$ components are specified. For $q \leq m$ it may be possible to assign an allowable eigenvector with these components specified exactly. For $q > m$ an allowable eigenvector which is in some sense 'closest' to the desired eigenvector must be selected. This technique of shaping the modal structure will be referred to in the work which follows as *direct eigenstructure assignment.* For a given problem only a few of the components of a desired eigenvector form v_i^d are usually specified (Harvey & Stein, 1978); the rest may be arbitrary. To account for this, a row reordering operation is computed such that

$$\tilde{R}v_i^d = \begin{bmatrix} \tilde{v}_i^s \\ \tilde{v}_i^u \end{bmatrix} \tag{4.55}$$

where $\tilde{v}_i^s \in \mathbb{R}^q$ is a vector of specified components and the vector $\tilde{v}_i^u \in \mathbb{R}^{n-q}$ represents unspecified elements. The desired form of equation (4.55) may not reside in the prescribed subspace as required by Theorem 4.1. It is thus necessary to determine a vector δ_i which corresponds to the projection of the desired eigenvector components $\tilde{v}_i^s$ into the allowable eigenvector subspace. To this end, apply the row reordering operation $\tilde{R}$ to the allowable eigenvector subspace $N(\lambda_i)$ and let

$$\tilde{R}N(\lambda_i) = \begin{bmatrix} N_1 \\ N_2 \end{bmatrix} \tag{4.56}$$

where $N_1 \in \mathbb{R}^{q \times m}$. The required δ_i is thus determined from the linear equation

$$N_1 \delta_i = \tilde{v}_i^s \tag{4.57}$$

Whether this equation has a solution will depend on the size of q relative to m and on the rank of N_1. It can be verified that if $q \leq m$ and N_1 is full rank, the equation always has an exact solution and thus the $q \leq m$ components of the eigenvector can be assigned exactly. If N_1 is rank deficient for the chosen $q \leq m$ components, the designer should consider interchanging some of the specified desired nonzero eigenvector entries with the unspecified components to render N_1 full rank. In this way $q \leq m$ components can again be specified exactly. In the case when $q > m$, equation (4.57) does not have an exact solution unless

$$\text{rank}\begin{bmatrix} N_1 & \tilde{v}_i^s \end{bmatrix} = \text{rank}\begin{bmatrix} N_1 \end{bmatrix}$$

However, the choice

$$\delta_i = (N_1^{\mathrm{T}} N_1)^{-1} N_1^{\mathrm{T}} \tilde{v}_i^s \tag{4.58}$$

minimises the error between the desired and allowable components in a least-squares sense. The allowable eigenvector is then determined as

$$v_i = N(\lambda_i)\delta_i \tag{4.59}$$

It should be noted that S is not uniquely determined from the selected eigenvectors since the equation

$$SV = 0 \qquad V = \begin{bmatrix} v_1 & \dots & v_{n-m} \end{bmatrix} \tag{4.60}$$

has m^2 degrees of freedom. If the eigenvector matrix of equation (4.60) results from a modal shaping of the system response, the null space dynamics determined by S will exhibit the required response shaping. In the mfile fragment which follows the null space of $G(\lambda_i)$, which in turn determines $N(\lambda_i)$, is obtained numerically using the singular value decomposition approach described in Appendix A.2.4. The hyperplane equation (4.60) is solved in a similar way to obtain the matrix S.

Once a controller is found to drive the system state and thereafter maintain it within the null space of S, the technique of VSCS is seen to have the potential to force a nonlinear or perturbed system to exhibit a required modal structure when in the sliding mode. The MATLAB commands given below allow the designer to assign desired real eigenvalues and eigenvectors to the ideal sliding mode dynamics. Appendix B contains MATLAB commands which enable the assignment of an eigenstructure that includes complex conjugate assignment.

mfile: hyperplane design using eigenstructure assignment

```
% For the system pair (A,B), the vector lambda defines the n-m desired
% sliding mode poles. The columns of the nx(n-m) matrix specpos are
% used to distinguish vector entries which are arbitrary (denoted by
% 0 in the relevant matrix entry) from entries which are to be
% specified (denoted by 1 in the relevant matrix entry). The columns
% of the nx(n-m) matrix specent specify any desired eigenvector entries.

[nn,mm]=size(B);

for i=1:(nn-mm),

   % Determine the space in which the eigenvector corresponding
   % to a specific desired pole must lie
   glambda=[lambda(i)*eye(nn)-A B];
   [u,v,w]=svd(glambda);
   nlambda=w(1:nn,nn+1:nn+mm);

   % Find subvector of specified entries
   despos=find(specpos(:,i));
   numspec=length(despos);
   n1=[]; desent=[];
   for j=1:numspec,
     n1(j,:)=nlambda(despos(j),:);
     desent(j)=specent(despos(j),i);
   end

   % Perform least-squares projection
   delta=n1\desent';
   V(:,i)=nlambda*delta;

end

% Use a singular value decomposition to determine the switching
% function matrix from the selected eigenvectors
[u,v,w]=svd(V);
S=u(:,nn-mm+1:nn)';
```

4.4 INCORPORATION OF A TRACKING REQUIREMENT

Up to now only regulation problems have been considered where the objective is to drive the states to zero. The possible ways in which a tracking requirement can be incorporated in the design framework will now be explored. It should be recalled that when in the sliding mode a system completely rejects any signals which satisfy the matching conditions. The inclusion of a tracking requirement is thus nontrivial and must always be achieved in such a way as to ensure that the command is not wholly or partly rejected by the sliding system.

4.4.1 A Model-Reference Approach

The model-following design objective is to develop a control scheme which forces the plant dynamics to follow the dynamics of an ideal model. The controller should thus force the error between the plant and the model states to zero as time tends to infinity. This will ensure that the plant output follows the model output faithfully.

Model-following control systems theory developed because of the difficulties encountered in the direct design of multivariable control systems using linear optimal control techniques. The two major problems concerning linear optimal control arise because of difficulties in specifying design objectives in terms of a performance index and also due to the large variations in plant parameters which may occur. A linear model-following approach to the control system design problem will avoid the difficulty of performance specification because the model will specify the design objectives; the controller is required to minimise the tracking error between the model and the plant. The problem of parameter variations still remains, however. This has led to the development of the so-called adaptive model-following control schemes, which are required to maintain a high performance in the presence of uncertainty and disturbances. There are two main design methods; the first is based on the hyperstability concept whilst the second uses Lyapunov methods. Both techniques are concerned with guaranteeing that the error between the model and plant states tends to zero, although neither prescribes the transient behaviour of this error. It is shown that an adaptive model-following control system can be designed as a variable structure system by applying the theory of variable structure systems such that a sliding mode exists. In this way the transient response of the error dynamics can be prescribed. The approach does not require any global convergence properties and so is well suited to uncertain, time-varying plants. In addition, adaption is achieved using signal synthesis in a fixed parameter structure and so no parameter identification algorithms are required.

Consider the linear time invariant multivariable plant

$$\dot{x}(t) = Ax(t) + Bu(t) \tag{4.61}$$

and corresponding ideal model

$$\dot{w}(t) = A_m w(t) + B_m r(t) \tag{4.62}$$

where $x \in \mathbb{R}^n$ and $w \in \mathbb{R}^n$ are the state vectors of the plant and model respectively, $u \in \mathbb{R}^m$ is the control vector, $r \in \mathbb{R}^r$ is an input vector and A, B, A_m and B_m are compatibly dimensioned matrices. It is assumed that the pair (A, B) is controllable and that the ideal model is stable so that the eigenvalues of A_m have negative real parts. Define a tracking error state e as the difference between the plant and model state responses:

$$e(t) = x(t) - w(t) \tag{4.63}$$

This error is required to tend asymptotically to zero. Differentiating the error equation (4.63) with respect to time yields

$$\dot{e}(t) = \dot{x}(t) - \dot{w}(t) \tag{4.64}$$

The dynamics of the model-following error system can now be determined directly from equations (4.61) and (4.62)

$$\dot{e}(t) = Ax(t) - A_m w(t) + Bu(t) - B_m r(t) \tag{4.65}$$

Adding and subtracting a term Aw in equation (4.65) yields

$$\dot{e}(t) = Ae(t) + (A - A_m)w(t) + Bu(t) - B_m r(t) \tag{4.66}$$

It is evident that for any given plant and model, a perfect model-following system may be impossible to achieve. A sufficient condition is that all orders of time derivatives of the error are zero at any time t. Commencing with the zeroth derivative, it follows that

$$w(t) = x(t) \tag{4.67}$$

Suppose some arbitrary term, feeding forward the model states, is added to the control action so that (4.61) becomes

$$\dot{x}(t) = Ax(t) + B(u(t) + Gw(t)) \tag{4.68}$$

for some arbitrary gain matrix G. If the first derivative of the error is to be zero then, using equations (4.68) and (4.62), the equation

$$Ax(t) + Bu(t) + BGx(t) = A_m w(t) + B_m r(t) \tag{4.69}$$

must hold. Solving equation (4.69) in order to obtain an expression for the control gives

$$u(t) = B^\dagger (A_m w(t) + B_m r(t) - Ax(t) - BGx(t))$$

where $B^\dagger$ denotes the Moore–Penrose pseudo-inverse of the matrix B. Substituting this expression for the control into (4.69) and rearranging yields

$$Ax(t) + BB^\dagger (A_m w(t) + B_m r(t) - Ax(t) - BGx(t)) + BGx(t) - A_m w(t) - B_m r(t) = 0$$

Note that $BB^\dagger B = B$ and the expression above simplifies to become

$$(BB^\dagger - I)A_m w(t) - (BB^\dagger - I)Ax(t) + (BB^\dagger - I)B_m r(t) = 0 \tag{4.70}$$

It is thus seen that direct use of the model states w through the control loop has no effect on the condition for model-following. Noting equation (4.67), equation (4.70) is satisfied for all x, w and r if

$$\begin{aligned} (BB^\dagger - I)(A - A_m) &= 0 && (4.71)\\ (BB^\dagger - I)B_m &= 0 && (4.72) \end{aligned}$$

If equations (4.71) to (4.72) are to hold for an arbitrary time t, all higher-order derivatives of the error will also be zero. Consider a control law with the structure

$$u(t) = u_1(t) + u_2(t) \tag{4.73}$$

where

$$\begin{aligned} u_1(t) &= -Ke(t) && (4.74)\\ u_2(t) &= B^\dagger (A_m - A)x(t) + B^\dagger B_m r(t) && (4.75) \end{aligned}$$

Substituting the control law (4.73) into equation (4.66) and assuming conditions (4.71) and (4.72) hold, then

$$\dot{e}(t) = (A_m - BK)e(t) \tag{4.76}$$

If (A_m, B) is a controllable pair, the closed-loop matrix $A_m - BK$ can have an arbitrary set of eigenvalues through suitable choice of K and in this way the error settling rates can be controlled. Equations (4.71) and (4.72) are the conditions for a perfect following and equation (4.73) is the control law for implementing it.

An equivalent test for (4.71) to (4.72) is available using a well-known theorem from linear algebra.

Theorem 4.2 *For the system of simultaneous equations denoted by*

$$HD = E \tag{4.77}$$

a solution for D exists if and only if rank$[\,H \quad E\,]$ = rank$[\,H\,]$

Proof
See for example Strang (1988). ■

If the following rank conditions hold

$$\text{rank}\begin{bmatrix} B & A_m - A \end{bmatrix} = \text{rank}[\,B\,] \tag{4.78}$$

$$\text{rank}\begin{bmatrix} B & B_m \end{bmatrix} = \text{rank}[\,B\,] \tag{4.79}$$

it follows from Theorem 4.2 that there exist compatibly dimensioned matrices F and G such that

$$BF = A_m - A \tag{4.80}$$

$$BG = B_m \tag{4.81}$$

It follows that the control law defined by equations (4.73) and (4.74) with u_2 expressed as

$$u_2(t) = Fx(t) + Gr(t) \tag{4.82}$$

will also achieve perfect following. However, it should be noted that equation (4.75) will use minimum control effort due to the use of the pseudo-inverse.

The design of variable structure model-following control systems will now be considered. The required system performance is usually specified via the model, and the controller synthesis is to minimise the error between the model states and the controlled plant. The associated control component (4.74) must provide robustness to parameter variations and other external uncertainties which occur in the plant. Consider the time invariant multivariable plant described by equation (4.61). The plant output is required to follow the output of the model (4.62). The dynamics of the model-following error system can be obtained in a suitable form from equation (4.65) by the addition and subtraction of a term $A_m x$, yielding

$$\dot{e}(t) = A_m e(t) + (A - A_m)x(t) + Bu(t) - B_m r(t) \tag{4.83}$$

The design objective is to choose a hyperplane matrix S and related discontinuous control law such that the error state attains a sliding mode. Define an error dependent switching function

$$s(e) = Se \tag{4.84}$$

which gives rise to a hyperplane in the error space

$$\mathcal{S}_e = \{e \in \mathbb{R}^n \ : \ Se = 0\} \tag{4.85}$$

During the sliding mode the error state will satisfy the equation

$$Se(t) = 0 \tag{4.86}$$

Differentiating this equation with respect to time and substituting from equation (4.83) yields

$$\begin{aligned} S\dot{e}(t) &= S(A_m e(t) + (A - A_m)x(t) + Bu(t) - B_m r(t)) \\ &= 0 \end{aligned} \tag{4.87}$$

Assuming that the matrix product SB is chosen to be nonsingular, the equivalent control when sliding can be determined from equation (4.87) as

$$u_{eq} = -(SB)^{-1} S(A_m e(t) + (A - A_m)x(t) - B_m r(t))$$

Substituting this equivalent control into the model-following error system (4.83) gives

$$\dot{e}(t) = (I - B(SB)^{-1}S)(A_m e(t) + (A - A_m)x(t) - B_m r(t)) \tag{4.88}$$

Suppose A, B, A_m and B_m satisfy the perfect model-matching conditions described by equations (4.80) and (4.81). Comparing these equations with the sliding mode invariance conditions in Theorem 3.1, it is seen that the two coincide. Therefore, if x and r are considered to be disturbances to the error dynamics, then the perfect model-matching conditions guarantee that the behaviour of the variable structure system in the sliding mode is insensitive to these disturbances. Equation (4.88) thus reduces to

$$\begin{aligned} \dot{e}(t) &= (I - B(SB)^{-1}S)A_m e(t) \\ &= A_{eq} e(t) \end{aligned} \tag{4.89}$$

In the sliding mode, the error system is formally equivalent to a system of order $n - m$ and the response will be determined by the $n - m$ nonzero eigenvalues of the system (4.89). Given a matrix pair (A_m, B) which is stabilisable, a hyperplane matrix S can be found such that the error tends to zero with increasing time. This hyperplane matrix can be designed by any of the techniques described in Section 4.2. Sufficiently fast error decay when sliding can be ensured by placing the $n - m$ eigenvalues sufficiently far enough into the left-hand half of the complex plane. Having chosen a suitable hyperplane matrix, a unit vector nonlinearity control structure (3.83) to (3.85) can be determined to drive the error vector into the null space of S and thereafter maintain it within this sliding subspace. The m range space eigenvalues, associated with the design matrix Φ, are chosen to ensure that the error state approaches the sliding surface in a suitably short time. Let $P_2 \in \mathbb{R}^{m \times m}$ denote a symmetric positive definite matrix satisfying the Lyapunov equation

$$P_2 \Phi + \Phi^{\mathrm{T}} P_2 = -I \tag{4.90}$$

The error feedback component u_1 from (4.73) is then defined to be $u_l + u_n$ where

$$u_l(t) = -(SB)^{-1}\left(SA_m - \Phi S\right)e(t) \tag{4.91}$$

and

$$u_n(t) = -\rho(t,e)(SB)^{-1}\frac{P_2 s(t)}{\|P_2 s(t)\|} \qquad \text{for } s(t) \neq 0 \tag{4.92}$$

The scalar function $\rho(t, e)$, which premultiplies the unit vector component, depends only on the magnitude of the uncertainty and can be derived from arguments similar to those in Section 3.6. The complete model-following variable structure control scheme thus has the form

$$u(t) = u_l(t) + u_n(t) + u_2(t) \tag{4.93}$$

where u_2 is defined in (4.82). Note that this component is not essential as the contributions of x and r to the error dynamics can be viewed as matched uncertainty which the component u_n can overcome. However, as they constitute known signals, it is sensible to tailor the distribution of control effort in this way.

4.4.2 An Integral Action Approach

Consider initially the development of a tracking control law for the nominal linear system

$$\dot{x}(t) = Ax(t) + Bu(t) \tag{4.94}$$

$$y(t) = Cx(t) \tag{4.95}$$

which is assumed to be square. In addition, for convenience, assume the matrix pair (A, B) is in regular form. The control law described here utilises an integral action methodology. Consider the introduction of additional states $x_r \in \mathbb{R}^p$ satisfying

$$\dot{x}_r(t) = r(t) - y(t) \tag{4.96}$$

where the differentiable signal $r(t)$ satisfies

$$\dot{r}(t) = \Gamma\left(r(t) - R\right) \tag{4.97}$$

with $\Gamma \in \mathbb{R}^{p\times p}$ a stable design matrix and R a constant demand vector. Augment the states with the integral action states and define

$$\tilde{x} = \begin{bmatrix} x_r \\ x \end{bmatrix} \tag{4.98}$$

and partition the augmented states as

$$\tilde{x} = \begin{bmatrix} x_1 \\ x_2 \end{bmatrix} \tag{4.99}$$

where $x_1 \in \mathbb{R}^n$ and $x_2 \in \mathbb{R}^m$. The (augmented) nominal system can then be conveniently written in the form

$$\dot{x}_1(t) = \tilde{A}_{11}x_1(t) + \tilde{A}_{12}x_2(t) + B_r r(t) \tag{4.100}$$

$$\dot{x}_2(t) = \tilde{A}_{21}x_1(t) + A_{22}x_2(t) + B_2 u(t) \tag{4.101}$$

where

$$\begin{bmatrix} \tilde{A}_{11} & \tilde{A}_{12} \\ \tilde{A}_{21} & A_{22} \end{bmatrix} \triangleq \left[\begin{array}{cc|c} 0 & -C_1 & -C_2 \\ 0 & A_{11} & A_{12} \\ \hline 0 & A_{21} & A_{22} \end{array}\right] \tag{4.102}$$

and the gain on the 'demand' signal $r(t)$ is given by

$$B_r = \begin{bmatrix} I_p \\ 0 \end{bmatrix} \tag{4.103}$$

The proposed controller seeks to induce a sliding motion on the surface

$$\mathcal{S} = \{\tilde{x} \in \mathbb{R}^{n+p} \ : \ S\tilde{x} = S_r r\} \tag{4.104}$$

where $S \in \mathbb{R}^{m\times(n+p)}$ and $S_r \in \mathbb{R}^{p\times p}$ are design parameters which govern the reduced-order motion. Partition the hyperplane system matrix as that

$$S = \begin{bmatrix} \overset{n}{\leftrightarrow} & \overset{p}{\leftrightarrow} \\ S_1 & S_2 \end{bmatrix} \tag{4.105}$$

and assume $S_2 = \Lambda B_2^{-1}$ where Λ is a nonsingular diagonal design matrix. If a controller exists which induces an ideal sliding motion on $\mathcal{S}$ then the ideal sliding motion is given by

$$\dot{x}_1(t) = (\tilde{A}_{11} - \tilde{A}_{12}M)x_1(t) + \left(\tilde{A}_{12}S_2^{-1}S_r + B_r\right) r(t) \tag{4.106}$$

where $M \triangleq S_2^{-1}S_1$. In order for the hyperplane design methods described earlier to be valid, it is necessary for the matrix pair $(\tilde{A}_{11}, \tilde{A}_{12})$ to be completely controllable. Necessary conditions on the original system are given by the following result.

Lemma 4.1 *If (A, B, C) is completely controllable and has no invariant zeros at the origin then the matrix pair $(\tilde{A}_{11}, \tilde{A}_{12})$ is completely controllable.*

Proof
Denote Rosenbrock's system matrix by

$$P(z) = \begin{bmatrix} zI - A & B \\ -C & 0 \end{bmatrix}$$

The invariant zeros of the triple (A, B, C) are given by

$$\{z \in \mathbb{C} \ : \ \det P(z) = 0\}$$

Therefore the system has zeros at the origin if and only if $\det P(0) = 0$. Because (A, B, C) is already in regular form and B_2 is nonsingular

$$\begin{aligned} \det P(0) = 0 \quad &\Leftrightarrow \quad \det \begin{bmatrix} -C & 0 \\ -A & B \end{bmatrix} = 0 \\ &\Leftrightarrow \quad \det \begin{bmatrix} -C_1 & -C_2 & 0 \\ -A_{11} & -A_{12} & 0 \\ -A_{21} & -A_{22} & B_2 \end{bmatrix} = 0 \\ &\Leftrightarrow \quad \det \begin{bmatrix} C_1 & C_2 \\ A_{11} & A_{12} \end{bmatrix} = 0 \end{aligned}$$

Utilising the PBH rank test, the pair $(\tilde{A}_{11}, \tilde{A}_{12})$ is controllable if and only if

$$\operatorname{rank}\begin{bmatrix} zI_p & C_1 & -C_2 \\ 0 & zI - A_{11} & A_{12} \end{bmatrix} = n \qquad \text{for all } z \in \mathbb{C} \tag{4.107}$$

If $z = 0$ then

$$\begin{aligned}
\operatorname{rank}\begin{bmatrix} zI_p & C_1 & -C_2 \\ 0 & zI - A_{11} & A_{12} \end{bmatrix} = n \quad &\Leftrightarrow \quad \det\begin{bmatrix} C_1 & -C_2 \\ -A_{11} & A_{12} \end{bmatrix} \neq 0 \\
&\Leftrightarrow \quad \det\begin{bmatrix} C_1 & C_2 \\ A_{11} & A_{12} \end{bmatrix} \neq 0 \\
&\Leftrightarrow \quad (A, B, C) \text{ has no zeros at the origin}
\end{aligned}$$

Otherwise $z \neq 0$ and

$$\operatorname{rank}\begin{bmatrix} zI_p & C_1 & C_2 \\ 0 & zI - A_{11} & A_{12} \end{bmatrix} = n \quad \Leftrightarrow \quad \operatorname{rank}\begin{bmatrix} zI - A_{11} & A_{12} \end{bmatrix} \;=\; n - p$$

However, from Proposition 3.3, (A, B) is controllable if and only if (A_{11}, A_{12}) is controllable and therefore by assumption

$$\operatorname{rank}\begin{bmatrix} zI - A_{11} & A_{12} \end{bmatrix} \;=\; n - p \qquad \text{for all } z$$

from the PBH rank test applied to (A_{11}, A_{12}). Therefore assertion (4.107) is true and $(\tilde{A}_{11}, \tilde{A}_{12})$ is controllable.

■

The development that follows mirrors the approach in Section 3.6. Specifically, define

$$T_s = \begin{bmatrix} I_n & 0 \\ S_1 & S_2 \end{bmatrix} \tag{4.108}$$

which is nonsingular because by construction S_2 is nonsingular. Let

$$\begin{bmatrix} x_1 \\ s \end{bmatrix} \triangleq T_s \begin{bmatrix} x_1 \\ x_2 \end{bmatrix}$$

then with respect to these new coordinates the nominal system can be written as

$$\dot{x}_1(t) = \bar{A}_{11} x_1(t) + \bar{A}_{12} s(t) + B_r r(t) \tag{4.109}$$

$$\dot{s}(t) = S_2 \bar{A}_{21} x_1(t) + S_2 \bar{A}_{22} S_2^{-1} s(t) + \Lambda u(t) + S_1 B_r r(t) \tag{4.110}$$

where $\bar{A}_{11} = \tilde{A}_{11} - \tilde{A}_{12} M$, $\bar{A}_{21} = M \bar{A}_{11} + \tilde{A}_{21} - A_{22} M$, $\bar{A}_{22} = M \tilde{A}_{12} + A_{22}$ and for convenience $\bar{A}_{12} = \tilde{A}_{12} S_2^{-1}$. Define a linear operator $u_L(\cdot)$ as

$$u_L(x_1, s, r) = \Lambda^{-1}\big(-S_2 \bar{A}_{21} x_1 + (\Phi - S_2 \bar{A}_{22} S_2^{-1}) s - (\Phi S_r + S_1 B_r) r + S_r \dot{r}\big) \tag{4.111}$$

where Φ is any stable design matrix. The overall control law is then given by

$$u = u_L(x_1, s, r) + u_N(s, r) \tag{4.112}$$

where the discontinuous vector

$$u_N(s,r) = \begin{cases} -\rho_c(u_L,y)\Lambda^{-1}\frac{\bar{P}_2(s-S_r r)}{\|\bar{P}_2(s-S_r r)\|} & \text{if } s \neq S_r r \\ 0 & \text{otherwise} \end{cases} \tag{4.113}$$

where $\bar{P}_2$ is a symmetric positive definite matrix satisfying

$$\bar{P}_2\Phi + \Phi^{\mathrm{T}}\bar{P}_2 = -I \tag{4.114}$$

The positive scalar function which multiplies the unit vector component can be obtained from arguments similar to those in Section 3.6. It follows that, in terms of the original coordinates

$$u_L(\tilde{x},r) = L\tilde{x} + L_r r + L_{\dot{r}}\dot{r} \tag{4.115}$$

with gains defined as

$$L = -\Lambda^{-1}(S\tilde{A} - \Phi S) \tag{4.116}$$

$$L_r = -\Lambda^{-1}(\Phi S_r + S_1 B_r) \tag{4.117}$$

$$L_{\dot{r}} = \Lambda^{-1}S_r \tag{4.118}$$

where the augmented system matrix

$$\tilde{A} \triangleq \begin{bmatrix} \tilde{A}_{11} & \tilde{A}_{12} \\ \tilde{A}_{21} & A_{22} \end{bmatrix} \tag{4.119}$$

4.5 DESIGN STUDY: PITCH-POINTING FLIGHT CONTROLLER

The configuration of the aircraft has changed very little in principle since the beginning of aviation; almost all aircraft have a standard arrangement of rudder, aileron and elevator control surfaces. However, it can easily be shown (Huenecke, 1987) that this design has considerable disadvantages for particular aircraft manoeuvres. Consider the situation illustrated in Figure 4.1.

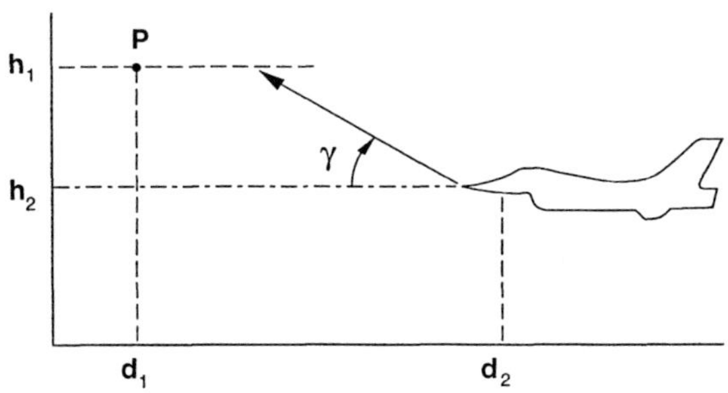

Figure 4.1: Aircraft manoeuvre

An aircraft is flying on a straight track. The pilot wants to point the aircraft towards the point P. Note that there is a longitudinal distance separating the aircraft from P as well as a vertical separation. The aircraft must either increase its altitude or pitch by the angle γ. A conventional aircraft will first of all use the elevator, after which a rotation about the lateral axis will be initiated. The required movement in the vertical direction can now be accomplished; the path alteration will therefore take the form of an oscillation.

If the combined rotational and translational motion could be successfully decoupled, the pilot would have a much more straightforward flying task. Consider the following two configurations for decoupling of the longitudinal motions. Firstly, the pilot commands the pitch angle θ whilst maintaining zero perturbation of the flight path angle γ. This configuration is illustrated in Figure 4.2 and will be referred to as *pitch-pointing control.* For the second configuration, the pilot commands the flight path angle γ whilst maintaining zero perturbation of the pitch angle θ. This so-called *vertical translation* control is illustrated in Figure 4.3.

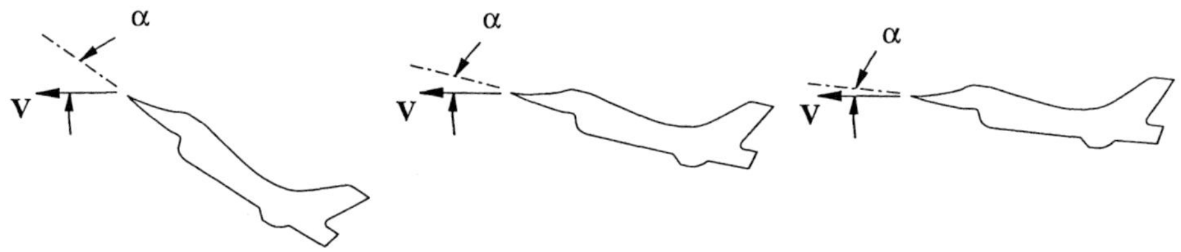

Figure 4.2: Pitch-pointing manoeuvre

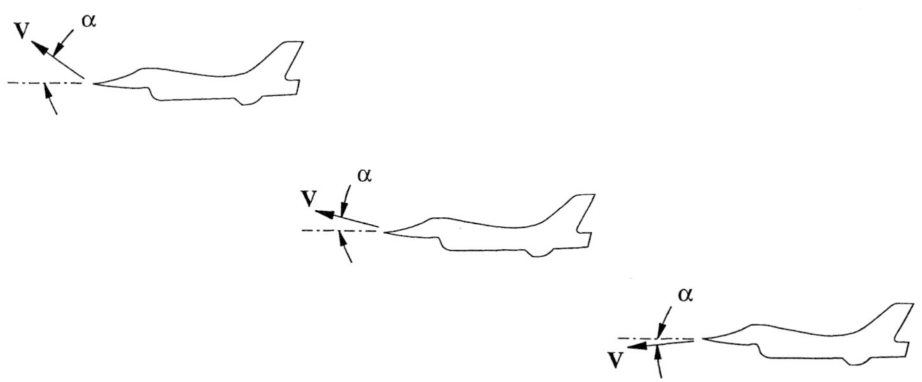

Figure 4.3: Vertical translation manoeuvre

Such advanced longitudinal control modes can be realised by using flaps or canard control surfaces in conjunction with the elevator. In this way, for example, a change of flight path without rotation can be achieved by directly influencing the lift acting upon the vehicle. A computer-based automatic flight control system is required for the coordinated actuation of the control elements. Smith & Anderson (1982) applied optimal control theory for the design of such decoupled controllers. However, it is extremely difficult to translate specifications for piloted vehicles into the required form for an optimal control problem; for example, the required damping, frequency and decoupling are not easily incorporated into a quadratic performance index, which results in quadratic weight adjustments by trial and error. This can be computationally expensive, time-consuming and does not utilise the engineer's knowledge of the aircraft system. An alternative design procedure has been proposed by Ridgely *et al.* (1982). This process uses high gain error-actuated control in conjunction with singular perturbation methods. The method does allow modal considerations, but has been found to yield both poles and feedback gains in excess of 1000. The possibilities of solving the control problem using the techniques developed in this chapter will now be explored.

Consider the longitudinal motion model of a conventional combat aircraft. The lateral and longitudinal motions are assumed decoupled for the control system design and a linearisation at the trim flight condition corresponding to an altitude of 3048 m and 0.77 Mach number yields

$$\dot{x}(t) = Ax(t) + Bu(t) \tag{4.120}$$

where

$$x(t) = \begin{bmatrix} \theta \\ q \\ \alpha \\ \eta \\ \delta \end{bmatrix} \begin{matrix} \text{pitch angle (rad)} \\ \text{pitch rate (rad/sec)} \\ \text{angle of attack (rad)} \\ \text{elevator deflection (rad)} \\ \text{flap deflection (rad)} \end{matrix} \tag{4.121}$$

and

$$u(t) = \begin{bmatrix} \eta_c \\ \delta_c \end{bmatrix} \begin{matrix} \text{elevator command (rad)} \\ \text{flap command (rad)} \end{matrix} \tag{4.122}$$

and the linearised plant matrices have the form

$$A = \begin{bmatrix} 0 & 1 & 0 & 0 & 0 \\ 0 & -1.99 & -13.41 & -18.95 & -3.60 \\ 0 & 1.00 & -1.74 & -0.08 & -0.59 \\ 0 & 0 & 0 & -20 & 0 \\ 0 & 0 & 0 & 0 & -20 \end{bmatrix} \tag{4.123}$$

and

$$B^{\mathrm{T}} = \begin{bmatrix} 0 & 0 & 0 & 20 & 0 \\ 0 & 0 & 0 & 0 & 20 \end{bmatrix} \tag{4.124}$$

The eigenvalues of the open-loop system are given by

$$\begin{matrix} -1.864 \pm j3.660 & \text{short period mode} \\ 0 & \text{pitch attitude mode} \\ -20 & \text{elevator actuator mode} \\ -20 & \text{flap actuator mode} \end{matrix}$$

The performance of the flight path angle is of direct interest and so in the state-space representation the pitch angle state is replaced by the flight path angle using the relationship

$$\theta = \gamma + \alpha \tag{4.125}$$

The state-space representation in equations (4.120) to (4.124) is therefore modified. The state vector (4.121) becomes

$$x^{\mathrm{T}} = (\gamma, q, \alpha, \eta, \delta) \tag{4.126}$$

The first row of the state matrix A in (4.123) becomes

$$\begin{bmatrix} 0 & 0 & 1.74 & 0.08 & 0.59 \end{bmatrix} \tag{4.127}$$

The remaining rows of the A matrix are unchanged, as are the B and u formulations.

4.5.1 Model-Reference Design

Here a design method using direct eigenstructure assignment to obtain the required decoupling, damping and rise time of the system response together with a set of feed-forward gains which are computed to ensure steady-state tracking of a step command from the pilot will be used to prescribe an ideal model. A variable structure model-following control system will then be derived.

In order to provide rapid manoeuvring, precision tracking and precise flight path control, the following desired closed-loop eigenvalues are chosen

$$\begin{array}{lcll} \lambda_{1,2} & = & -5.6 \pm 4.2j & \text{short period mode} \\ \lambda_3 & = & -1.0 & \text{flight path mode} \\ \lambda_4 & = & -20 & \text{elevator actuator mode} \\ \lambda_5 & = & -20 & \text{flap actuator mode} \end{array}$$

The desired short period frequency and damping are thus chosen to be $\zeta = 0.8$ and $\omega_n = 7\,\text{rad/sec}$. The desired eigenvectors are chosen to decouple the short period and flight path modes. The desired eigenvectors corresponding to the short period mode are thus selected to have no contribution from the first state, which corresponds to the flight path angle, but are allowed contributions from the pitching motions of the aircraft. The desired eigenvector corresponding to the flight path mode is chosen to have the opposite characteristics. These choices should prevent a pitch attitude command from causing significant change in the flight path angle; the performance specification is thus to minimise the coupling between pitch rate and flight path angle. The desired and achievable eigenvectors are

$$\begin{aligned} V^d &= \begin{bmatrix} v_1^d & v_2^d & v_3^d & v_4^d & v_5^d \end{bmatrix} \\ &= \begin{bmatrix} 0 & 0 & 1 & v_{14} & v_{15} \\ 1 + v_{22}j & 1 - v_{22}j & 0 & v_{24} & v_{25} \\ v_{31} + j & v_{31} - j & 0 & v_{34} & v_{35} \\ v_{41} + v_{42}j & v_{41} - v_{42}j & v_{43} & 1 & v_{45} \\ v_{51} + v_{52}j & v_{51} - v_{52}j & v_{53} & v_{54} & 1 \end{bmatrix} \end{aligned} \tag{4.128}$$

where the v_{ij} elements are unrestricted and

$$\begin{aligned} V^a &= \begin{bmatrix} v_1^a & v_2^a & v_3^a & v_4^a & v_5^a \end{bmatrix} \\ &= \begin{bmatrix} 0 & 0 & 0.67 & 0.01 & -0.03 \\ 1.00 - 9.50j & 1.00 + 9.50j & -0.33 & 0.97 & 0.11 \\ -0.93 + j & -0.93 - j & -0.33 & -0.06 & 0.03 \\ -1.82 - 2.24j & -1.82 + 2.24j & 0.29 & 1.00 & -0.11 \\ 2.97 - 2.62j & 2.97 + 2.62j & -0.18 & -0.21 & 1.00 \end{bmatrix} \end{aligned} \tag{4.129}$$

The resulting full-state feedback matrix F is computed from (4.37) to be

$$F = \begin{bmatrix} 1.9310 & 0.5743 & 2.1142 & -0.4750 & -0.1066 \\ -1.9441 & -0.2241 & -3.1057 & 0.0693 & 0.0514 \end{bmatrix} \tag{4.130}$$

It is now necessary to compute the feed-forward gains which ensure steady-state tracking of a step command of flight path angle γ or pitch angle θ. The outputs of interest are given by

$$\begin{aligned} y(t) &= \begin{bmatrix} \theta(t) \\ \gamma(t) \end{bmatrix} = \begin{bmatrix} 1 & 0 & 1 & 0 & 0 \\ 1 & 0 & 0 & 0 & 0 \end{bmatrix} x(t) \\ &= Hx(t) \end{aligned} \tag{4.131}$$

The inverse of the steady-state gain for the triple $(A + BF, B, H)$ will provide an appropriate feed-forward injection signal

$$Gr(t) = -(H(A+BF)^{-1}B)^{-1}r(t) \tag{4.132}$$

$$= \begin{bmatrix} -2.6495 & 0.7184 \\ 0.3382 & 1.6059 \end{bmatrix} \begin{bmatrix} \theta_c(t) \\ \gamma_c(t) \end{bmatrix} \tag{4.133}$$

where θ_c and γ_c are the pitch and flight path angle commands at the pilot's seat respectively. If $\gamma_c = 0$ a pitch-pointing manoeuvre will result, whereas if $\theta_c = 0$ a vertical translation manoeuvre results. It should be noted that, using this control type, it is possible to command θ and γ simultaneously if required. The closed-loop system can thus be expressed in the general form

$$\dot{x}(t) = Ax(t) + Bu(t) \tag{4.134}$$

$$u(t) = Fx(t) + Gr(t) \tag{4.135}$$

In order to define an ideal model with the required response characteristics, let

$$\dot{w}(t) = A_m w(t) + B_m r(t) \tag{4.136}$$

where

$$A_m = A + BF \tag{4.137}$$

$$B_m = BG \tag{4.138}$$

A required model structure has now been precisely defined; development of a non-linear control system can now be considered. The matrix pair (A_m, B) is taken and a corresponding hyperplane designed using the quadratic minimisation technique developed in Section 4.2.2, where the performance index is given by

$$Q = \mathrm{diag}(10, 5, 5, 20, 20) \tag{4.139}$$

This choice is based upon the assumptions that the deflections of the elevator and flaps of the actual system should follow the model deflections closely and that fast response is required to errors in desired flight path angle. The following hyperplane matrix S is obtained:

$$S = \begin{bmatrix} 0.7025 & 0.4209 & 0.1894 & -1 & 0 \\ -0.0810 & 0.0635 & 0.0267 & 0 & -1 \end{bmatrix} \tag{4.140}$$

The eigenvalues of the range space dynamics, linked with the design of the linear control component (4.91), are chosen to be -20 and -20. This ensures rapid approach to the sliding subspace whilst accommodating the bandwidths of the elevator and flaps. The linear component may be expressed in terms of the physically meaningful error coordinate system as

$$\begin{aligned} u_l(t) &= Le(t) \\ &= \begin{bmatrix} -1.2285 & -0.1858 & -2.1624 & 0.0782 & 0.0460 \\ 1.8631 & 0.2827 & 3.0804 & -0.1299 & -0.0660 \end{bmatrix} e(t) \end{aligned} \tag{4.141}$$

The parameters of the nonlinear control component (4.92) are given by

$$P_2 = \begin{bmatrix} 0.025 & 0 \\ 0 & 0.025 \end{bmatrix} \tag{4.142}$$

where the gain ρ is set equal to unity. This control scheme is augmented to form the full model-following control law of equation (4.93). The step response of the aircraft system when pitch-pointing is shown in Figure 4.4. Excellent decoupling of the pitch and flight path channels is exhibited.

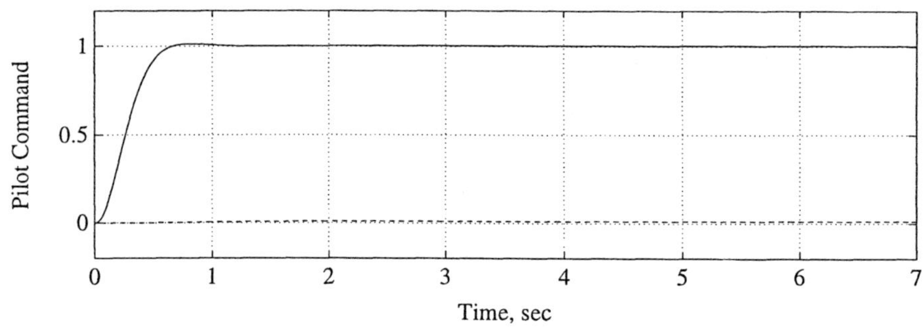

Figure 4.4: Pitch-pointing manoeuvre with model-reference VSCS

4.5.2 Integral Action Based Design

Here the pilot's command is incorporated using the alternative integral action approach. The output equation is taken to be defined by the matrix H in equation (4.131). The design matrix Γ in equation (4.97) is given by

$$\Gamma = \begin{bmatrix} -0.9 & 0 \\ 0 & -0.7 \end{bmatrix} \tag{4.143}$$

and the poles of the ideal sliding mode dynamics have been chosen to be

$$\{-5.6 \pm 4.2j, -1, -0.4, -0.7\}$$

The corresponding desired eigenvectors which define the required decoupling are selected as

$$V^d = \begin{bmatrix} v_1^d & v_2^d & v_3^d & v_4^d & v_5^d \end{bmatrix}$$

$$= \begin{bmatrix} v_{11}+v_{12}j & v_{11}-v_{12}j & v_{13} & v_{14} & v_{15} \\ v_{21}+v_{22}j & v_{21}-v_{22}j & v_{23} & v_{24} & v_{25} \\ 0 & 0 & 1 & 0 & 1 \\ 1+v_{42}j & 1-v_{42}j & 0 & 1 & 0 \\ v_{51}+j & v_{51}-j & 0 & v_{54} & 0 \\ v_{61}+v_{62}j & v_{61}-v_{62}j & v_{63} & v_{64} & v_{65} \\ v_{71}+v_{72}j & v_{71}-v_{72}j & v_{73} & v_{74} & v_{75} \end{bmatrix} \quad (4.144)$$

The corresponding switching surface is determined as

$$S = \begin{bmatrix} 0.0531 & 0.0137 & -0.1436 & -0.0260 & -0.1373 & 0.0500 & 0 \\ -0.0072 & -0.0612 & 0.1635 & 0.0035 & 0.1661 & 0 & 0.0500 \end{bmatrix} \quad (4.145)$$

From equations (4.115) to (4.118) the controller gains are given by

$$L = \begin{bmatrix} -1.0616 & -0.2743 & 2.9379 & 0.6060 & 2.4605 & -0.4927 & -0.0900 \\ 0.1440 & 1.2236 & -3.3390 & -0.2296 & -3.2769 & 0.0671 & 0.0142 \end{bmatrix}$$

$$L_r = \begin{bmatrix} -2.9499 & 0.0120 \\ 0.4000 & 2.9391 \end{bmatrix}$$

$$L_{\dot{r}} = \begin{bmatrix} -0.1448 & 0.0013 \\ 0.0196 & 0.1439 \end{bmatrix} \quad (4.146)$$

where the eigenvalues of the range space dynamics are taken to be -20 and -20. The step response of the aircraft system when undergoing a vertical translation manoeuvre is shown in Figure 4.5. Excellent decoupling of the pitch and flight path channels is again exhibited.

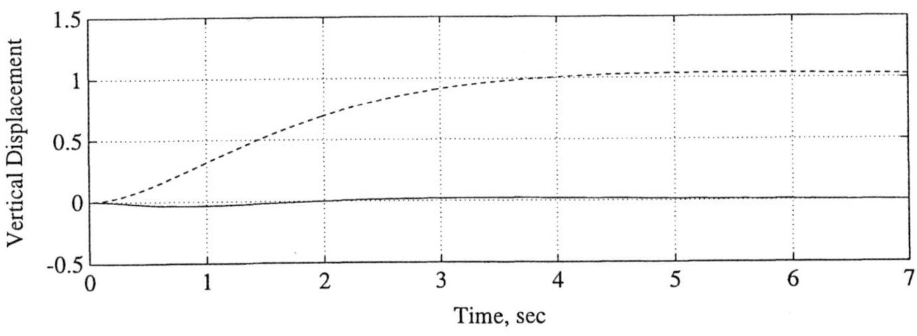

Figure 4.5: Vertical translation manoeuvre with integral action VSCS

4.6 SUMMARY

This chapter has developed design algorithms appropriate for the construction of sliding mode control schemes for uncertain systems where all of the states are measurable. One particular algorithm has been presented which seeks to minimise the effects of unmatched parameter variations, and in this way it has been shown that the sliding system cannot only reject a class of uncertainty completely but can also be made maximally robust to the remaining unmatched uncertainty. Appropriate methodologies have been described to solve servomechanism and other tracking problems. Much of the work has been illustrated in the development of an advanced aircraft flight control system. In later chapters industrial application of these results will be explored and the further development of the design algorithms to encompass the output feedback case will be considered.

4.7 NOTES AND REFERENCES

The regular form used for sliding mode design is closely related to the controllability canonical form for a multivariable linear system due to Kwakernaak & Sivan (1972).

The robust eigenstructure assignment procedure used in Section 4.2.1 is based upon the work of Kautsky, Nichols and co-workers (Kautsky *et al.*, 1985; Kautsky & Nichols, 1983; Nichols, 1987). The quadratic minimisation approach of Utkin & Young (1978) is also described in Zinober & Dorling (1990). The latter also considers eigenstructure assignment procedures for hyperplane design. Further approaches for solving the existence problem are available in the contributions of Woodham & Zinober (1993) and Zinober (1994), which describe procedures that force the reduced-order eigenvalues to lie in specified regions of the complex plane.

Alternative methods for achieving robust model-reference control in the presence of uncertainty and disturbances use the hyperstability concept or Lyapunov techniques. Further details can be found in the following references: Landau (1974), Landau & Courtoil (1974), Landau (1979), Butchart & Shakcloth (1966) and also Monopoli *et al.* (1968).

For further details on the conditions for perfect model-following, see the contributions of Erzberger (1968), Chan (1973) and Corless *et al.* (1985).

VSCS model-following control schemes are considered in the papers by Young (1977), Zinober *et al.* (1982), Balestrino *et al.* (1984), Dorling & Zinober (1985) and Zinober (1984).

The integral action control law in this chapter is based on the work of Davies & Spurgeon (1993).

Broussard's method (O'Brien & Broussard, 1978; Broussard & O'Brien, 1980) can be useful to compute the feed-forward control effort that is required to ensure steady-state tracking of more complex signals in model-reference frameworks.

Further details of the variable structure model-following aircraft control system can be found in the paper by Mudge & Patton (1988).

Chapter 5

Sliding Mode Controllers Using Output Information

5.1 INTRODUCTION

In the previous chapter, arguably the most stringent constraint related to the assumption that all the states of the system to be regulated were available to the controller. In most practical situations this is not the case. In some circumstances it is impossible or prohibitively expensive to measure all of the process variables. Alternatively, the system may be so complex that a system identification approach must be adopted to obtain a reasonable model. In this case the states will usually have no physical meaning and thus cannot be measured. This chapter considers the problem of designing the sliding surface and the variable structure control law in such a way that only output information is required.

The majority of this chapter is devoted to developing a framework for designing a regulator for uncertain systems where only output information is available to the controller. A particular canonical form, which may be viewed as a special case of the regular form introduced in Section 3.4, will be shown to be important in the design procedure. At the end of the chapter, it will be shown that these ideas can be incorporated into a model-reference structure to generate an output tracking controller. The control law that will be developed is similar to the unit vector structure employed in the previous chapter except that only output information is used.

5.2 PROBLEM FORMULATION

Consider an uncertain dynamical system of the form

$$\begin{aligned} \dot{x}(t) &= Ax(t) + Bu(t) + f(t,x,u) \\ y(t) &= Cx(t) \end{aligned} \tag{5.1}$$

where $x \in \mathbb{R}^n$, $u \in \mathbb{R}^m$ and $y \in \mathbb{R}^p$ with $m \leq p < n$. Assume that the nominal linear system (A, B, C) is known and that the input and output matrices B and C are both of full rank. The unknown function $f : \mathbb{R}_+ \times \mathbb{R}^n \times \mathbb{R}^m \to \mathbb{R}^n$, which represents the system nonlinearities plus any model uncertainties in the system, is assumed to satisfy the matching condition

$$f(t,x,u) = B\xi(t,x,u) \tag{5.2}$$

where the bounded function $\xi : \mathbb{R}_+ \times \mathbb{R}^n \times \mathbb{R}^m \to \mathbb{R}^m$ satisfies

$$\|\xi(t,x,u)\| < k_1\|u\| + \alpha(t,y) \tag{5.3}$$

for some known function $\alpha : \mathbb{R}_+ \times \mathbb{R}^p \to \mathbb{R}_+$ and positive constant $k_1 < 1$.

The intention is to develop a control law which induces an ideal sliding motion on the surface

$$\mathcal{S} = \{x \in \mathbb{R}^n \;\; : FCx = 0\} \tag{5.4}$$

for some selected matrix $F \in \mathbb{R}^{m\times p}$. A control law of the form

$$u(t) = Gy(t) - \nu_y \tag{5.5}$$

will be sought where G is a fixed gain matrix and the discontinuous vector

$$\nu_y = \begin{cases} \rho(t,y)\frac{Fy(t)}{\|Fy(t)\|} & \text{if } Fy \neq 0 \\ 0 & \text{otherwise} \end{cases} \tag{5.6}$$

where $\rho(t,y)$ is some positive scalar function of the outputs.

Before establishing a general framework, it is instructive to first consider the special case when the number of inputs equals the number of outputs.

5.3 A SPECIAL CASE: SQUARE PLANTS

Consider the special case when the number of inputs equals the number of outputs: such plants will be referred to as *square*. Consider first the choice of hyperplane to guarantee a stable reduced-order motion.

From Section 3.3, in order for a unique equivalent control to exist, $\det(FCB) \neq 0$. Because by assumption $m = p$, it follows that F is a square matrix and therefore

$$\det(FCB) = \det(F)\ \det(CB)$$

which implies that for a unique equivalent control to exist both $\det(CB) \neq 0$ and $\det(F) \neq 0$. Note that the term 'CB' is a Markov parameter[1] and is independent of any similarity transformation. Another direct consequence is that F acts only as a scaling of the hyperplane and hence

$$\mathcal{S} = \{x \in \mathbb{R}^n \;\; : \;\; Cx = 0\} = \mathcal{N}(C)$$

Using Proposition 3.4 it follows that the poles of the reduced-order sliding motion are the invariant zeros of the triple (A, B, C). Consequently, in this special case there is no freedom of choice in the selection of the reduced-order motion and it can be concluded that necessary conditions for a stable sliding motion are that

[1]The Markov parameters CA^iB for $i = 0, 1, 2, \ldots$ are the coefficients of the powers of $1/s$ in the expansion of $C(sI - A)^{-1}B$; for further details see for example Chen (1984).

- $\det(CB) \neq 0$
- invariant zeros of (A, B, C) are in $\mathbb{C}_-$.

All the preceding analysis of course depends on the existence of a control law which induces sliding. A natural starting point is to examine the effect that considering outputs only has on the control structure developed in Section 3.6. The nonlinear component in (3.85) depends only on the switching function and thus poses no problem. The linear component, however, was shown to be

$$u_l(t) = -(SB)^{-1}SAx(t) + (SB)^{-1}\Phi Sx(t) \tag{5.7}$$

which, in this particular case, becomes

$$u_l(t) = -(FCB)^{-1}FCAx(t) + (FCB)^{-1}\Phi FCx(t) \tag{5.8}$$

It is clear that unless by some good fortune $CA = MC$ for some $M \in \mathbb{R}^{m\times m}$ then (5.8) cannot be realised. In the remainder of this section a different approach will be explored.

Assume that (5.1) is expressed in regular form and the triple (A, B, C) is partitioned as

$$A = \begin{bmatrix} A_{11} & A_{12} \\ A_{21} & A_{22} \end{bmatrix} \quad B = \begin{bmatrix} 0 \\ B_2 \end{bmatrix} \quad C = \begin{bmatrix} C_1 & C_2 \end{bmatrix} \tag{5.9}$$

where $A_{22}, B_2 \in \mathbb{R}^{m\times m}$ and B_2 is nonsingular. Also assume that $\det(CB) \neq 0$ and the invariant zeros of (A, B, C) have negative real parts. Consider the change of coordinates $x \mapsto \bar{T}x$ where

$$\bar{T} = \begin{bmatrix} I & 0 \\ C_1 & C_2 \end{bmatrix} \tag{5.10}$$

Since $CB = C_2B_2$ and both $\det(CB) \neq 0$ and $\det(B_2) \neq 0$ it follows $\det C_2 \neq 0$ and the transformation matrix in (5.10) is nonsingular. In the new coordinate system

$$\bar{A} = \begin{bmatrix} \bar{A}_{11} & \bar{A}_{12} \\ \bar{A}_{21} & \bar{A}_{22} \end{bmatrix} \qquad \bar{B} = \begin{bmatrix} 0 \\ B_2 \end{bmatrix} \qquad F\bar{C} = \begin{bmatrix} 0 & F \end{bmatrix} \tag{5.11}$$

where $\bar{A}_{11} = A_{11} - A_{12}C_2^{-1}C_1$.

Lemma 5.1 *The invariant zeros of $(\bar{A}, \bar{B}, \bar{C})$ are the eigenvalues of $\bar{A}_{11}$.*

Proof
This can be shown using a slight modification to the arguments in the proof of Proposition 3.4.

∎

Let P be a symmetric positive definite matrix partitioned conformably with the matrices in (5.11) so that

$$P = \begin{bmatrix} P_1 & 0 \\ 0 & P_2 \end{bmatrix} \tag{5.12}$$

where the symmetric positive definite sub-block P_2 is a design matrix and the symmetric positive definite sub-block P_1 satisfies the Lyapunov equation

$$P_1\bar{A}_{11} + \bar{A}_{11}^{\mathrm{T}}P_1 = -Q_1 \tag{5.13}$$

for some symmetric positive definite matrix Q_1. If

$$F \triangleq B_2^{\mathrm{T}}P_2 \tag{5.14}$$

then the matrix P satisfies the structural constraint

$$P\bar{B} = \bar{C}^{\mathrm{T}}F^{\mathrm{T}} \tag{5.15}$$

For notational convenience let

$$Q_2 \triangleq P_1\bar{A}_{12} + \bar{A}_{21}^{\mathrm{T}}P_2 \tag{5.16}$$

$$Q_3 \triangleq P_2\bar{A}_{22} + \bar{A}_{22}^{\mathrm{T}}P_2 \tag{5.17}$$

and define

$$\gamma_0 \triangleq \tfrac{1}{2}\lambda_{max}\left((F^{-1})^{\mathrm{T}}(Q_3 + Q_2^{\mathrm{T}}Q_1^{-1}Q_2)F^{-1}\right) \tag{5.18}$$

This scalar is well defined since the matrix on the right is symmetric and therefore has no complex eigenvalues.

Lemma 5.2 *The symmetric matrix* $L(\gamma) \triangleq PA_0 + A_0^{\mathrm{T}}P$ *where* $A_0 = \bar{A} - \gamma\bar{B}F\bar{C}$ *is negative definite if and only if* $\gamma > \gamma_0$.

Proof

Using the definition of A_0 and the structural constraint (5.15) it follows that

$$\begin{aligned} L(\gamma) \triangleq PA_0 + A_0^{\mathrm{T}}P &= P\bar{A} + \bar{A}^{\mathrm{T}}P - \gamma P\bar{B}F\bar{C} - \gamma\bar{C}^{\mathrm{T}}F^{\mathrm{T}}\bar{B}^{\mathrm{T}}P \\ &= P\bar{A} + \bar{A}^{\mathrm{T}}P - 2\gamma(F\bar{C})^{\mathrm{T}}F\bar{C} \end{aligned}$$

Using the partitions of $\bar{A}, F\bar{C}$ and P from equations (5.11) and (5.12) respectively and the definitions of Q_2 and Q_3 it follows that

$$L(\gamma) = \begin{bmatrix} -Q_1 & Q_2 \\ Q_2^{\mathrm{T}} & Q_3 - 2\gamma F^{\mathrm{T}}F \end{bmatrix}$$

Using the properties of symmetric matrices (see Appendix A.2.7), and the fact that Q_1 is positive definite, it follows that

$$\begin{aligned} L(\gamma) < 0 &\Leftrightarrow Q_3 - 2\gamma F^{\mathrm{T}}F + Q_2^{\mathrm{T}}Q_1^{-1}Q_2 < 0 \\ &\Leftrightarrow 2\gamma F^{\mathrm{T}}F > Q_3 + Q_2^{\mathrm{T}}Q_1^{-1}Q_2 \\ &\Leftrightarrow 2\gamma I_m > (F^{-1})^{\mathrm{T}}(Q_3 + Q_2^{\mathrm{T}}Q_1^{-1}Q_2)F^{-1} \\ &\Leftrightarrow 2\gamma > \lambda_{max}\left((F^{-1})^{\mathrm{T}}(Q_3 + Q_2^{\mathrm{T}}Q_1^{-1}Q_2)F^{-1}\right) \end{aligned}$$

and the proof is complete. ■

Define a variable structure control law, depending only on outputs, by

$$u(t) = -\gamma Fy(t) - \nu_y \tag{5.19}$$

where $\gamma > \gamma_0$ and ν_y is the discontinuous vector given by

$$\nu_y = \begin{cases} \rho(t,y)\frac{Fy(t)}{\|Fy(t)\|} & \text{if } Fy \neq 0 \\ 0 & \text{otherwise} \end{cases} \tag{5.20}$$

and $\rho(t,y)$ is the positive scalar function

$$\rho(t,y) = (k_1\gamma\|Fy\| + \alpha(t,y) + \gamma_2)\,/\,(1-k_1) \tag{5.21}$$

where γ_2 is a positive design scalar which will be shown to define the region in which sliding takes place. In the new coordinate system, the uncertain system (5.1) can be written as

$$\dot{\bar{x}}(t) = \bar{A}\bar{x}(t) + \bar{B}(u(t) + \xi(t,x,u)) \tag{5.22}$$

Proposition 5.1 *The variable structure control law above quadratically stabilises the uncertain system given in equation (5.22).*

Proof
Consider as a candidate Lyapunov function the positive definite expression

$$V(\bar{x}) \triangleq \bar{x}^{\mathrm{T}}P\bar{x} \tag{5.23}$$

Taking derivatives along the system trajectory, and using the structural constraint from equation (5.15) gives

$$\begin{aligned} \dot{V} &= \bar{x}^{\mathrm{T}}\left(\bar{A}^{\mathrm{T}}P + P\bar{A} - 2\gamma(F\bar{C})^{\mathrm{T}}F\bar{C}\right)\bar{x} + 2\bar{x}^{\mathrm{T}}P\bar{B}(\xi - \nu_y) \\ &= \bar{x}^{\mathrm{T}}L(\gamma)\bar{x} + 2y^{\mathrm{T}}F^{\mathrm{T}}(\xi - \nu_y) \\ &\leq \bar{x}^{\mathrm{T}}L(\gamma)\bar{x} - 2y^{\mathrm{T}}F^{\mathrm{T}}\nu_y + 2\|Fy\|\|\xi\| \\ &= \bar{x}^{\mathrm{T}}L(\gamma)\bar{x} - 2\rho(t,y)\|Fy\| + 2\|Fy\|\|\xi\| \\ &< \bar{x}^{\mathrm{T}}L(\gamma)\bar{x} - 2\|Fy\|\left(\rho(t,y) - k_1\|u\| - \alpha(t,y)\right) \end{aligned}$$

But by definition

$$\rho(t,y) = (k_1\gamma\|Fy\| + \alpha(t,y) + \gamma_2)\,/(1-k_1)$$

and so by rearranging

$$\begin{aligned} \rho(t,y) &= k_1\rho(t,y) + k_1\gamma\|Fy\| + \alpha(t,y) + \gamma_2 \\ &\geq k_1(\|\nu_y\| + \gamma\|Fy\|) + \alpha(t,y) + \gamma_2 \\ &\geq k_1\|u\| + \alpha(t,y) + \gamma_2 \end{aligned} \tag{5.24}$$

Using (5.24) in the inequality for the Lyapunov derivative

$$\dot{V} < \bar{x}^{\mathrm{T}}L(\gamma)\bar{x} - 2\gamma_2\|Fy\| \;<\; 0 \qquad \text{if } \bar{x} \neq 0 \text{ and } \gamma > \gamma_0$$

and therefore the system is quadratically stable. ∎

Corollary 5.1 *An ideal sliding motion takes place on the surface* $\mathcal{S}$ *in the domain* $\Omega = \{\bar{x} \in \mathbb{R}^n \ : \ \|B_2^{-1} A_0^L \bar{x}\| < \gamma_2 - \eta\}$ *where the matrix* A_0^L *represents the last* m *rows of the closed-loop matrix* A_0 *and* η *is a small scalar satisfying* $0 < \eta < \gamma_2$.

Proof
Let the switching function $s(\bar{x}) = F\bar{C}\bar{x}$ then from equation (5.22) it follows that

$$\dot{s} = F\bar{C}A_0\bar{x} + FB_2\,(\xi - \nu_y)$$

Let $V_c : \mathbb{R}^m \rightarrow \mathbb{R}$ be defined by

$$V_c(s) = 2s^{\mathrm{T}}(F^{-1})^{\mathrm{T}} P_2 F^{-1} s$$

Using the fact that $F^{\mathrm{T}} = P_2 B_2$ it follows that $(F^{-1})^{\mathrm{T}} P_2 F^{-1} F\bar{C}A_0 = B_2^{-1} A_0^L$ and arguing as in Proposition 5.1 it can be verified that

$$\dot{V}_c = 2s^{\mathrm{T}} B_2^{-1} A_0^L \bar{x} + 2s^{\mathrm{T}}\,(\xi - \nu) \ \leq \ 2\|s\|\|B_2^{-1} A_0^L \bar{x}\| - 2\gamma_2\|s\| < -2\eta\|s\| \qquad (5.25)$$

if $\bar{x} \in \Omega$. From Proposition 5.1 the states $\bar{x}(t)$ are quadratically stable and so there exists a t_0 such that $\bar{x}(t) \in \Omega$ for all $t > t_0$. Consequently (5.25) holds for all $t > t_0$, and arguing as in Section 3.6.1, a sliding motion will be attained in finite time. ∎

Remark
From the statement of Corollary 5.1 it is clear that the parameter γ_2 defines the region Ω in which sliding takes place. As a result, outside of this region the system trajectories may pierce the sliding surface without sliding taking place and the control law will behave in a similar fashion to a relay. Because quadratic stability has been demonstrated, eventually the trajectories will enter Ω and sliding will take place.

A solution to the special case of square systems has been presented. The next section aims to provide a framework for the general case when the number of outputs is greater than (or equal to) the number of inputs.

5.4 A GENERAL FRAMEWORK

Consider the system in (5.1) and assume that $p \geq m$ and $\text{rank}(CB) = m$. The reason for imposing this rank restriction at the outset is that for a unique equivalent control to exist, the matrix $FCB \in \mathbb{R}^{m \times m}$ must have full rank. It is well known that

$$\text{rank}(FCB) \leq \min\{\text{rank}(F),\ \text{rank}(CB)\}$$

and so in order for FBC to have full rank both F and CB must have rank m. The matrix F is a design parameter and therefore by choice can be chosen to be of full rank. A necessary condition therefore for the matrix FBC to be full rank is that $\text{rank}(CB) = m$.

The first problem which must be considered is how to choose F so that the associated sliding motion is stable. A control law analogous to (5.19) to (5.21) will then be used to guarantee the existence of a sliding motion.

5.4.1 Hyperplane Design

In view of the fact that the outputs will be considered, it is convenient to introduce a coordinate transformation to make the last p states of the system the outputs. Define

$$T_c = \begin{bmatrix} N_c^{\mathrm{T}} \\ C \end{bmatrix} \tag{5.26}$$

where $N_c \in \mathbb{R}^{n\times(n-p)}$ and the columns span the null space of C. The coordinate transformation $x \mapsto T_c x$ is nonsingular by construction and, as a result, in the new coordinate system

$$C = \begin{bmatrix} 0 & I_p \end{bmatrix}$$

From this starting point a special case of regular form will be established: suppose in the new coordinate system

$$B = \begin{bmatrix} B_{c1} \\ B_{c2} \end{bmatrix} \begin{matrix} \updownarrow n-p \\ \updownarrow p \end{matrix}$$

Then $CB = B_{c2}$ and so by assumption $\mathrm{rank}(B_{c2}) = m$. Hence the left pseudo-inverse

$$B_{c2}^{\dagger} = (B_{c2}^{\mathrm{T}} B_{c2})^{-1} B_{c2}^{\mathrm{T}}$$

is well defined and there exists an orthogonal matrix $T \in \mathbb{R}^{p\times p}$ such that

$$T^{\mathrm{T}} B_{c2} = \begin{bmatrix} 0 \\ B_2 \end{bmatrix} \tag{5.27}$$

where $B_2 \in \mathbb{R}^{m\times m}$ is nonsingular. Consequently, the coordinate transformation $x \mapsto T_b x$ where

$$T_b = \begin{bmatrix} I_{n-p} & -B_{c1} B_{c2}^{\dagger} \\ 0 & T^{\mathrm{T}} \end{bmatrix} \tag{5.28}$$

is nonsingular, and with respect to the new coordinates the triple (A, B, C) is in the form

$$A = \begin{bmatrix} A_{11} & A_{12} \\ A_{21} & A_{22} \end{bmatrix} \qquad B = \begin{bmatrix} 0 \\ B_2 \end{bmatrix} \qquad C = \begin{bmatrix} 0 & T \end{bmatrix} \tag{5.29}$$

where $A_{11} \in \mathbb{R}^{(n-m)\times(n-m)}$ and the remaining sub-blocks in the system matrix are partitioned accordingly.

Let

$$\begin{bmatrix} \overset{p-m}{\leftrightarrow} \\ F_1 \end{bmatrix.\!\!\begin{matrix} \overset{m}{\leftrightarrow} \\ F_2 \end{matrix} = FT$$

where T is the matrix from equation (5.27). As a result

$$FC = \begin{bmatrix} F_1 C_1 & F_2 \end{bmatrix} \tag{5.30}$$

where

$$C_1 \triangleq \begin{bmatrix} 0_{(p-m)\times(n-p)} & I_{(p-m)} \end{bmatrix} \tag{5.31}$$

Therefore $FCB = F_2B_2$ and the square matrix F_2 is nonsingular. By assumption the uncertainty is matched and therefore the sliding motion is independent of the uncertainty. In addition, because the canonical form in (5.29) can be viewed as a special case of the regular form normally used in sliding mode controller design, the reduced-order sliding motion is governed by a free motion with system matrix

$$A_{11}^s \triangleq A_{11} - A_{12}F_2^{-1}F_1C_1 \tag{5.32}$$

which must therefore be stable. If $K \in \mathbb{R}^{m\times(p-m)}$ is defined as $K = F_2^{-1}F_1$ then

$$A_{11}^s = A_{11} - A_{12}KC_1 \tag{5.33}$$

and the problem of hyperplane design is equivalent to a *static output feedback problem* for the system (A_{11}, A_{12}, C_1). In order to utilise the existing literature it is necessary that the pair (A_{11}, A_{12}) is controllable and (A_{11}, C_1) is observable. The former has already been established (Proposition 3.3); the observability of (A_{11}, C_1) is not so straightforward. An appropriate way to make progress is as follows.

Partition the submatrix A_{11} so that

$$A_{11} = \begin{bmatrix} A_{1111} & A_{1112} \\ A_{1121} & A_{1122} \end{bmatrix} \tag{5.34}$$

where $A_{1111} \in \mathbb{R}^{(n-p)\times(n-p)}$ and suppose the matrix pair (A_{1111}, A_{1121}) is observable.

It follows that

$$\begin{aligned}\operatorname{rank}\begin{bmatrix} zI - A_{11} \\ C_1 \end{bmatrix} &= \operatorname{rank}\begin{bmatrix} zI - A_{1111} & A_{1112} \\ A_{1121} & zI - A_{1122} \\ 0 & I_{p-m} \end{bmatrix} \\ &= \operatorname{rank}\begin{bmatrix} zI - A_{1111} \\ A_{1121}^o \end{bmatrix} + (p-m) \qquad \text{for all } z \in \mathbb{C}\end{aligned}$$

and hence from the PBH rank test, and using the fact that (A_{1111}, A_{1121}) is observable, it follows that

$$\operatorname{rank}\begin{bmatrix} zI - A_{11} \\ C_1 \end{bmatrix} = n - m$$

for all $z \in \mathbb{C}$ and hence (A_{11}, C_1) is observable.

If the pair (A_{1111}, A_{1121}) is not observable then there exists a $T_{obs} \in \mathbb{R}^{(n-p)\times(n-p)}$ which puts the pair into the following observability canonical form[2]:

$$T_{obs}A_{1111}T_{obs}^{-1} = \begin{bmatrix} A_{11}^o & A_{12}^o \\ 0 & A_{22}^o \end{bmatrix} \quad \text{and} \quad A_{1121}T_{obs}^{-1} = \begin{bmatrix} 0 & A_{21}^o \end{bmatrix}$$

where $A_{11}^o \in \mathbb{R}^{r\times r}$, $A_{21}^o \in \mathbb{R}^{(p-m)\times(n-p-r)}$, the pair (A_{22}^o, A_{21}^o) is completely observable and $r \geq 0$ represents the number of unobservable states of (A_{1111}, A_{1121}).

[2]For further details see Chen (1984).

The transformation T_{obs} can be embedded in a new state transformation matrix

$$T_a = \begin{bmatrix} T_{obs} & 0 \\ 0 & I_p \end{bmatrix} \qquad (5.35)$$

which, when used in conjunction with T_c and T_b from equations (5.26) and (5.28), generates the canonical form described below.

Lemma 5.3 *Let (A, B, C) be a linear system with $p > m$ and rank$(CB) = m$. Then a change of coordinates exists so that the system triple with respect to the new coordinates has the following structure:*

(*a*) *The system matrix can be written as*

$$A = \begin{bmatrix} A_{11} & A_{12} \\ A_{21} & A_{22} \end{bmatrix} \quad \text{where } A_{11} \in \mathbb{R}^{(n-m)\times(n-m)} \qquad (5.36)$$

and the sub-block A_{11} when partitioned has the structure

$$A_{11} = \left[\begin{array}{cc|c} A_{11}^o & A_{12}^o & \multirow{2}{*}{A_{12}^m} \\ 0 & A_{22}^o & \\ \hline 0 & A_{21}^o & A_{22}^m \end{array}\right] \qquad (5.37)$$

where $A_{11}^o \in \mathbb{R}^{r\times r}$, $A_{22}^o \in \mathbb{R}^{(n-p-r)\times(n-p-r)}$ and $A_{21}^o \in \mathbb{R}^{(p-m)\times(n-p-r)}$ for some $r \geq 0$ and the pair (A_{22}^o, A_{21}^o) is completely observable.

(*b*) *The input distribution matrix has the form*

$$B = \begin{bmatrix} 0 \\ B_2 \end{bmatrix} \qquad (5.38)$$

where $B_2 \in \mathbb{R}^{m\times m}$ and is nonsingular.

(*c*) *The output distribution matrix has the form*

$$C = \begin{bmatrix} 0 & T \end{bmatrix} \qquad (5.39)$$

where $T \in \mathbb{R}^{p\times p}$ and is orthogonal.

MATLAB commands to generate the appropriate canonical form are given below. This form will be fundamental for the sliding mode output feedback and observer-based design methods which are to be developed.

mfile: establishes the canonical form in Lemma 5.3

```
% Establish the number of inputs and outputs
[nn,mm]=size(B);
[pp,nn]=size(C);

% Change coordinates so the output distribution matrix is [0 I]
nc = null(C);
Tc=[nc'; C];
```

```
Ac=Tc*A*inv(Tc);
Bc=Tc*B;
Cc=C*inv(Tc);

% Partition the input distribution matrix conformably
Bc1=Bc(1:nn-pp,:);
Bc2=Bc(nn-pp+1:nn,:);

% Finds a tranformation to bring about a special structure in
% the input and output distribution matrices
[T,temp]=qr(Bc2);
T=(flipud(T'))';
clear temp
Tb=[eye(nn-pp) -Bc1*inv(Bc2'*Bc2)*Bc2'; zeros(pp,nn-pp) T'];

Aa=Tb*Ac*inv(Tb);                  % In this new coordinate system
Ba=Tb*Bc;                          % we have C=[0 T] and B=[0 B2']'
Ca=Cc*inv(Tb);                     % A has no special structure yet

A1111=Aa(1:nn-pp,1:nn-pp);
A1121=Aa(nn-pp+1:nn-mm,1:nn-pp);

% The aim is to put (A1111,A1121) in the observability canonical form
[Ab,Bb,Cb,Tobs,k]=obsvf(A1111,zeros(nn-pp,1),A1121,1000*eps);

% r is the dimension of the unobservable subspace and
% the number of invariant zeros of the system (A,B,C)
r=nn-pp-sum(k);
fprintf('Dimension of the unobservable subspace is %.0f \n',r);

Ta=[Tobs zeros(nn-pp,pp);zeros(pp,nn-pp) eye(pp)];

Af=Ta*Aa*inv(Ta);
Bf=Ta*Ba;
Cf=Ca*inv(Ta);
```

In the case where $r > 0$, the intention is to construct a new system $(\tilde{A}_{11}, \tilde{B}_1, \tilde{C}_1)$ which is both controllable and observable with the property that

$$\lambda(A_{11}^s) = \lambda(A_{11}^o) \cup \lambda(\tilde{A}_{11} - \tilde{B}_1 K \tilde{C}_1)$$

To this end, partition the matrices A_{12} and A_{12}^m as

$$A_{12} = \begin{bmatrix} A_{121} \\ A_{122} \end{bmatrix} \quad \text{and} \quad A_{12}^m = \begin{bmatrix} A_{121}^m \\ A_{122}^m \end{bmatrix} \tag{5.40}$$

where $A_{122} \in \mathbb{R}^{(n-m-r)\times m}$ and $A_{122}^m \in \mathbb{R}^{(n-p-r)\times(p-m)}$ and form a new subsystem represented by the triple $(\tilde{A}_{11}, A_{122}, \tilde{C}_1)$ where

$$\tilde{A}_{11} \triangleq \begin{bmatrix} A_{22}^o & A_{122}^m \\ A_{21}^o & A_{22}^m \end{bmatrix} \qquad \tilde{C}_1 \triangleq \begin{bmatrix} 0_{(p-m)\times(n-p-r)} & I_{(p-m)} \end{bmatrix} \tag{5.41}$$

The following lemma can be established.

Lemma 5.4 *The spectrum of* A_{11}^s *decomposes as*

$$\lambda(A_{11} - A_{12}KC_1) = \lambda(A_{11}^o) \cup \lambda(\tilde{A}_{11} - A_{122}K\tilde{C}_1)$$

Proof
Using the partition in equation (5.37), by definition

$$\begin{aligned} A_{11}^s = A_{11} - A_{12}KC_1 &= \left[\begin{array}{cc|c} A_{11}^o & A_{12}^o & \multirow{2}{*}{A_{12}^m} \\ 0 & A_{22}^o & \\ \hline 0 & A_{21}^o & A_{22}^m \end{array}\right] - \left[\begin{array}{cc} 0_{(n-m)\times(n-p)} & A_{12}K \end{array}\right] \\ &= \left[\begin{array}{cc} A_{11}^o & [A_{12}^o \ A_{121}^m] \\ 0 & \tilde{A}_{11} \end{array}\right] - \left[\begin{array}{cc} 0 & A_{121}K \\ 0 & A_{122}K \end{array}\right] \\ &= \left[\begin{array}{cc} A_{11}^o & \left[\begin{array}{c|c} A_{12}^o & A_{121}^m - A_{121}K \end{array}\right] \\ 0 & \tilde{A}_{11} - A_{122}K\tilde{C}_1 \end{array}\right] \end{aligned}$$

Therefore $\lambda(A_{11}^s) = \lambda(A_{11}^o) \cup \lambda(\tilde{A}_{11} - A_{122}K\tilde{C}_1)$ as claimed.

■

Lemma 5.5 *The spectrum of* A_{11}^o *represents the invariant zeros of* (A, B, C).

Proof
Let $P(z)$ denote Rosenbrock's system matrix

$$P(z) = \left[\begin{array}{cc} zI - A & B \\ -C & 0 \end{array}\right]$$

The invariant zeros of (A, B, C) are defined to be

$$\{z \in \mathbb{C} \ : \ P(z) \text{ loses normal rank}\}$$

By assumption (A, B, C) is in the canonical form of Lemma 5.3 and so

$$\begin{aligned} P(z) \text{ loses rank} &\Leftrightarrow \left[\begin{array}{ccc} zI - A_{11} & -A_{12} & 0 \\ -A_{21} & zI - A_{22} & B_2 \\ [0 \ \ -T_1] & -T_2 & 0 \end{array}\right] \text{ loses rank} \\ &\Leftrightarrow \left[\begin{array}{cc} zI - A_{11} & -A_{12} \\ [0 \ \ -T_1] & -T_2 \end{array}\right] \text{ loses rank} \end{aligned}$$

since $\det(B_2) \neq 0$ where $T_1 \in \mathbb{R}^{(p-m)\times m}$ and $T_2 \in \mathbb{R}^{m\times m}$ represent a partition of the matrix T. Substituting for A_{11} from (5.37) and repartitioning gives

$$\left[\begin{array}{cc} zI - A_{11} & -A_{12} \\ [0 \ \ -T_1] & -T_2 \end{array}\right] \equiv \left[\begin{array}{cc|c} zI - A_{11}^o & -A_{12}^o & \\ 0 & zI - A_{22}^o & * \\ 0 & A_{21}^o & \\ \hline 0 & 0 & -T \end{array}\right]$$

where $*$ represents a matrix sub-block which plays no part in the analysis. Because T is full rank, the expression on the right, and hence $P(z)$, loses rank if and only if

$$\begin{bmatrix} zI - A_{11}^o & -A_{12}^o \\ 0 & zI - A_{22}^o \\ 0 & -A_{21}^o \end{bmatrix} \text{ loses rank}$$

By construction the pair (A_{22}^o, A_{21}^o) is completely observable and therefore from the PBH observability rank test

$$\text{rank}\begin{bmatrix} zI - A_{22}^o \\ -A_{21}^o \end{bmatrix} = n - p - r \quad \text{for all } z \in \mathbb{C}$$

Therefore

$$P(z) \text{ loses rank} \quad \Leftrightarrow \quad \det(zI - A_{11}^o) = 0$$

and so the zeros of (A, B, C) are the eigenvalues of A_{11}^o ■

It follows directly that for a stable sliding motion, the invariant zeros of the system (A, B, C) must lie in the open left-half plane and the triple $(\tilde{A}_{11}, A_{122}, \tilde{C}_1)$ must be stabilisable with respect to output feedback.[3]

The matrix A_{122} is not necessarily full rank. Suppose $\text{rank}(A_{122}) = m'$ then it is possible to construct a matrix of elementary column operations $T_{m'} \in \mathbb{R}^{m \times m}$ such that

$$A_{122}T_{m'} = \begin{bmatrix} \tilde{B}_1 & 0 \end{bmatrix} \tag{5.42}$$

where $\tilde{B}_1 \in \mathbb{R}^{(n-m-r)\times m'}$ and is of full rank. If $K_{m'} = T_{m'}^{-1}K$ and $K_{m'}$ is partitioned compatibly as

$$K_{m'} = \begin{bmatrix} K_1 \\ K_2 \end{bmatrix} \begin{matrix} \updownarrow m' \\ \updownarrow m-m' \end{matrix}$$

then

$$\tilde{A}_{11} - A_{122}K\tilde{C}_1 = \tilde{A}_{11} - \begin{bmatrix} \tilde{B}_1 & 0 \end{bmatrix} K_{m'}\tilde{C}_1 = \tilde{A}_{11} - \tilde{B}_1K_1\tilde{C}_1$$

and $(\tilde{A}_{11}, A_{122}, \tilde{C}_1)$ is stabilisable by output feedback if and only if $(\tilde{A}_{11}, \tilde{B}_1, \tilde{C}_1)$ is stabilisable by output feedback. The reason for the somewhat tortuous argument to establish the subsystem $(\tilde{A}_{11}, \tilde{B}_1, \tilde{C}_1)$ is that in order to use standard output feedback results, the triple must be controllable, observable and satisfy the Kimura–Davison conditions (see Section 2.3.4), which in this case amount to

$$m' + p + r \geq n + 1 \tag{5.43}$$

Lemma 5.6 *The pair* $(\tilde{A}_{11}, \tilde{B}_1)$ *is completely controllable and* $(\tilde{A}_{11}, \tilde{C}_1)$ *is completely observable.*

Proof
Because the pair (A, B) is in the canonical form of Lemma 5.3, which is a special

[3]The phrase ***stabilisable with respect to output feedback*** will indicate that for the linear system (A, B, C) there exists a fixed gain K such that the matrix $A - BKC$ is stable.

case of the 'regular form', the pair (A, B) is completely controllable if and only if the pair (A_{11}, A_{12}) is completely controllable. Therefore from the PBH rank test

$$\text{rank}\left[\begin{array}{cc} zI - A_{11} & A_{12} \end{array}\right] = n - m \qquad \text{for all } z \in \mathbb{C}$$

Substituting for A_{11} from (5.37) and A_{12} from (5.40) gives

$$\text{rank}\left[\begin{array}{ccc} zI - A_{11}^o & -[A_{12}^o \;\; A_{121}^m] & A_{121} \\ 0 & zI - \tilde{A}_{11} & A_{122} \end{array}\right] = n - m \qquad \text{for all } z \in \mathbb{C}$$

This implies

$$\text{rank}\left[\begin{array}{cc} zI - \tilde{A}_{11} & A_{122} \end{array}\right] = n - m - r \qquad \text{for all } z \in \mathbb{C}$$

and therefore $(\tilde{A}_{11}, A_{122})$ is completely controllable by the PBH test. By construction $(\tilde{A}_{11}, A_{122})$ is controllable if and only if $(\tilde{A}_{11}, \tilde{B}_1)$ is controllable, so the first part of the lemma is established. Applying the PBH observability rank test to the pair $(\tilde{A}_{11}, \tilde{C}_1)$

$$\begin{aligned} \text{rank}\left[\begin{array}{c} zI - \tilde{A}_{11} \\ \tilde{C}_1 \end{array}\right] &= \text{rank}\left[\begin{array}{cc} zI - A_{22}^o & -A_{122}^m \\ -A_{21}^o & zI - A_{22}^m \\ 0 & I_{p-m} \end{array}\right] \\ &= \text{rank}\left[\begin{array}{c} zI - A_{22}^o \\ -A_{21}^o \end{array}\right] + (p - m) \qquad \text{for all } z \in \mathbb{C} \end{aligned}$$

Now (A_{22}^o, A_{21}^o) is observable and so by the PBH observability test

$$\text{rank}\left[\begin{array}{c} zI - A_{22}^o \\ -A_{21}^o \end{array}\right] = n - p - r \qquad \text{for all } z \in \mathbb{C}$$

and therefore

$$\text{rank}\left[\begin{array}{c} zI - \tilde{A}_1 \\ \tilde{C}_1 \end{array}\right] = n - m - r \qquad \text{for all } z \in \mathbb{C}$$

hence $(\tilde{A}_{11}, \tilde{C}_1)$ is completely observable. ■

The rest of this section considers the problem of constructing an appropriate control law to induce sliding.

5.4.2 Control Law Synthesis

Assume there exists a $K_1 \in \mathbb{R}^{m' \times (p-m)}$ such that $\tilde{A}_{11} - \tilde{B}_1 K_1 \tilde{C}_1$ is stable. Let

$$K = T_{m'} \left[\begin{array}{c} K_1 \\ K_2 \end{array}\right] \tag{5.44}$$

where $K_2 \in \mathbb{R}^{(m-m') \times (p-m)}$ and is arbitrary and the matrix $T_{m'} \in \mathbb{R}^{m \times m}$ is defined in equation (5.42). Then providing any invariant zeros are stable, it follows that

$$\lambda(A_{11} - A_{12} K C_1) = \lambda(A_{11}^o) \cup \lambda(\tilde{A}_{11} - \tilde{B}_1 K_1 \tilde{C}_1)$$

and so the matrix $A_{11} - A_{12}KC_1$ is stable. Choose

$$F = F_2 \begin{bmatrix} K & I_m \end{bmatrix} T^{\mathrm{T}}$$

where $F_2 \in \mathrm{IR}^{m \times m}$ is nonsingular and will be defined later. Introduce a nonsingular state transformation $x \mapsto \bar{T}x$ where

$$\bar{T} = \begin{bmatrix} I_{(n-m)} & 0 \\ KC_1 & I_m \end{bmatrix} \tag{5.45}$$

and C_1 is defined in (5.31). In this new coordinate system the system triple $(\bar{A}, \bar{B}, F\bar{C})$ has the property that

$$\bar{A} = \begin{bmatrix} \bar{A}_{11} & \bar{A}_{12} \\ \bar{A}_{21} & \bar{A}_{22} \end{bmatrix} \qquad \bar{B} = \begin{bmatrix} 0 \\ B_2 \end{bmatrix} \qquad F\bar{C} = \begin{bmatrix} 0 & F_2 \end{bmatrix} \tag{5.46}$$

where $\bar{A}_{11} = A_{11} - A_{12}KC_1$ and is therefore stable. This is the same structure as considered in (5.11) of Section 5.3. An alternative description is that by an appropriate choice of F a new square system $(\bar{A}, \bar{B}, F\bar{C})$ has been synthesised which is minimum phase and relative degree 1. Consequently, the control structure described in Section 5.3 can be utilised to induce sliding on $\mathcal{S}$. Let the control law be given by

$$u(t) = -\gamma F y(t) - \nu_y \tag{5.47}$$

where

$$\nu_y = \begin{cases} \rho(t,y)\frac{Fy(t)}{\|Fy(t)\|} & \text{if } Fy \neq 0 \\ 0 & \text{otherwise} \end{cases} \tag{5.48}$$

and

$$\gamma > \tfrac{1}{2}\lambda_{max}\left((F_2^{-1})^{\mathrm{T}}(Q_3 + Q_2^{\mathrm{T}} Q_1^{-1} Q_2)F_2^{-1}\right) \tag{5.49}$$

where the Q_1, Q_2 and Q_3 satisfy expressions (5.13), (5.16) and (5.17) and $\rho(t,y)$ is defined in (5.21). Arguing as in Section 5.3 the uncertain system is quadratically stable and an ideal sliding motion is induced on $\mathcal{S}$.

The results of this section can be summarised as follows.

Proposition 5.2 *There exists a matrix F defining a surface $\mathcal{S}$ which provides a stable sliding motion with a unique equivalent control if and only if*

- *the invariant zeros of (A, B, C) lie in $\mathbb{C}_-$*
- *the triple $(\tilde{A}_{11}, \tilde{B}_1, \tilde{C}_1)$ is stabilisable with respect to output feedback.*

5.4.3 Example 1

Consider the nominal linear system

$$A = \begin{bmatrix} 0 & 1 & 0 \\ 0 & 0 & 1 \\ -1 & \frac{1}{3} & -1 \end{bmatrix} \qquad B = \begin{bmatrix} 0 \\ 1 \\ 1 \end{bmatrix} \qquad C = \begin{bmatrix} 1 & \frac{8}{3} & 1 \\ 4 & \frac{2}{3} & -2 \end{bmatrix} \tag{5.50}$$

In the canonical form of Lemma 5.3 the system becomes

$$A = \begin{bmatrix} -1.5816 & 0.0192 & 0.1457 \\ 1.4071 & 0.3845 & -1.7080 \\ 0.2953 & 0.3400 & 0.1971 \end{bmatrix} \qquad B = \begin{bmatrix} 0 \\ 0 \\ -3.9016 \end{bmatrix}$$

and

$$C = \begin{bmatrix} 0 & 0.3417 & -0.9398 \\ 0 & 0.9398 & 0.3417 \end{bmatrix}$$

It can be verified that the one-dimensional matrix $B_2 = -3.9016$, the orthogonal matrix

$$T = \begin{bmatrix} 0.3417 & -0.9398 \\ 0.9398 & 0.3417 \end{bmatrix}$$

and the triple $(\tilde{A}_{11}, \tilde{B}_1, \tilde{C}_1)$ is given by

$$\tilde{A}_{11} = \begin{bmatrix} -1.5816 & 0.0192 \\ 1.4071 & 0.3845 \end{bmatrix} \qquad \tilde{B}_1 = \begin{bmatrix} 0.1457 \\ -1.7080 \end{bmatrix} \qquad \tilde{C}_1 = \begin{bmatrix} 0 & 1 \end{bmatrix}$$

Here $r = 0$ hence the original system does not possess any invariant zeros. Arbitrary placement of the poles of $\tilde{A}_{11} - \tilde{B}_1 K_1 \tilde{C}_1$ is not possible since only a single scalar is available as design freedom. For the single-input single-output system $(\tilde{A}_{11}, \tilde{B}_1, \tilde{C}_1)$ the variation in the poles of $\tilde{A}_{11} - \tilde{B}_1 K_1 \tilde{C}_1$ with respect to K_1 can be examined by *root locus* techniques; see for example Ogata (1997). In this case if the gain matrix $K = K_1 = -1.0556$ then $\lambda(\tilde{A}_{11} - \tilde{B}_1 K \tilde{C}_1) = \{-1, -2\}$, from which

$$\begin{aligned} F &= F_2 \begin{bmatrix} K & 1 \end{bmatrix} T^{\mathrm{T}} \\ &= F_2 \begin{bmatrix} -1.3005 & -0.6503 \end{bmatrix} \end{aligned} \tag{5.51}$$

where F_2 is a nonzero scalar which will be computed later. Transforming the system into the canonical form using $\bar{T}$ defined in (5.45) generates

$$\bar{A}_{11} = \begin{bmatrix} -1.5816 & 0.1729 \\ 1.4071 & -1.4184 \end{bmatrix}$$

where $\lambda(\bar{A}_{11}) = \{-1, -2\}$ by construction. It can be verified that

$$P_1 = \begin{bmatrix} 0.3368 & 0.1891 \\ 0.1891 & 0.5401 \end{bmatrix}$$

is a Lyapunov matrix for $\bar{A}_{11}$ and that if $P_2 = 1$ then the parameter $F_2 = -3.9016$. It can be checked that $\gamma_0 = 0.2452$ and substituting for F_2 in (5.51) gives

$$F = \begin{bmatrix} 5.0741 & 2.5370 \end{bmatrix}$$

The following closed-loop simulation represents the regulation of the initial states $[1\ \ 0\ \ 0]$ to the origin. Figure 5.1 represents a plot of the switching function versus time. The hyperplane is not globally attractive since at approximately 0.3 second it is pierced and a sliding motion cannot be maintained. Only after approximately 1 sec is sliding established. Figure 5.2 shows the decay of the states to the origin.

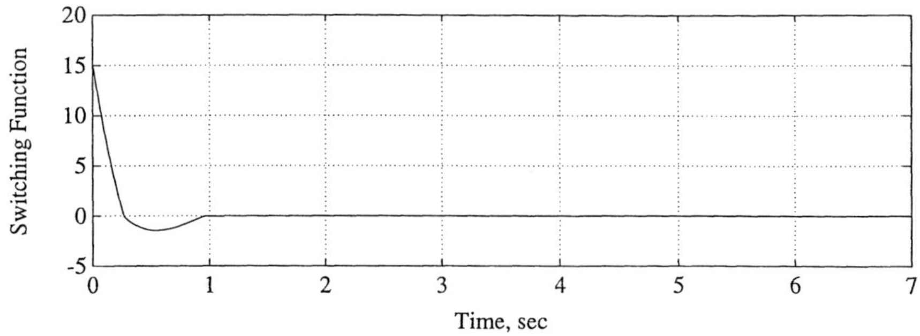

Figure 5.1: Switching function versus time

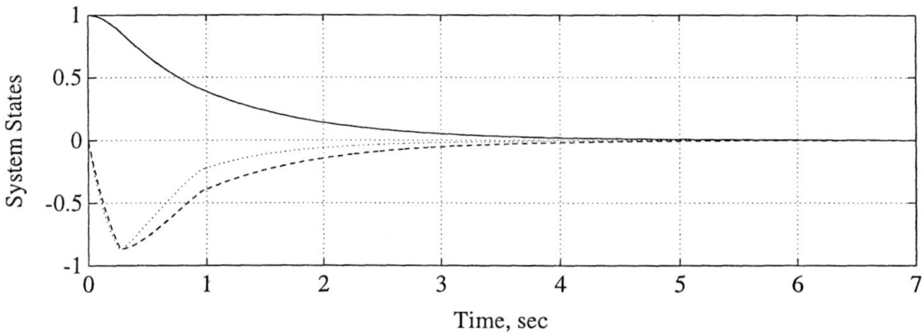

Figure 5.2: Evolution of system states with respect to time

5.5 DYNAMIC COMPENSATION

In the analysis encapsulated in Proposition 5.2, it was assumed that the triple $(\tilde{A}_{11}, \tilde{B}_1, \tilde{C}_1)$ was stabilisable with respect to output feedback. This property can be guaranteed if the so-called Kimura–Davison condition holds.[4] If it is not possible to synthesise a K_1 to stabilise the triple $(\tilde{A}_{11}, \tilde{B}_1, \tilde{C}_1)$ then it is natural to explore the effect of introducing a *compensator* – i.e. a dynamical system driven by the output of the plant – to introduce extra dynamics to provide additional degrees of freedom.

Consider the uncertain system from equation (5.1) together with a compensator given by

$$\dot{x}_c(t) = Hx_c(t) + Dy(t) \tag{5.52}$$

where the matrices $H \in \mathbb{R}^{q \times q}$ and $D \in \mathbb{R}^{q \times p}$ are to be determined. Define a new hyperplane in the augmented state space, formed from the plant and compensator state spaces, as

$$\mathcal{S}_c = \{(x, x_c) \in \mathbb{R}^{n+q} \ : \ F_c x_c + FCx = 0\} \tag{5.53}$$

[4]It is quite possible of course for the triple $(\tilde{A}_{11}, \tilde{B}_1, \tilde{C}_1)$ to be output feedback stabilisable if the Kimura–Davison constraint does not hold, since it represents only a sufficient condition; for instance, see Section 5.4.3.

where $F_c \in \mathbb{R}^{m\times q}$ and $F \in \mathbb{R}^{m\times p}$. As in Section 5.4, assume that the nominal linear system (A, B, C) is in the canonical form of Lemma 5.3 and partition the matrix FT, where T is the orthogonal matrix from the output distribution matrix, as

$$\begin{array}{cc} \overset{p-m}{\leftrightarrow} & \overset{m}{\leftrightarrow} \\ \left[\; F_1 \right. & \left. F_2 \; \right] \end{array} = FT$$

In an analogous way define $D_1 \in \mathbb{R}^{q\times(p-m)}$ and $D_2 \in \mathbb{R}^{q\times m}$ as

$$\left[\begin{array}{cc} D_1 & D_2 \end{array} \right] = DT \tag{5.54}$$

If the states of the uncertain system are partitioned as

$$x = \left[\begin{array}{c} x_1 \\ x_2 \end{array} \right] \begin{array}{l} \updownarrow n-m \\ \updownarrow m \end{array} \tag{5.55}$$

then the compensator can be written as

$$\dot{x}_c(t) = Hx_c(t) + D_1C_1x_1(t) + D_2x_2(t) \tag{5.56}$$

where C_1 is defined in equation (5.31). Assume that a control action exists which forces and maintains motion on the hyperplane $\mathcal{S}_c$ given in (5.53). As usual, in order for a unique equivalent control to exist, the square matrix F_2 must be invertible. By writing $K = F_2^{-1}F_1$ and defining $K_c = F_2^{-1}F_c$ then the system matrix governing the reduced-order sliding motion, obtained by eliminating the coordinates x_2, can be written as

$$\dot{x}_1(t) = (A_{11} - A_{12}KC_1)x_1(t) - A_{12}K_cx_c(t) \tag{5.57}$$

$$\dot{x}_c(t) = (D_1 - D_2K)C_1x_1(t) + (H - D_2K_c)x_c(t) \tag{5.58}$$

From the above equations it is clear that the introduction of the compensator has introduced more design freedom. Unfortunately, the invariant zeros of the uncertain system are still embedded in the dynamics, since from the definition of the partition of A_{12} given in (5.40) and from the partitioned form of $A_{11} - A_{12}KC_1$ given in the proof of Lemma 5.4, it follows that

$$\left[\begin{array}{cc} A_{11} - A_{12}KC_1 & -A_{12}K_c \\ (D_1 - D_2K)C_1 & H - D_2K_c \end{array} \right] = \left[\begin{array}{ccc} A_{11}^o & [A_{12}^o \mid A_{121}^m - A_{121}K] & -A_{121}K_c \\ 0 & \tilde{A}_{11} - A_{122}K\tilde{C}_1 & -A_{122}K_c \\ 0 & (D_1 - D_2K)\tilde{C}_1 & H - D_2K_c \end{array} \right]$$

As in the uncompensated case, it is necessary for the eigenvalues of A_{11}^o to have negative real parts. The design problem becomes one of selecting a compensator, represented by the matrices D_1, D_2 and H, and a hyperplane represented by the matrices K and K_c so that the matrix

$$A_c \triangleq \left[\begin{array}{cc} \tilde{A}_{11} - A_{122}K\tilde{C}_1 & -A_{122}K_c \\ (D_1 - D_2K)\tilde{C}_1 & H - D_2K_c \end{array} \right] \tag{5.59}$$

is stable. Again if there is rank deficiency in the matrix A_{122} the problem is over-parametrised. As in Section 5.4, suppose $\text{rank}(A_{112}) = m' < m$ and let $T_{m'} \in \mathbb{R}^{m\times m}$ be a matrix of elementary column operations such that

$$A_{122}T_{m'} = \left[\begin{array}{cc} \tilde{B}_1 & 0 \end{array} \right]$$

where $\tilde{B}_1 \in \mathbb{R}^{(n-m-r)\times m'}$ and is of full rank. Define partitions of the transformed hyperplane matrices as

$$T_{m'}^{-1}K = \begin{bmatrix} K_1 \\ K_2 \end{bmatrix} \begin{matrix} \updownarrow m' \\ \updownarrow m-m' \end{matrix} \qquad \text{and} \qquad T_{m'}^{-1}K_c = \begin{bmatrix} K_{c1} \\ K_{c2} \end{bmatrix} \begin{matrix} \updownarrow m' \\ \updownarrow m-m' \end{matrix}$$

then it follows that

$$A_c = \begin{bmatrix} \tilde{A}_{11} - \tilde{B}_1 K_1 \tilde{C}_1 & -\tilde{B}_1 K_{c1} \\ (D_1 - D_2 K)\tilde{C}_1 & H - D_2 K_c \end{bmatrix} \tag{5.60}$$

As before, the matrix given in (5.60) will be written as the result of an output feedback problem for a certain system triple. Unfortunately, a degree of over-parametrisation is still present in (5.60), which for simplicity will be removed by defining

$$\tilde{D}_1 \triangleq D_1 - D_2 K \qquad \text{and} \qquad \tilde{H} \triangleq H - D_2 K_c \tag{5.61}$$

This is comparable to the situation which occurred in the uncompensated case where K_2 was found to have no effect on $\tilde{A}_{11} - \tilde{B}_1 K \tilde{C}_1$. The key observation is that equation (5.60) can now be written as

$$\begin{bmatrix} \tilde{A}_{11} - \tilde{B}_1 K_1 \tilde{C}_1 & -B_1 K_{c1} \\ \tilde{D}_1 \tilde{C}_1 & \tilde{H} \end{bmatrix} = \begin{bmatrix} \tilde{A}_{11} & 0 \\ 0 & 0 \end{bmatrix} - \begin{bmatrix} \tilde{B}_1 & 0 \\ 0 & -I_q \end{bmatrix} \begin{bmatrix} K_1 & K_{c1} \\ \tilde{D}_1 & \tilde{H} \end{bmatrix} \begin{bmatrix} \tilde{C}_1 & 0 \\ 0 & I_q \end{bmatrix}$$

Thus by defining

$$A_q = \begin{bmatrix} \tilde{A}_{11} & 0 \\ 0 & 0_{q\times q} \end{bmatrix} \quad B_q = \begin{bmatrix} \tilde{B}_1 & 0 \\ 0 & -I_q \end{bmatrix} \quad C_q = \begin{bmatrix} \tilde{C}_1 & 0 \\ 0 & I_q \end{bmatrix}$$

then the parameters $K_1, K_{c1}, \tilde{D}_1$ and $\tilde{H}$ can be obtained from output feedback pole placement of the triple (A_q, B_q, C_q). In order to use standard output feedback results it is necessary for the triple (A_q, B_q, C_q) to be both controllable and observable.

Lemma 5.7 *The matrix pairs (A_q, B_q) and (A_q, C_q) are controllable and observable respectively.*

Proof
From the definition of (A_q, B_q) it follows that

$$\text{rank}\begin{bmatrix} zI - A_q & B_q \end{bmatrix} = \text{rank}\begin{bmatrix} zI - \tilde{A}_{11} & \tilde{B}_1 \end{bmatrix} + q$$

for all $z \in \mathbb{C}$. From Lemma 5.6 the pair $(\tilde{A}_{11}, \tilde{B}_1)$ is controllable and therefore from the PBH rank test (A_q, B_q) is controllable, as claimed. Using the fact that from Lemma 5.6 pair $(\tilde{A}_{11}, \tilde{C}_1)$ is observable, a similar argument proves that (A_q, C_q) is observable. ■

The Kimura–Davison conditions for the triple (A_q, B_q, C_q) amount to requiring that

$$m' + q + p \geq n - r + 1 \tag{5.62}$$

Thus for a large enough q, the Kimura–Davison conditions can always be satisfied and the static output feedback method can be employed.[5]

[5] An example of such an approach is given in Bag *et al.* (1997).

5.6 DYNAMIC COMPENSATION (OBSERVER BASED)

Rather than formulating the hyperplane design problem as a static output feedback problem, this section explores an observer-based methodology. Consider the compensator defined in (5.52) then, as in the previous section, eliminating any invariant zeros, the assignable dynamics of the sliding motion are given by the system matrix

$$A_c = \begin{bmatrix} \tilde{A}_{11} - A_{122}K\tilde{C}_1 & -A_{122}K_c \\ (D_1 - D_2K)\tilde{C}_1 & H - D_2K_c \end{bmatrix} \tag{5.63}$$

This section provides an alternative method for choosing appropriate compensator variables H, D_1 and D_2, and the hyperplane matrix gains K and K_c.

Consider the triple $(\tilde{A}_{11}, A_{122}, \tilde{C}_1)$ where by definition

$$\tilde{A}_{11} = \begin{bmatrix} A_{22}^o & A_{122}^m \\ A_{21}^o & A_{22}^m \end{bmatrix} \qquad \tilde{C}_1 = \begin{bmatrix} 0_{(p-m)\times(n-p-r)} & I_{(p-m)} \end{bmatrix} \tag{5.64}$$

and the pair (A_{22}^o, A_{21}^o) is observable. Recall that the structure of the output distribution matrix above was used as a starting point for the development of a reduced-order Luenberger observer in Section 2.3.5. Consequently, if the input distribution matrix is partitioned conformably so that

$$A_{122} = \begin{bmatrix} A_{1221} \\ A_{1222} \end{bmatrix} \begin{matrix} \updownarrow n-p-r \\ \updownarrow p-m \end{matrix} \tag{5.65}$$

then a reduced-order observer for the fictitious system

$$\begin{aligned} \dot{\tilde{x}}(t) &= \tilde{A}_{11}\tilde{x}(t) + A_{122}\tilde{u}(t) \\ \tilde{y}(t) &= \tilde{C}_1\tilde{x}(t) \end{aligned} \tag{5.66}$$

is given by

$$\dot{z} = (A_{22}^o + L^oA_{21}^o)z + (A_{122}^m + L^oA_{22}^m - (A_{22}^o + L^oA_{21}^o)L^o)\,\tilde{y} + (A_{1221} + L^oA_{1222})\tilde{u} \tag{5.67}$$

where $L^o \in \mathbb{R}^{(n-p-r)\times(p-m)}$ is any gain matrix so that $A_{22}^o + L^oA_{21}^o$ is stable. Let $\mathcal{K}$ be any state feedback matrix for the controllable pair $(\tilde{A}_{11}, A_{122})$ so that $\tilde{A}_{11} - A_{122}\mathcal{K}$ is stable, and partition the state feedback matrix so that

$$\begin{matrix} \overset{n-r-p}{\leftrightarrow} & \overset{p-m}{\leftrightarrow} \\ [\,\mathcal{K}_1 & \mathcal{K}_2\,] \end{matrix} = \mathcal{K}$$

Then arguing as in Section 2.3.5 the state feedback law can be implemented using the observer states and the outputs in the form

$$\tilde{u} = -\mathcal{K}_1 z - (\mathcal{K}_2 - \mathcal{K}_1 L^o)\tilde{y} \tag{5.68}$$

and the closed-loop system comprising (5.66) and (5.67) is stable. Define

$$\begin{aligned} H &= A_{22}^o + L^oA_{21}^o && (5.69) \\ D_1 &= A_{122}^m + L^oA_{22}^m - (A_{22}^o + L^oA_{21}^o)L^o && (5.70) \\ D_2 &= A_{1221} + L^oA_{1222} && (5.71) \\ K &= \mathcal{K}_2 - \mathcal{K}_1 L^o && (5.72) \\ K_c &= \mathcal{K}_1 && (5.73) \end{aligned}$$

then equation (5.67) can be written

$$\dot{z}(t) = Hz(t) + D_1\tilde{y} + D_2\tilde{u} \tag{5.74}$$

where

$$\tilde{u} = -K_c z(t) - K\tilde{y} \tag{5.75}$$

It can easily be verified that the closed-loop system formed from (5.66) and (5.67) is given by

$$\begin{bmatrix} \dot{\tilde{x}}(t) \\ \dot{z}(t) \end{bmatrix} = \begin{bmatrix} \tilde{A}_{11} - A_{122}K\tilde{C}_1 & -A_{122}K_c \\ (D_1 - D_2K)\tilde{C}_1 & H - D_2K_c \end{bmatrix} \begin{bmatrix} \tilde{x}(t) \\ z(t) \end{bmatrix} \tag{5.76}$$

and from the separation principle the closed-loop poles are $\lambda(H) \cup \lambda(\tilde{A}_{11} - A_{122}\mathcal{K})$. The system matrix associated with (5.76) is identical to the system matrix of the reduced-order sliding motion given in (5.59). Therefore the choice of compensator matrices in (5.69) to (5.71) and the hyperplane matrices (5.72) and (5.73) give rise to a stable sliding mode.

MATLAB commands to generate the compensator matrices are given below.

mfile: generates a compensator based on a reduced-order observer

```
% Assumes the triple (A,B,C) is in the canonical form of Lemma 5.3
% p1 is an nn-pp-r vector containing the desired poles of (A22o+Lo A21o)
% p2 is an nn-mm-r vector containing the desired poles of (A11tilde-A122 K)
% p3 is an mm vector representing the poles of the range space dynamics

A11o=Af(1:r,1:r);
A12o=Af(1:r,r+1:nn-pp);
A22o=Af(r+1:nn-pp,r+1:nn-pp);
A21o=Af(nn-pp+1:nn-mm,r+1:nn-pp);
A121m=Af(1:r,nn-pp+1:nn-mm);                  % Extracts the various sub-blocks
A122m=Af(r+1:nn-pp,nn-pp+1:nn-mm);            % from the system matrix A under
A22m=Af(nn-pp+1:nn-mm,nn-pp+1:nn-mm);         % the assumption that a coordinate
A121=Af(1:r,nn-mm+1:nn);                      % transformation has been used to
A122=Af(r+1:nn-mm,nn-mm+1:nn);                % obtain the canonical form of
A211=Af(nn-mm+1:nn,1:r);                      % Lemma 5.3
A212=Af(nn-mm+1:nn,r+1:nn-pp);
A213=Af(nn-mm+1:nn,nn-pp+1:nn-mm);
A22=Af(nn-mm+1:nn,nn-mm+1:nn);

A11tilde=[A22o A122m; A21o A22m];
A1221=A122(1:nn-pp-r,:);
A1222=A122(nn-pp-r+1:nn-mm-r,:);

Lo=vplace(A22o',-A21o',p1);
Lo=Lo';
calK=vplace(A11tilde,A122,p2);
K1=calK(:,1:nn-r-pp);
K2=calK(:,nn-pp-r+1:nn-mm-r);

H=A22o+Lo*A21o;
```

```
D1=A122m+Lo*A22m-H*Lo;
D2=A1221+Lo*A1222;
K=K2-K1*Lo;
Kc=K1;

Hhat=[A11o A12o; zeros(nn-pp-r,r) H];
Dhat=[A121m-A12o*Lo A121; D1 D2]*T';
S2=eye(mm);                              % For simplicity
S=S2*[zeros(mm,r) Kc K eye(mm)];

Ahat=[A11o A12o A121m-A12o*Lo A121;  zeros(nn-pp-r,r) H D1 D2;
     zeros(pp-mm,r) A21o A22m-A21o*Lo A1222; A211 A212 A213-A212*Lo A22]

Phi=diag(p3);
Lambda=S*Bf;
L=-inv(Lambda)*S*Ahat + inv(Lambda)*Phi*S
P=lyap(Phi',eye(mm));
```

The control laws, based on the plant outputs described earlier in this chapter, tend to produce 'high gain' controllers. The next section considers a control law which utilises the plant outputs and the compensator states and is more akin to the state feedback controllers of Section 3.6.

5.6.1 Control Law Construction

Assume that there are r (stable) invariant zeros and partition the state vector x_1 as in (5.55) so that

$$x_1 = \begin{bmatrix} x_r \\ x_{11} \\ x_{12} \end{bmatrix} \begin{matrix} \updownarrow r \\ \updownarrow n-p-r \\ \updownarrow p-m \end{matrix}$$

As a result, the (original) compensator can be written as

$$\dot{x}_c(t) = Hx_c(t) + D_1 x_{12}(t) + D_2 x_2(t) \tag{5.77}$$

Define a new dynamical system by

$$\dot{z}_r(t) = A_{11}^o z_r(t) + A_{12}^o x_c(t) + (A_{121}^m - A_{12}^o L^o) x_{12}(t) + A_{121} x_2(t) \tag{5.78}$$

and augment (5.77) with (5.78) to form a new compensator

$$\dot{\hat{x}}_c(t) = \hat{H}\hat{x}_c(t) + \hat{D}y(t) \tag{5.79}$$

where

$$\hat{H} \triangleq \begin{bmatrix} A_{11}^o & A_{12}^o \\ 0 & H \end{bmatrix} \qquad \text{and} \qquad \hat{D} \triangleq \begin{bmatrix} (A_{121}^m - A_{12}^o L^o) & A_{121} \\ D_1 & D_2 \end{bmatrix} T^{\mathrm{T}}$$

Using the partitions (5.36), (5.37), (5.40) and (5.65), the original dynamics can be written as

$$\dot{x}_r(t) = A_{11}^o x_r(t) + A_{12}^o x_{11}(t) + A_{121}^m x_{12}(t) + A_{121} x_2(t) \tag{5.80}$$

$$\dot{x}_{11}(t) = A^o_{22}x_{11}(t) + A^m_{122}x_{12}(t) + A_{1221}x_2(t) \tag{5.81}$$
$$\dot{x}_{12}(t) = A^o_{21}x_{11}(t) + A^m_{22}x_{12}(t) + A_{1222}x_2(t) \tag{5.82}$$
$$\dot{x}_2(t) = A_{211}x_r(t) + A_{212}x_{11}(t) + A_{213}x_{12}(t) + A_{22}x_2(t) + B_2(u(t) + \xi(t,x)) \tag{5.83}$$

where the lower left sub-block of A from (5.36) has been partitioned so that

$$\begin{array}{ccccc} \overset{r}{\leftrightarrow} & \overset{n-p-r}{\leftrightarrow} & \overset{p-m}{\leftrightarrow} & & \\ [\ A_{211} & A_{212} & A_{213}\ ] & = & A_{21} \end{array} \tag{5.84}$$

Define two error states

$$e_r = z_r - x_r \tag{5.85}$$

and

$$e_c = x_c - x_{11} - L^o x_{12} \tag{5.86}$$

then straightforward algebra reveals

$$\dot{e}_r(t) = A^o_{11}e_r(t) + A^o_{12}e_c \tag{5.87}$$

and also

$$\dot{e}_c(t) = He_c(t) \tag{5.88}$$

These stable error systems result from the fact that, by construction, the compensator states x_c and z_r are observations of $x_{11} + L^o x_{12}$ and x_r respectively. Define a state matrix

$$\hat{x} = \begin{bmatrix} z_r \\ x_c \\ x_{12} \\ x_2 \end{bmatrix} \tag{5.89}$$

then standard algebra reveals

$$\dot{\hat{x}}(t) = \hat{A}\hat{x}(t) - \hat{A}_e\hat{e}(t) + B(u(t) + \xi(t,x,u)) \tag{5.90}$$

where

$$\hat{A} = \begin{bmatrix} A^o_{11} & A^o_{12} & A^m_{121} - A^o_{12}L^o & A_{121} \\ 0 & H & D_1 & D_2 \\ 0 & A^o_{21} & A^m_{22} - A^o_{21}L^o & A_{1222} \\ A_{211} & A_{212} & A_{213} - A_{212}L^o & A_{22} \end{bmatrix} \quad \text{and} \quad \hat{A}_e = \begin{bmatrix} 0 & 0 \\ 0 & 0 \\ 0 & A^o_{21} \\ A_{211} & A_{212} \end{bmatrix}$$

and the augmented error state

$$\hat{e} = \begin{bmatrix} e_r \\ e_c \end{bmatrix} \tag{5.91}$$

Note that the triple $(\hat{A}, B, C)$ can be obtained from the canonical form (A, B, C) via a similarity transformation. Thus the original system together with the compensator can be written as

$$\dot{\hat{e}}(t) = \hat{H}\hat{e}(t) \tag{5.92}$$
$$\dot{\hat{x}}(t) = \hat{A}\hat{x}(t) - \hat{A}_e\hat{e}(t) + B(u(t) + \xi(t,x,u)) \tag{5.93}$$

Note also that the sliding surface $\mathcal{S}_c$ can be written as

$$\{\hat{x} \in \mathbb{R}^n \ : \ S\hat{x} = 0\}$$

where

$$S = F_2 \begin{bmatrix} 0_{m\times r} & K_c & K & I_m \end{bmatrix} \tag{5.94}$$

Define a switching function

$$s(t) = S\hat{x}(t) \tag{5.95}$$

and define a linear feedback component

$$u_l(t) = -\Lambda^{-1} S\hat{A}\hat{x}(t) + \Lambda^{-1}\Phi S\hat{x}(t) \tag{5.96}$$

where $\Lambda = SB$ and $\Phi \in \mathbb{R}^{m\times m}$ is a stable design matrix. Let P be the unique positive definite solution to the Lyapunov equation

$$P\Phi + \Phi^{\mathrm{T}} P = -I \tag{5.97}$$

A control law to induce a sliding motion on the sliding surface $\mathcal{S}_c$ is given by

$$u(t) = u_l(t) - \nu_y \tag{5.98}$$

where

$$\nu_y = \begin{cases} \rho(t,y)\Lambda^{-1}\frac{Ps(t)}{\|Ps(t)\|} & \text{if } s(t) \neq 0 \\ 0 & \text{otherwise} \end{cases} \tag{5.99}$$

and $\rho(\cdot)$ is the positive scalar function

$$\rho(t,y) = (k_1\|\Lambda\|\|u_l(t)\| + \|\Lambda\|\alpha(t,y) + \gamma_2) \,/\, (1 - k_1\kappa(\Lambda)) \tag{5.100}$$

where γ_2 is a small positive constant.

Proposition 5.3 *The control law defined in (5.96) to (5.100) induces a sliding motion on the sliding surface $\mathcal{S}_c$.*

Proof
By definition $s(t) = S\hat{x}(t)$ which implies $\dot{s}(t) = S\dot{\hat{x}}(t)$ and therefore from (5.93) it follows that

$$\begin{aligned} \dot{s}(t) &= S\hat{A}\hat{x}(t) - S\hat{A}_e(t)\hat{e}(t) + \Lambda u(t) + \Lambda\xi(t,x,u) & (5.101) \\ &= \Phi s(t) - S\hat{A}_e\hat{e}(t) + \Lambda(\xi(t,x,u) - \rho(t,y)\nu_y) & (5.102) \end{aligned}$$

where the control law from (5.96) and (5.98) has been substituted into (5.101) and the expression simplified.

Let $V(s) = s^{\mathrm{T}} Ps$ and therefore

$$\begin{aligned} \dot{V} &= -\|s\|^2 - 2s^{\mathrm{T}} PS\hat{A}_e\hat{e} + 2s^{\mathrm{T}} P\Lambda(\xi - \rho\nu_y) \\ &\leq -\|s\|^2 + 2\|Ps\|\|S\hat{A}_e\hat{e}\| + 2\|Ps\|\|\Lambda\|\|\xi\| - 2\rho\|Ps\| & (5.103) \end{aligned}$$

From the uncertainty structure in (5.3) and the expression for the control law from (5.98), it can be shown that

$$\|\Lambda\|\|\xi\| - \rho \leq \|\Lambda\|(k_1\|u_l\| + k_1\|\Lambda^{-1}\|\rho + \alpha) - \rho = -\gamma_2 \tag{5.104}$$

where the equality on the right comes from rearranging (5.100). Therefore from inequalities (5.103) and (5.104) it follows that

$$\dot{V} \leq -\|s\|^2 + 2\|Ps\|\|S\hat{A}_e\hat{e}\| - 2\gamma_2\|Ps\|$$

Consequently, in the domain $\Omega = \{(\hat{x}, \hat{e}) \ : \ \|S\hat{A}_e\hat{e}\| \leq \gamma_2 - \eta\}$ where η is a (small) scalar satisfying $0 < \eta < \gamma_2$ it follows that

$$\dot{V} \leq -\|s\|^2 - 2\eta\|Ps\| \leq -2\eta\sqrt{\lambda_{min}(P)}\sqrt{V} \tag{5.105}$$

Because $\hat{e}(t)$ satisfies the unforced stable system in (5.92), $\hat{e}(t)$ enters Ω in finite time and remains there. Consequently, the inequality in (5.105) demonstrates that an ideal sliding motion is induced on $\mathcal{S}_c$ in finite time.

■

Remarks

The control law in (5.96) to (5.100) is effectively a state feedback controller since the components z_r and x_c are estimates of the true states x_r and x_{11} (up to a coordinate transformation). Indeed, viewed in this way, the control law can be thought of as being equivalent to the state feedback controller in Section 3.6 and not 'high gain' in nature.

5.6.2 Design Example 1

Consider the nominal linear system

$$A = \begin{bmatrix} -2 & 1 & 0 & 0 \\ 0 & 0 & 4 & 1 \\ 0 & 1 & 0 & 0 \\ 1 & -6 & -9 & -2 \end{bmatrix} \quad B = \begin{bmatrix} 0 \\ 0 \\ 0 \\ 1 \end{bmatrix} \quad C = \begin{bmatrix} 0 & 0 & 0 & 1 \\ 0 & 0 & 1 & 0 \end{bmatrix} \tag{5.106}$$

This system is already in the canonical form of Lemma 5.3 and thus

$$A_{11} = \begin{bmatrix} A_{11}^o & [A_{12}^o \ \ A_{121}^m] \\ 0 & \tilde{A}_{11} \end{bmatrix} = \left[\begin{array}{c|cc} -2 & 1 & 0 \\ \hline 0 & 0 & 4 \\ 0 & 1 & 0 \end{array}\right] \quad A_{12} = \begin{bmatrix} A_{121} \\ A_{122} \end{bmatrix} = \left[\begin{array}{c} 0 \\ \hline 1 \\ 0 \end{array}\right]$$

In terms of the compensator design, the triple of interest is given by

$$\tilde{A}_{11} = \begin{bmatrix} 0 & 4 \\ 1 & 0 \end{bmatrix} \quad A_{122} = \begin{bmatrix} 1 \\ 0 \end{bmatrix} \quad C_1 = \begin{bmatrix} 0 & 1 \end{bmatrix} \tag{5.107}$$

It can be shown by direct computation that for $K = k$

$$\lambda(\tilde{A}_{11} - A_{122}KC_1) = \{\pm\sqrt{4 - k^2}\}$$

and so the triple $(\tilde{A}_{11}, A_{122}, C_1)$ is not stabilisable by static output feedback and a compensator-based approach must be employed. It follows from (5.107) that

$$\begin{bmatrix} A_{22}^o & A_{122}^m \\ A_{21}^o & A_{22}^m \end{bmatrix} = \begin{bmatrix} 0 & 4 \\ 1 & 0 \end{bmatrix}$$

and so from equations (5.69) to (5.71) an appropriate parametrisation for the compensator is

$$H = L^o \quad D_1 = 4 - (L^o)^2 \quad D_2 = 1$$

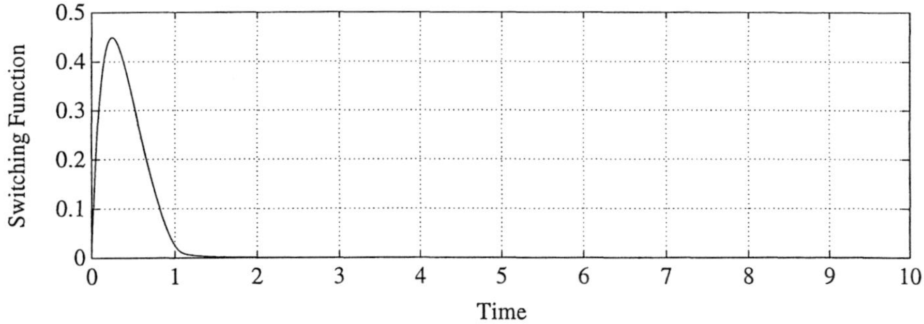

Figure 5.3: Switching function versus time

where L^o is any negative scalar which will appear as one of the eigenvalues of (5.59). In the simulation which follows $L^o = -2.5$ and $\lambda(\tilde{A}_{11} - A_{122}\mathcal{K}) = \{-1, -1.5\}$. Since the system has an invariant zero at -2, the sliding motion will have poles at $\{-1, -1.5, -2\}$. The pole represented by Φ which governs the range space dynamics has been chosen to be -5. For simplicity the scaling factor for the sliding surface is $F_2 = 1$. All the available degrees of freedom have now been assigned. Figure 5.3 is a plot of the switching function against time; it can be seen that sliding occurs after approximately 1 second. Figure 5.4 shows the evolution of the states against time. Initially the states of the compensator have been set to zero. The states of the system have a nonzero initial condition which needs to be regulated to zero.

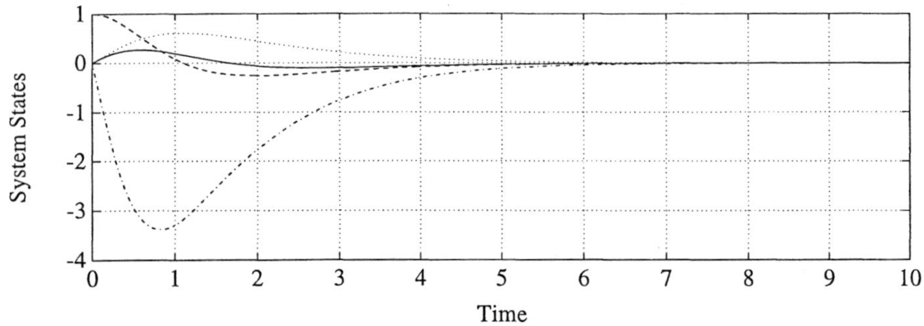

Figure 5.4: Evolution of the system states

Figures 5.5 and 5.6 show the evolution of the error states e_c and e_r. Initially e_c is nonzero since the state x_{11} was given a nonzero initial condition. As indicated in equation (5.88), this error system is completely decoupled and decays away to zero at a rate governed by the invariant zero (Figure 5.5). The error states e_r, shown in Figure 5.6, although initially zero, are coupled to the state e_c as shown in equation (5.87). However, this too decays asymptotically to zero in accordance with the theory. Notice from Figure 5.3 that, although the states initially lie on the sliding surface, a sliding motion is not maintained; this is due to the fact that the error term $\hat{e}$ is initially too large. A sliding motion occurs after approximately 1 second, by which time the error $\hat{e}$ has decayed sufficiently.

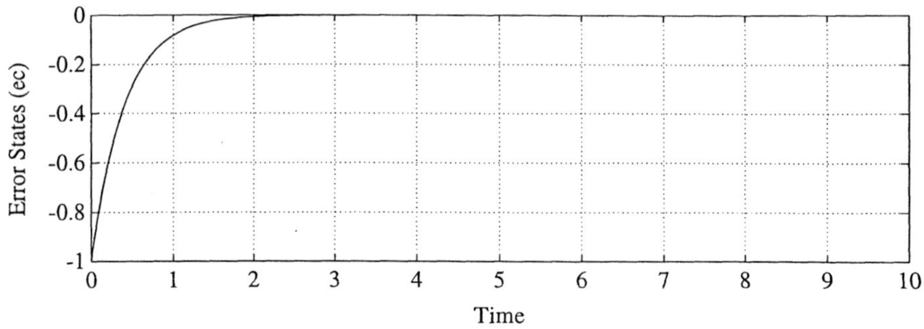

Figure 5.5: Evolution of the error states e_c

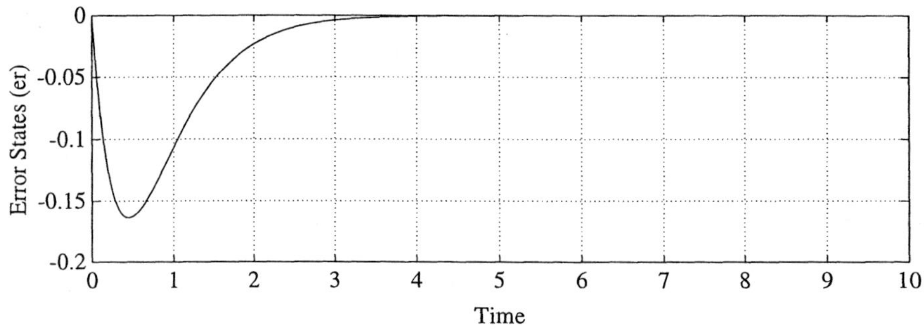

Figure 5.6: Evolution of the errors states e_r

5.6.3 Design Example 2: Inverted Pendulum

Consider the inverted pendulum with a cart shown in Figure 5.7. Assume that the pendulum rotates in the vertical plane and the cart is to be manipulated so that the pendulum remains in an upright position. The cart is linked by a transmission belt to a drive wheel which is driven by a DC motor.

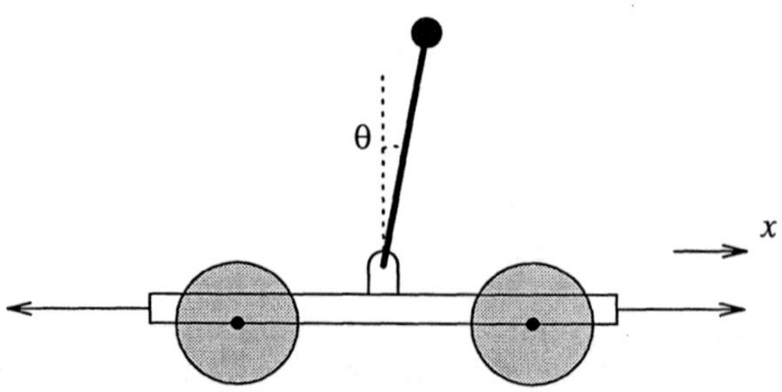

Figure 5.7: Schematic of inverted pendulum

The equations of motion are

$$(M+m)\ddot{x} + F_x\dot{x} + ml(\ddot{\theta}\cos\theta - \dot{\theta}^2\sin\theta) = u \quad (5.108)$$
$$J\ddot{\theta} + F_\theta\dot{\theta} - mlg\sin\theta + ml\ddot{x}\cos\theta = 0 \quad (5.109)$$

where the values of the physical parameters used are given in Table 5.1.

Table 5.1: Model parameters for the inverted pendulum

M	(kg)	3.2	F_x	(kg/sec)	6.2
m	(kg)	0.535	F_θ	(kg m^2)	0.009
J	(kg m^2)	0.062	g	(m/sec^2)	9.807
l	(m)	0.365			

A linearisation of the system in (5.108) and (5.109) has been made about the equilibrium point at the origin. Using x, θ, $\dot{x}$ and $\dot{\theta}$ as system states, and assuming that only θ, x and $\dot{x}$ are available as measured outputs, results in the triple

$$A = \begin{bmatrix} 0 & 0 & 1 & 0 \\ 0 & 0 & 0 & 1 \\ 0 & -1.9333 & -1.9872 & 0.0091 \\ 0 & 36.9771 & 6.2589 & -0.1738 \end{bmatrix} \qquad B = \begin{bmatrix} 0 \\ 0 \\ 0.3205 \\ -1.0095 \end{bmatrix}$$
$$C = \begin{bmatrix} 1 & 0 & 0 & 0 \\ 0 & 1 & 0 & 0 \\ 0 & 0 & 1 & 0 \end{bmatrix} \quad (5.110)$$

It can be easily verified that the Markov parameter CB is full rank and in the canonical form of Lemma 5.3

$$A = \begin{bmatrix} -0.1452 & 0 & 30.8881 & -0.4572 \\ 0 & 0 & 0 & 1.0000 \\ 1.0000 & 0 & 0 & 3.1496 \\ -0.0091 & 0 & 1.9333 & -2.0158 \end{bmatrix} \quad B = \begin{bmatrix} 0 \\ 0 \\ 0 \\ -0.3205 \end{bmatrix} \quad (5.111)$$

From the system matrix above

$$A_{22}^o = -0.1452 \quad \text{and} \quad A_{21}^o = \begin{bmatrix} 0 \\ 1 \end{bmatrix}$$

which are clearly an observable pair. Therefore, $r = 0$ and the triple (A, B, C) does not possess invariant zeros. As a consequence, in this example,

$$\tilde{A}_{11} = \begin{bmatrix} -0.1452 & 0 & 30.8881 \\ 0 & 0 & 0 \\ 1.0000 & 0 & 0 \end{bmatrix} \qquad A_{122} = \begin{bmatrix} -0.4572 \\ 1.0000 \\ 3.1496 \end{bmatrix}$$

The theory requires a state feedback gain $\mathcal{K}$ to be chosen so that $\tilde{A}_{11} - A_{122}\mathcal{K}$ is stable. In the absence of any initial mismatch between the compensator states and the states of the system, $e_c(t) = 0$ for all time, and the reduced-order sliding motion will be governed by this system matrix. If the canonical form in (5.111) is viewed as regular form then the quadratic minimisation method described in Section 4.2.2 can be used to optimally select $\mathcal{K}$.

If, in the notation of Section 4.2.2, the positive definite matrix Q in the cost function is chosen to be diag(10, 1, 1, 0.1), then the matrices $\hat{Q} = \text{diag}(0.052, 10, 1)$ and $Q_{22} = 1.9920$ can be identified. Solving the Riccati equation yields

$$\mathcal{K} = \begin{bmatrix} 0.8269 & -2.2406 & 4.5352 \end{bmatrix}$$

This gives a reduced-order motion with poles $\{-4.3241 \pm 1.7852j,\ -3.1623\}$. The range space dynamic matrix Φ has been chosen as -6 and the compensator gain matrix $L^o = -9.8548$, which results in $H = -10$ and

$$D = \begin{bmatrix} 0 & -67.6603 & 31.4960 \end{bmatrix}$$

In the simulations which follow, the scaling matrix $F_2 = I$ and $\rho = 1$. The following plots are from a simulation of the controller on the nominal linearisation. The initial conditions are all zero except for θ, which has been chosen to represent an initial displacement of 0.1 radian from the vertical. The initial condition for the compensator is -0.9855 which guarantees that, in this nominal situation, $e_c(t) = 0$ for all time.

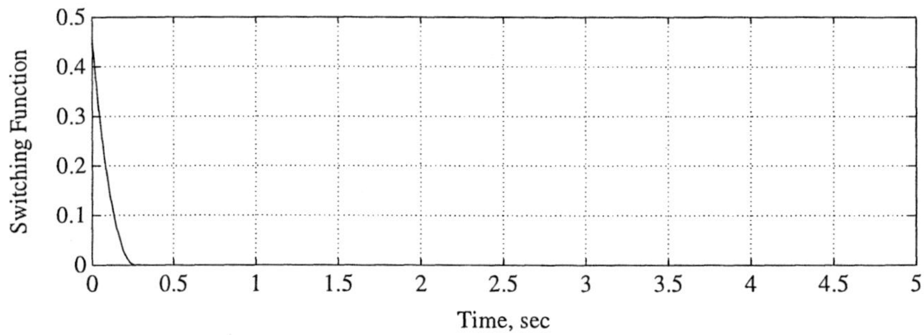

Figure 5.8: Switching function versus time

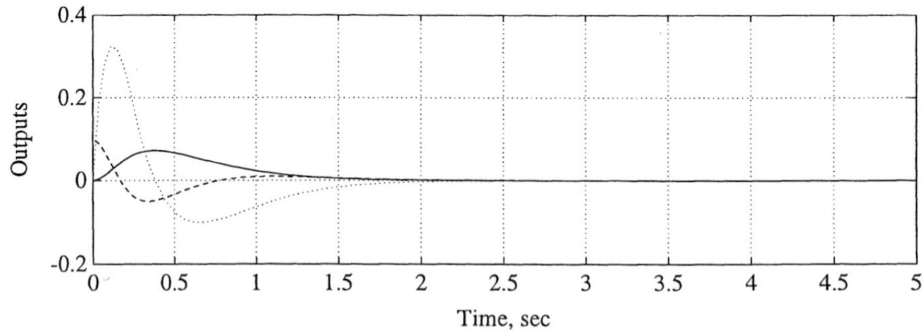

Figure 5.9: System outputs versus time

Figure 5.8 is a plot of the switching function against time and shows that within 0.25 second a sliding motion is established. It also demonstrates the advantage of setting the initial conditions of the compensator appropriately since a sliding motion is sustained at the first instant when $s(t) = 0$; this should be compared with Figure 5.3 in the previous example. Figure 5.9 plots the evolution of the system outputs. The dashed line represents the angular position θ and shows that the settling time is approximately 2 seconds.

The following results are obtained from using the controller on the nonlinear equations of motion. In this situation both matched and unmatched nonlinearities are present and so the theory described earlier is not technically valid. In practice, however, a good response is obtained. The following simulations use the same initial conditions as Figures 5.8 and 5.9.

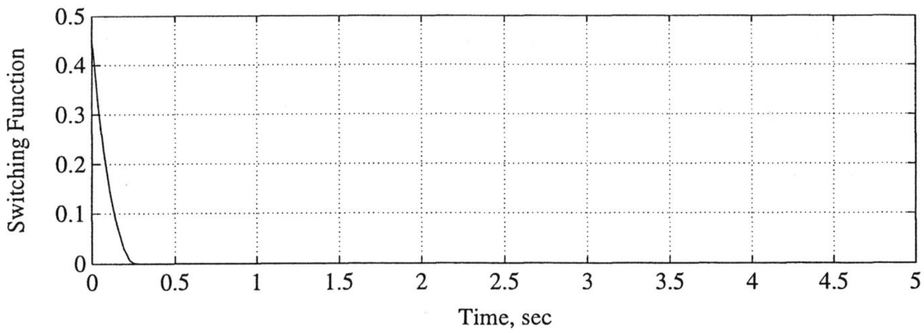

Figure 5.10: Switching function versus time

Figure 5.10 is a plot of the switching function against time and shows that within 0.25 second a sliding motion is established. The output responses shown in Figure 5.11 are identical to those of Figure 5.9. Again the dashed line represents angular position.

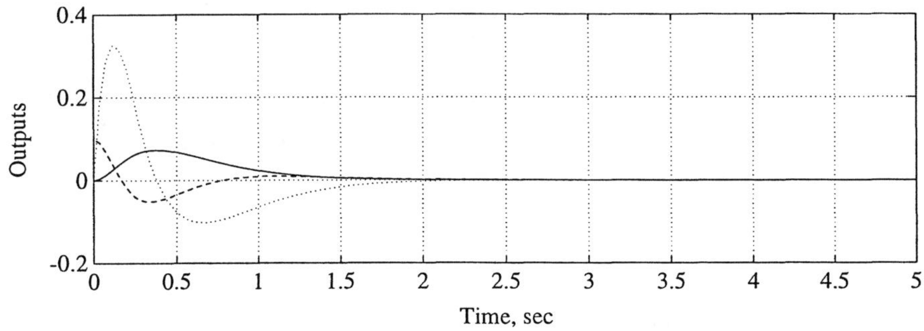

Figure 5.11: Nonlinear system outputs

5.7 A MODEL-REFERENCE SYSTEM USING ONLY OUTPUTS

In Chapter 4 a model-following strategy was employed to incorporate a tracking requirement under the assumption that all the states were available to the controller. In this section, a model-following problem is considered with the proviso that only output information is available. The intention is to formulate a control law using the model states and the output information from the plant which forces the discrepancy between the states of the plant and the model to zero. This section will demonstrate a link between such a problem and the output regulation problem that has occupied the majority of the chapter.

Consider the uncertain system described in Section 5.2 and define a new dynamical system, the ideal model, according to

$$\dot{x}_m(t) = A_m x_m(t) + B_m r(t) \tag{5.112}$$

$$y_m(t) = C x_m(t) \tag{5.113}$$

where $x_m \in \mathbb{R}^n$ are the model states and $r(t)$ are the reference inputs. Here, as in Section 4.4, it will be assumed that

$$A_m = A + BL_x \quad \text{and} \quad B_m = BL_r \tag{5.114}$$

for some appropriate fixed gains L_x and L_r and that the model state-space matrix A_m is stable. Together with the matching assumption from Section 5.1, this guarantees that the conditions for perfect model-following are satisfied. Formally it is desired to synthesise a control law so that the error

$$e(t) \triangleq x(t) - x_m(t) \tag{5.115}$$

is quadratically stable.

Using equations (5.1) and (5.112) to (5.113) the error system satisfies

$$\dot{e}(t) = Ae(t) + B\left(u(t) + \xi(t,x,u) - L_x x_m(t) - L_r r(t)\right) \tag{5.116}$$

If the proposed control law has the form

$$u(t) = u_f(t) + u_e(t) \tag{5.117}$$

where $u_e(t)$ is a control component law depending only on $e_y = Ce$ and the feedforward term $u_f(t) = L_x x_m(t) + L_r r(t)$ then

$$\dot{e}(t) = Ae(t) + B\left(u_e(t) + \xi(t,x,u)\right) \tag{5.118}$$

The problem is therefore to construct a control law $u_e(t)$ using only output information to stabilise the uncertain error system (5.118). This, of course, is the regulator problem considered in the rest of this chapter and hence all the results obtained there are applicable.

5.7.1 Aircraft Example

Consider the following fifth-order linear system representing the lateral axis model of an L-1011 fixed-wing aircraft with the actuator dynamics removed. The state vector is represented by

$$x = \begin{bmatrix} \phi \\ r \\ p \\ \beta \\ x_5 \end{bmatrix} \quad \begin{matrix} \text{bank angle (rad)} \\ \text{yaw rate (rad/sec)} \\ \text{roll rate (rad/sec)} \\ \text{sideslip angle (rad)} \\ \text{washed-out filter state} \end{matrix}$$

and the system triple (A, B, C) is given by

$$A = \begin{bmatrix} 0 & 0 & 1.0000 & 0 & 0 \\ 0 & -0.1540 & -0.0042 & 1.5400 & 0 \\ 0 & 0.2490 & -1.0000 & -5.2000 & 0 \\ 0.0386 & -0.9960 & -0.0003 & -0.1170 & 0 \\ 0 & 0.5000 & 0 & 0 & -0.5000 \end{bmatrix}$$

$$B = \begin{bmatrix} 0 & 0 \\ -0.7440 & -0.0320 \\ 0.3370 & -1.1200 \\ 0.0200 & 0 \\ 0 & 0 \end{bmatrix} \qquad C = \begin{bmatrix} 0 & 1 & 0 & 0 & -1 \\ 0 & 0 & 1 & 0 & 0 \\ 0 & 0 & 0 & 1 & 0 \\ 1 & 0 & 0 & 0 & 0 \end{bmatrix}$$

The inputs are

$$u = \begin{bmatrix} \delta_r \\ \delta_a \end{bmatrix} \quad \begin{matrix} \text{rudder deflection (rad)} \\ \text{aileron deflection (rad)} \end{matrix}$$

and the outputs

$$y = \begin{bmatrix} r_{wo} \\ p \\ \beta \\ \phi \end{bmatrix} \quad \begin{matrix} \text{washed-out yaw rate} \\ \text{roll rate (rad/sec)} \\ \text{sideslip angle (rad)} \\ \text{bank angle (rad)} \end{matrix}$$

An ideal model for this system has been developed using the eigenstructure assignment method described in Section 4.5.1. The state gain matrix

$$L_x = \begin{bmatrix} -0.3131 & 3.3211 & -0.1386 & -0.7379 & 4.1180 \\ 3.9524 & 5.6616 & 2.2906 & -65.6425 & -90.7262 \end{bmatrix} \tag{5.119}$$

provides an ideal model system matrix $A_m = A + BL_x$ with eigenvalues at

$$\{-0.05, -2 \pm 1.5j, -1.5 \pm 1.5j\}$$

This model essentially decouples the rolling and yawing motions of the aircraft. Clearly, since the system has only two inputs, all four outputs cannot be controlled independently. For this reason a matrix of controlled outputs

$$C_m = \begin{bmatrix} 1 & 0 & 0 & 0 & 0 \\ 0 & 0 & 0 & 1 & 0 \end{bmatrix} \tag{5.120}$$

has been introduced which identifies bank angle ϕ and sideslip angle β as the outputs of interest. (For control purposes the four signals in y will be fed back.) The feed-forward matrix L_r has been calculated as in Section 4.5.1 to ensure that the steady-state gain from the reference signal to the controlled outputs via the triple (A_m, BL_r, C_m) is unity.

It can easily be checked that the Markov parameter CB is of full rank and a realisation of (A, B, C) in the canonical form of Lemma 5.3 yields

$$\begin{bmatrix} A_{11} & A_{12} \\ A_{21} & A_{22} \end{bmatrix} = \left[\begin{array}{ccc|cc} -0.0133 & -0.0007 & -0.0172 & 0.2838 & 0.6266 \\ 0 & 0 & -0.0008 & -0.9110 & 0.4125 \\ 0.7071 & 0.0386 & -0.0858 & 0.4167 & 0.9210 \\ \hline 0.1227 & 0.0004 & 4.1018 & -0.8028 & 0.4471 \\ -0.1545 & 0.0009 & -3.5336 & 0.0580 & -0.8691 \end{array}\right] \tag{5.121}$$

Since $r = 0$ it follows that $\tilde{A}_{11} = A_{11}$ and $A_{122} = A_{12}$, hence

$$\begin{bmatrix} A^o_{22} & A^m_{122} \\ A^o_{21} & A^m_{22} \end{bmatrix} = \left[\begin{array}{c|cc} -0.0133 & -0.0007 & -0.0172 \\ \hline 0 & 0 & -0.0008 \\ 0.7071 & 0.0386 & -0.0858 \end{array}\right]$$

and

$$\begin{bmatrix} A_{1221} \\ A_{1222} \end{bmatrix} = \left[\begin{array}{cc} 0.2838 & 0.6266 \\ \hline -0.9110 & 0.4125 \\ 0.4167 & 0.9210 \end{array}\right]$$

Here the gain matrix L^o has been chosen so that $H = -5$, which results in

$$D = \begin{bmatrix} 5.5198 & 0.0243 & -34.8353 & -0.2729 \end{bmatrix}$$

Because in this example $\tilde{A}_{11} = A_{11}$ and $A_{122} = A_{12}$, the state feedback gain $\mathcal{K}$ can be chosen by applying the quadratic minimisation method of Section 4.2.2 as detailed in Section 5.6.3. In the notation of Section 4.2.2, the symmetric positive definite matrix Q in the cost function has been chosen to be diag$(5, 1, 1, 5, 5)$, which results in a switching function matrix

$$S = \begin{bmatrix} 0.4410 & -2.0310 & 3.9576 & 1.0000 & 0 \\ 0.9219 & 0.9368 & 8.4549 & 0 & 1.0000 \end{bmatrix}$$

and sliding mode poles at $\{-0.4470, -2.5061, -2.2372\}$. The range space dynamic matrix Φ has been chosen as diag$(-5, -5)$ and in the simulation which follows $\rho = 1$.

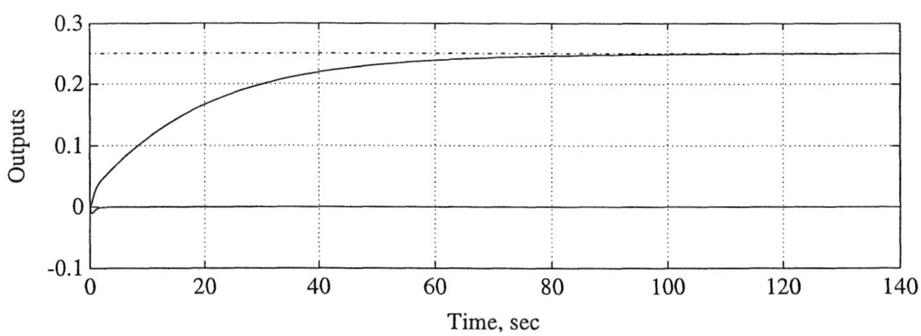

Figure 5.12: System and ideal model response

Figure 5.12 shows the response of the closed-loop system to a step change on bank angle. The bank angle plant response is identical to that of the ideal model. The plant exhibits an initial sideslip transient before attaining the perfect decoupling of the ideal model. In order to make the error system apparent, the initial state of the compensator has been chosen to be nonzero. Figure 5.13 shows the first 10 seconds of the simulation so that the plant and ideal model discrepancies can be observed. Figure 5.14 shows that a sliding motion in the error system occurs after approximately 1 second.

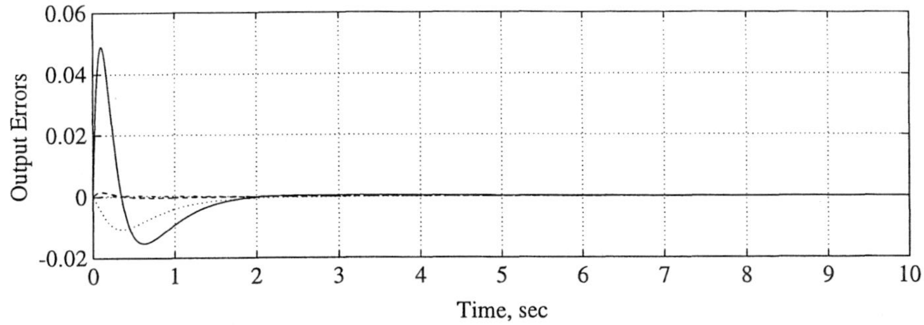

Figure 5.13: Evolution of the errors between plant and ideal model outputs e_y

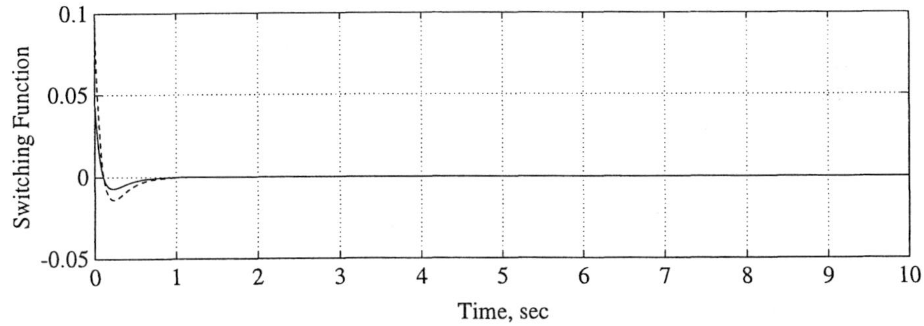

Figure 5.14: Evolution of the error-based switching function $s(t)$

5.8 SUMMARY

Design procedures have been presented to synthesise robust output feedback controllers for uncertain systems. The class of systems to which the results apply has been identified, and includes the requirement that the nominal linear system is minimum phase. Emphasis has been placed on the tractability of the associated design procedure. Two design methods have been employed for the design of the sliding surface: one utilises established static output feedback pole placement results, the other a reduced-order Luenberger observer approach. It has been demonstrated that all the assumptions imposed on the system pertain to the design of the sliding surface. The proposed controller which guarantees attainment of a sliding mode despite the presence of uncertainty, requires no additional assumptions. The last section demonstrated how the previous output feedback results can be viewed within the context of a model-reference approach.

5.9 NOTES AND REFERENCES

For linear systems with no uncertainty the output feedback problem has been investigated by White (1990) and by El-Khazali & DeCarlo (1991,1992). For uncertain linear systems an approach has recently been reported by Hui & Żak (1993) and

Żak & Hui (1993), who propose an algorithm for output dependent hyperplane design, which is based upon eigenvector methods. The conceptual approach of Żak & Hui (1993) is similar to that of El-Khazali & DeCarlo in that it relies on establishing an appropriate eigenstructure for the reduced-order sliding motion.

White (1990) considers the case when, in the notation of this chapter, FCB is rank deficient. In this situation an ideal sliding motion as given in Definition 3.1 cannot be guaranteed.

The compensator-based approach considered towards the end of the chapter is similar to the structure considered in El-Khazali & DeCarlo (1993). The framing of the problem as a static output feedback pole placement problem is described in detail in Bag *et al.* (1997) where a so-called normal matrix approach is employed to obtain the feedback gain. No single universally acknowledged algorithm for output feedback pole placement has emerged, although a useful recent survey of the state of the art is given in Syrmos *et al.* (1997).

The quadratic stability approach used in the early sections relies on constructing relative degree 1 minimum phase systems. It is well known that such systems are stabilisable via a linear static output feedback controller; see Gu (1990).

Work adopting a model-reference approach using only input/output information has appeared within the context of adaptive control where the variable structure ideas are utilised in the parameter adaptation mechanism. A survey of these methods appears in Hsu *et al.* (1994).

This chapter has not considered a class of controllers found in the literature which employ numerical methods to synthesise a linear static output feedback gain to ensure that a reachability condition similar to that of equation (3.122) is satisfied. For further details see Heck *et al.* (1995) and Bag *et al.* (1997).

The linear system in Example 1 of Section 5.4.3 is from Hui & Żak (1993).

The data for the inverted pendulum in Table 5.1 is from an experimental rig in the Department of Electronic Engineering at the University of Hull.

The lateral axis model of the L-1011 aircraft at cruise flight conditions in Section 5.7.1 appears in Sobel & Shapiro (1986). In Section 5.7.1, for control law design purposes, the states associated with the actuator dynamics have been removed.

Chapter 6

Sliding Mode Observers

6.1 INTRODUCTION

Many of the theoretical developments in the area of sliding mode control systems assume that the system state vector is available for use by the control scheme. In order to exploit these control strategies, a suitable estimate of the state vector may be constructed for use in the original control law. The idea of using a dynamical system to generate estimates of the system states – an *observer* – can be traced to Luenberger (1971), who proposed a method for linear systems which now bears his name and which was described in Section 2.3.5. In this approach, the observer system is driven by the control input and by the difference between the output of the observer and the output of the plant. The latter should ideally become zero. This naturally suggests the exploration of ideas to generate a sliding mode on the subspace for which the output error is zero. In particular, it is of interest to explore the possibilities of using sliding mode observers for *robust* state reconstruction.

6.2 SLIDING MODE OBSERVERS

Consider initially the linear system described by

$$\begin{aligned} \dot{x}(t) &= Ax(t) + Bu(t) \\ y(t) &= Cx(t) \end{aligned} \tag{6.1}$$

where $A \in \mathbb{R}^{n\times n}$, $B \in \mathbb{R}^{n\times m}$, $C \in \mathbb{R}^{p\times n}$ and $p \geq m$. Assume that the matrices B and C are of full rank and the pair (A, C) is observable.

6.2.1 An Utkin Observer

As the outputs are to be considered, it is logical to effect a change of coordinates so that the outputs appear as components of the states. As in Chapter 5, one possibility is to consider the transformation $x \mapsto T_c x$ where

$$T_c = \begin{bmatrix} N_c^{\mathrm{T}} \\ C \end{bmatrix} \tag{6.2}$$

where the columns of $N_c \in \mathbb{R}^{n\times(n-p)}$ span the null space of C. This transformation is nonsingular, and with respect to this new coordinate system, the new output distribution matrix is

$$CT_c^{-1} = \begin{bmatrix} 0 & I_p \end{bmatrix}$$

If the other system matrices are written as

$$T_cAT_c^{-1} = \begin{bmatrix} A_{11} & A_{12} \\ A_{21} & A_{22} \end{bmatrix} \quad \text{and} \quad T_cB = \begin{bmatrix} B_1 \\ B_2 \end{bmatrix}$$

then the nominal system can be written as

$$\dot{x}_1(t) = A_{11}x_1(t) + A_{12}y(t) + B_1u(t) \tag{6.3}$$
$$\dot{y}(t) = A_{21}x_1(t) + A_{22}y(t) + B_2u(t) \tag{6.4}$$

where

$$T_cx = \begin{bmatrix} x_1 \\ y \end{bmatrix} \begin{matrix} \updownarrow n-p \\ \updownarrow p \end{matrix}$$

The observer proposed by Utkin (1981) has the form

$$\dot{\hat{x}}_1(t) = A_{11}\hat{x}_1(t) + A_{12}\hat{y}(t) + B_1u(t) + L\nu \tag{6.5}$$
$$\dot{\hat{y}}(t) = A_{21}\hat{x}_1(t) + A_{22}\hat{y}(t) + B_2u(t) - \nu \tag{6.6}$$

where $(\hat{x}_1, \hat{y})$ represent the state estimates for (x_1, y), $L \in \mathbb{R}^{(n-p)\times p}$ is a constant feedback gain matrix and the discontinuous vector ν is defined componentwise by

$$\nu_i = M\,\mathrm{sgn}(\hat{y}_i - y_i) \tag{6.7}$$

where $M \in \mathbb{R}_+$. If the errors between the estimates and the true states are written as $e_1 = \hat{x}_1 - x_1$ and $e_y = \hat{y} - y$ then from equations (6.3) to (6.6) the following error system is obtained

$$\dot{e}_1(t) = A_{11}e_1(t) + A_{12}e_y(t) + L\nu \tag{6.8}$$
$$\dot{e}_y(t) = A_{21}e_1(t) + A_{22}e_y(t) - \nu \tag{6.9}$$

Since the pair (A, C) is observable, the pair (A_{11}, A_{21}) is also observable.[1] As a consequence, L can be chosen to make the spectrum of $A_{11}+LA_{21}$ lie in $\mathbb{C}_-$. Define a further change of coordinates, dependent on L, by

$$\tilde{T} = \begin{bmatrix} I_{n-p} & L \\ 0 & I_p \end{bmatrix} \tag{6.10}$$

and let $\tilde{e}_1 = e_1 + Ly$. The error system with respect to the new coordinates can be written as

$$\dot{\tilde{e}}_1(t) = \tilde{A}_{11}\tilde{e}_1(t) + \tilde{A}_{12}e_y(t) \tag{6.11}$$
$$\dot{e}_y(t) = A_{21}\tilde{e}_1(t) + \tilde{A}_{22}e_y(t) - \nu \tag{6.12}$$

where $\tilde{A}_{11} = A_{11} + LA_{21}$, $\tilde{A}_{12} = A_{12} + LA_{22} - \tilde{A}_{11}L$ and $\tilde{A}_{22} = A_{22} - A_{21}L$. It follows from (6.12) that in the domain

$$\Omega = \{(e_1, e_y) \;:\; \|A_{21}e_1\| + \tfrac{1}{2}\lambda_{max}(\tilde{A}_{22} + \tilde{A}_{22}^{\mathrm{T}})\|e_y\| < M - \eta\} \tag{6.13}$$

[1]This is the dual result to Proposition 3.3.

where $\eta < M$ is some small positive scalar, the reachability condition

$$e_y^{\mathrm{T}}\dot{e}_y < -\eta\|e_y\| \tag{6.14}$$

is satisfied. Consequently, an ideal sliding motion will take place on the surface

$$\mathcal{S}_o = \{(e_1, e_y) \ : \ e_y = 0\} \tag{6.15}$$

It follows that after some finite time t_s, for all subsequent time, $e_y = 0$ and $\dot{e}_y = 0$. Equation (6.11) then reduces to

$$\dot{\tilde{e}}_1(t) = \tilde{A}_{11}\tilde{e}_1(t) \tag{6.16}$$

which, by choice of L, represents a stable system and so $\tilde{e}_1 \to 0$ and consequently, $\hat{x}_1 \to x_1$ as $t \to \infty$. Equation (6.16) represents the reduced order sliding mode error dynamics.

6.2.2 Example 1

Consider the second-order linear system

$$\begin{aligned} \dot{x}(t) &= Ax(t) + Bu(t) \quad (6.17)\\ y(t) &= Cx(t) \quad (6.18) \end{aligned}$$

where

$$A = \begin{bmatrix} 0 & 1 \\ -2 & 0 \end{bmatrix} \quad B = \begin{bmatrix} 0 \\ 1 \end{bmatrix} \quad C = \begin{bmatrix} 1 & 1 \end{bmatrix}$$

which represents a simple harmonic oscillator. For simplicity assume $u = 0$ and consider the problem of designing a sliding mode observer. Define a nonsingular matrix

$$T_c = \begin{bmatrix} 1 & 0 \\ 1 & 1 \end{bmatrix} \tag{6.19}$$

and change coordinates according to $x \mapsto T_c x$ to give the system triple

$$T_c A T_c^{-1} = \begin{bmatrix} -1 & 1 \\ -3 & 1 \end{bmatrix} \quad T_c B = \begin{bmatrix} 0 \\ 1 \end{bmatrix} \quad C T_c^{-1} = \begin{bmatrix} 0 & 1 \end{bmatrix}$$

The system is now in the form of equations (6.3) and (6.4). An appropriate choice of gain in the observer given in (6.5) is $L = 0.5$ which results in an error system governed by $\tilde{A}_{11} = -2.5$.

In the simulations which follow, the scaling constant M in the discontinuous component in equation (6.7) has been set to unity. Figure 6.1 shows the state estimation errors $e_1(t)$ and $e_y(t)$ resulting from the initial conditions $e_1 = -1$ and $e_y = 0$. It can be seen that although the error system starts on the sliding surface $\mathcal{S}_o$, an ideal sliding motion cannot be maintained; only after approximately 1.2 seconds is sliding established. At this point in the time interval 1.2 to 3.0 seconds, $e_1(t)$ exhibits a first-order exponential decay to the origin. After 3.0 seconds almost perfect replication of the states takes place.

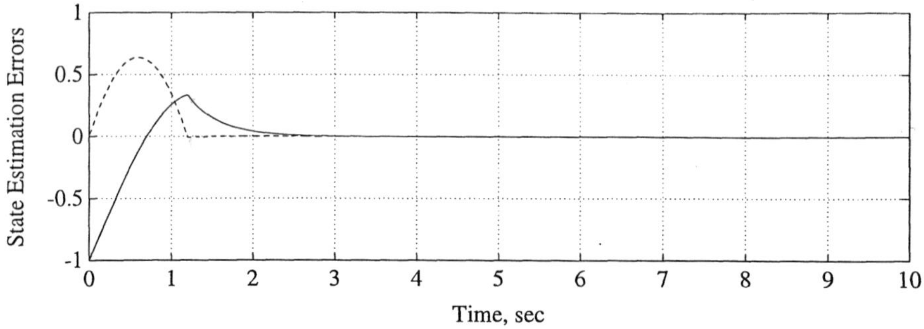

Figure 6.1: State estimation errors versus time

In the original coordinates it can be seen from Figure 6.2 that perfect tracking occurs after approximately 3 seconds. The dotted lines represent the true states and the solid line the estimates from the observer.

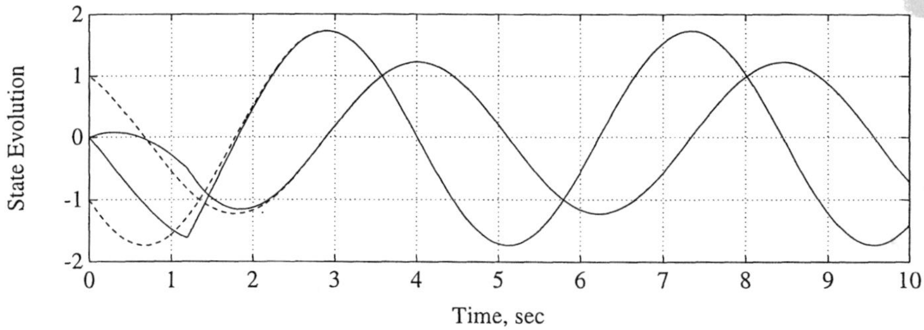

Figure 6.2: State evolution with respect to time

Finally Figure 6.3 shows the value of ν with respect to time and shows switching taking place from 1.2 seconds onwards.

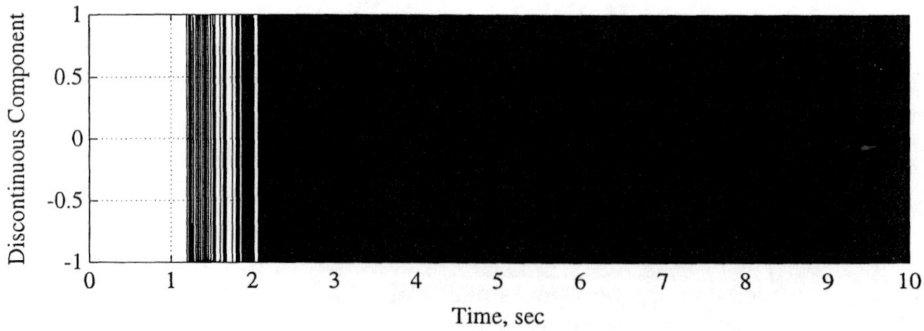

Figure 6.3: Discontinuous switching component

6.2.3 A Modification to Include a Linear Term

Consider the effect of adding a negative output error feedback term to equations (6.5) and (6.6) of the Utkin observer. This results in a new error system

$$\dot{\tilde{e}}_1(t) = \tilde{A}_{11}\tilde{e}_1(t) + \tilde{A}_{12}e_y(t) - G_1 e_y(t) \tag{6.20}$$
$$\dot{e}_y(t) = A_{21}\tilde{e}_1(t) + \tilde{A}_{22}e_y(t) - G_2 e_y(t) - \nu \tag{6.21}$$

By selecting $G_1 = \tilde{A}_{12}$ and $G_2 = \tilde{A}_{22} - A_{22}^s$, where A_{22}^s is any stable design matrix of appropriate dimension, then

$$\dot{\tilde{e}}_1(t) = \tilde{A}_{11}\tilde{e}_1(t) \tag{6.22}$$
$$\dot{e}_y(t) = A_{21}\tilde{e}_1(t) + A_{22}^s e_y(t) - \nu \tag{6.23}$$

In this form, the error system is asymptotically stable for $\nu \equiv 0$ because the poles of the combined system are given by $\lambda(\tilde{A}_{11}) \cup \lambda(A_{22}^s)$ and so lie in the open left half complex plane. In the original Utkin observer, the switching action ν was potentially required to make the error system stable. The addition of a Luenberger type gain matrix, feeding back the output error, yields the potential to provide robustness against certain classes of uncertainty by virtue of the discontinuous component ν.

Remark
It should be noted that thus far the only restriction imposed on the nominal linear system is that the pair (A, C) is observable.

6.2.4 A Walcott–Żak Observer

Consider now the uncertain system described by

$$\begin{aligned} \dot{x}(t) &= Ax(t) + Bu(t) + f(t,x,u) \\ y(t) &= Cx(t) \end{aligned} \tag{6.24}$$

where $A \in \mathbb{R}^{n\times n}$, $B \in \mathbb{R}^{n\times m}$, $C \in \mathbb{R}^{p\times n}$ and $p \geq m$; in addition the matrices B and C are assumed to be of full rank. The function $f : \mathbb{R}_+ \times \mathbb{R}^n \times \mathbb{R}^m \to \mathbb{R}^n$ is unknown and represents the system uncertainty. A natural problem to consider initially is the special case when the uncertainty is matched: suppose

$$f(t,x,u) = B\xi(t,x,u) \tag{6.25}$$

where the function $\xi : \mathbb{R}_+ \times \mathbb{R}^n \times R^m \to \mathbb{R}^m$ is unknown, but bounded, so that

$$\|\xi(t,x,u)\| \leq r_1\|u\| + \alpha(t,y) \tag{6.26}$$

where r_1 is a known scalar and $\alpha : \mathbb{R}_+ \times \mathbb{R}^p \to \mathbb{R}_+$ is a known function.

The problem to be considered involves estimating the states of the uncertain system given in (6.24) so that the error system

$$e(t) = \hat{x}(t) - x(t) \tag{6.27}$$

is quadratically stable despite the presence of the uncertainty. A conceptually appealing approach is adopted by Walcott & Żak (1987) who make the following assumption. There exists a $G \in \mathbb{R}^{n \times p}$ such that $A_0 = A - GC$ has stable eigenvalues and there exists a Lyapunov pair (P, Q) for A_0 such that the structural constraint

$$C^{\mathrm{T}} F^{\mathrm{T}} = PB \tag{6.28}$$

is satisfied for some $F \in \mathbb{R}^{m \times p}$.

Utilising this assumption, consider an observer of the form

$$\dot{\hat{x}}(t) = A\hat{x}(t) + Bu(t) - G(C\hat{x}(t) - y(t)) + P^{-1}C^{\mathrm{T}}F^{\mathrm{T}}\nu \tag{6.29}$$

where

$$\nu = \begin{cases} -\rho(t, y, u)\frac{FCe}{\|FCe\|} & \text{if } FCe \neq 0 \\ 0 & \text{otherwise} \end{cases} \tag{6.30}$$

and the scalar function $\rho(\cdot)$ is any function satisfying

$$\rho(t, y, u) \geq r_1\|u\| + \alpha(t, y) + \eta \tag{6.31}$$

for some positive scalar η. To prove that this observer guarantees quadratic stability of the error system, consider $V(e) = e^{\mathrm{T}}Pe$ as a candidate Lyapunov function. It follows from (6.24) and (6.29) that the error system satisfies

$$\dot{e}(t) = (A - GC)e(t) - B\xi(t, x, u) + B\nu \tag{6.32}$$

and therefore

$$\begin{aligned} \dot{V} &= e^{\mathrm{T}}(PA_0 + A_0P)e - 2e^{\mathrm{T}}PB\xi + 2e^{\mathrm{T}}PB\nu \\ &\leq -e^{\mathrm{T}}Qe - 2e^{\mathrm{T}}PB\xi - 2\rho(t, y, u)\|FCe\| \end{aligned} \tag{6.33}$$

Then using the structural constraint (6.28) it follows that

$$\begin{aligned} \dot{V} &\leq -e^{\mathrm{T}}Qe - 2e^{\mathrm{T}}C^{\mathrm{T}}F^{\mathrm{T}}\xi - 2\rho(t, y, u)\|FCe\| \\ &\leq -e^{\mathrm{T}}Qe - 2\|FCe\|(\rho(t, y, u) - \|\xi\|) \\ &\leq -e^{\mathrm{T}}Qe - 2\eta\|FCe\| \end{aligned} \tag{6.34}$$

Using an argument similar to that in Section 5.3 it can be shown that there exists a domain in which sliding motion is induced on the surface in the state error space given by

$$\mathcal{S}_{wz} = \{ e \in \mathbb{R}^n \ : \ FCe = 0 \} \tag{6.35}$$

This is a very appealing approach but relies on establishing whether there exists a gain matrix G such that, for the resulting closed-loop matrix A_0, there exists a Lyapunov matrix P for A_0 satisfying (6.28) for some $F \in \mathbb{R}^{m \times p}$. This represents a nontrivial problem. Walcott & Żak (1988) propose an algorithm for the design of P which can be summarised as follows:

(1) Choose the spectrum of A_0, and compute G accordingly.

(2) Solve the structural constraint *symbolically* to obtain an expression for P_F in terms of the entries of F, ensuring that P_F is symmetric.

(3) Compute $Q(P_F)$ symbolically in terms of the entries of P_F using the expression $Q(P_F) = -(P_F A_0 + A_0^{\mathrm{T}} P_F)$.

(4) Choose the elements of F and P_F to ensure $Q(P_F)$ is symmetric positive definite by checking that all the principal minors are positive.

From a computational point of view, the algebra associated with the manipulation and solution of the inequalities resulting from step 4 is impractical without the use of a symbolic manipulation package for systems of order 4 and above. Also no indication is given of the class of systems for which the algorithm will produce a successful design.

Remarks

- For a stable square system given by (A_0, B, FC) the existence of a Lyapunov matrix P which satisfies the structural constraint (6.28) is equivalent to the transfer function

$$G_F(s) = FC(sI - A_0)^{-1}B \tag{6.36}$$

being *strictly positive real.*[2] This observation does not offer a solution to the problem of finding a suitable G and F but it does give an indication of the class of systems for which one exists, since it is well known that positive real systems are both minimum phase and relative degree 1.

- The structural constraint in (6.28) is similar to the one encountered in Section 5.3 and, as such, a similar approach to the one described there will be adopted for the remainder of this chapter.

The next section considers an explicit solution to the problem of designing a sliding mode observer for an uncertain system.

6.3 SYNTHESIS OF A DISCONTINUOUS OBSERVER

Consider the dynamical system

$$\begin{aligned} \dot{x}(t) &= Ax(t) + Bu(t) + D\xi(t,x,u) \\ y(t) &= Cx(t) \end{aligned} \tag{6.37}$$

where $A \in \mathbb{R}^{n\times n}$, $B \in \mathbb{R}^{n\times m}$, $C \in \mathbb{R}^{p\times n}$ and $D \in \mathbb{R}^{n\times q}$ where $p \geq q$. Assume that the matrices B, C and D are full rank and the function $\xi : \mathbb{R}_+ \times \mathbb{R}^n \times \mathbb{R}^m \to \mathbb{R}^q$ is unknown but bounded so that

$$\|\xi(t,x,u)\| \leq r_1\|u\| + \alpha(t,y) \tag{6.38}$$

where r_1 is a known scalar and $\alpha : \mathbb{R}_+ \times \mathbb{R}^p \to \mathbb{R}_+$ is a known function.

[2] A transfer function $G(s)$ is defined to be *strictly positive real* if it is stable and the Hermitian matrix defined by $G(j\omega) + G^{\mathrm{T}}(-j\omega) > 0$ for all real ω.

6.3.1 A Canonical Form for Observer Design

Suppose that there exists a linear change of coordinates T_o so that the system can be written as

$$\begin{aligned}\dot{x}_1(t) &= \mathcal{A}_{11}x_1(t) + \mathcal{A}_{12}y(t) + \mathcal{B}_1u(t) \\ \dot{y}(t) &= \mathcal{A}_{21}x_1(t) + \mathcal{A}_{22}y(t) + \mathcal{B}_2u(t) + \mathcal{D}_2\xi\end{aligned} \tag{6.39}$$

where $x_1 \in \mathbb{R}^{(n-p)}$, $y \in \mathbb{R}^p$ and the matrix $\mathcal{A}_{11}$ has *stable* eigenvalues. Consider an observer of the form

$$\begin{aligned}\dot{\hat{x}}_1(t) &= \mathcal{A}_{11}\hat{x}_1(t) + \mathcal{A}_{12}\hat{y}(t) + \mathcal{B}_1u(t) - \mathcal{A}_{12}e_y(t) \\ \dot{\hat{y}}(t) &= \mathcal{A}_{21}\hat{x}_1(t) + \mathcal{A}_{22}\hat{y}(t) + \mathcal{B}_2u(t) - (\mathcal{A}_{22} - \mathcal{A}_{22}^s)e_y(t) + \nu\end{aligned} \tag{6.40}$$

where $\mathcal{A}_{22}^s$ is a stable design matrix and $e_y = \hat{y} - y$. Let $P_2 \in \mathbb{R}^{p\times p}$ be symmetric positive definite Lyapunov matrix for $\mathcal{A}_{22}^s$ then the discontinuous vector ν is defined by

$$\nu = \begin{cases} -\rho(t,y,u)\|\mathcal{D}_2\|\frac{P_2e_y}{\|P_2e_y\|} & \text{if } e_y \neq 0 \\ 0 & \text{otherwise}\end{cases} \tag{6.41}$$

where the scalar function $\rho : \mathbb{R}_+ \times \mathbb{R}^p \times \mathbb{R}^m \to \mathbb{R}_+$ satisfies

$$\rho(t,y,u) \geq r_1\|u\| + \alpha(t,y) + \gamma_o \tag{6.42}$$

and γ_o is a positive scalar. If the state estimation error $e_1 = \hat{x}_1 - x_1$, then it is straightforward to show

$$\dot{e}_1(t) = \mathcal{A}_{11}e_1(t) \tag{6.43}$$

$$\dot{e}_y(t) = \mathcal{A}_{21}e_1(t) + \mathcal{A}_{22}^s e_y(t) + \nu - \mathcal{D}_2\xi \tag{6.44}$$

The lower block triangular structure has been shown to occur quite naturally as a result of the state-space representation chosen and the output error feedback gains employed.

Proposition 6.1 *There exist a family of symmetric positive definite matrices P_2 such that the uncertain dynamical error system above is quadratically stable.*

Proof
Let $Q_1 \in \mathbb{R}^{(n-p)\times(n-p)}$ and $Q_2 \in \mathbb{R}^{p\times p}$ be symmetric positive definite design matrices and define $P_2 \in \mathbb{R}^{p\times p}$ to be the unique symmetric positive definite solution to the Lyapunov equation

$$P_2\mathcal{A}_{22}^s + (\mathcal{A}_{22}^s)^{\mathrm{T}}P_2 = -Q_2 \tag{6.45}$$

Using the computed value of P_2 define

$$\hat{Q} = \mathcal{A}_{21}^{\mathrm{T}}P_2Q_2^{-1}P_2\mathcal{A}_{21} + Q_1 \tag{6.46}$$

and notice that $\hat{Q} = \hat{Q}^{\mathrm{T}} > 0$. Let $P_1 \in \mathbb{R}^{(n-p)\times(n-p)}$ be the unique symmetric positive definite solution to the Lyapunov equation

$$P_1\mathcal{A}_{11} + \mathcal{A}_{11}^{\mathrm{T}}P_1 = -\hat{Q} \tag{6.47}$$

Consider the quadratic form given by

$$V(e_1, e_y) = e_1^{\mathrm{T}} P_1 e_1 + e_y^{\mathrm{T}} P_2 e_y \tag{6.48}$$

as a candidate Lyapunov function. The derivative along the system trajectory

$$\dot{V} = -e_1^{\mathrm{T}} \hat{Q} e_1 + e_1^{\mathrm{T}} \mathcal{A}_{21}^{\mathrm{T}} P_2 e_y + e_y^{\mathrm{T}} P_2 \mathcal{A}_{21} e_1 - e_y^{\mathrm{T}} Q_2 e_y + 2e_y^{\mathrm{T}} P_2 \nu - 2e_y^{\mathrm{T}} P_2 \mathcal{D}_2 \xi \tag{6.49}$$

It is easy to verify that

$$\begin{aligned}(e_y - Q_2^{-1} P_2 \mathcal{A}_{21} e_1)^{\mathrm{T}} Q_2 (e_y - Q_2^{-1} P_2 \mathcal{A}_{21} e_1) \equiv \\ e_y^{\mathrm{T}} Q_2 e_y - e_1^{\mathrm{T}} \mathcal{A}_{21}^{\mathrm{T}} P_2 e_y - e_y^{\mathrm{T}} P_2 \mathcal{A}_{21} e_1 + e_1^{\mathrm{T}} \mathcal{A}_{21}^{\mathrm{T}} P_2 Q_2^{-1} P_2 \mathcal{A}_{21} e_1\end{aligned} \tag{6.50}$$

Substituting the identity (6.50) into equation (6.49) and writing for notational convenience $(e_y - Q_2^{-1} P_2 \mathcal{A}_{21} e_1)$ as $\tilde{e}_y$ then

$$\begin{aligned}\dot{V} &= -e_1^{\mathrm{T}} \hat{Q} e_1 + e_1^{\mathrm{T}} \mathcal{A}_{21}^{\mathrm{T}} P_2 Q_2^{-1} P_2 \mathcal{A}_{21} e_1 - \tilde{e}_y^{\mathrm{T}} Q_2 \tilde{e}_y + 2e_y^{\mathrm{T}} P_2 \nu - 2e_y^{\mathrm{T}} P_2 \mathcal{D}_2 \xi \\ &= -e_1^{\mathrm{T}} Q_1 e_1 - \tilde{e}_y^{\mathrm{T}} Q_2 \tilde{e}_y + 2e_y^{\mathrm{T}} P_2 \nu - 2e_y^{\mathrm{T}} P_2 \mathcal{D}_2 \xi \\ &= -e_1^{\mathrm{T}} Q_1 e_1 - \tilde{e}_y^{\mathrm{T}} Q_2 \tilde{e}_y - 2\rho(t, y, u) \|\mathcal{D}_2\| \|P_2 e_y\| - 2e_y^{\mathrm{T}} P_2 \mathcal{D}_2 \xi\end{aligned}$$

Using the uncertainty bound (6.38) and the bound for $\rho(\cdot)$ from equation (6.42) in the inequality above

$$\begin{aligned}\dot{V} &\leq -e_1^{\mathrm{T}} Q_1 e_1 - \tilde{e}_y^{\mathrm{T}} Q_2 \tilde{e}_y - 2\rho(t, y, u) \|\mathcal{D}_2\| \|P_2 e_y\| + 2\|\mathcal{D}_2\| (r_1 \|u\| + \alpha(y)) \|P_2 e_y\| \\ &\leq -e_1^{\mathrm{T}} Q_1 e_1 - \tilde{e}_y^{\mathrm{T}} Q_2 \tilde{e}_y - 2\gamma_o \|\mathcal{D}_2\| \|P_2 e_y\| \\ &< 0 \qquad \text{for } (e_1, e_y) \neq 0\end{aligned}$$

and hence the error system is quadratically stable. ■

Consider the hyperplane in the error space given by

$$\mathcal{S}_o = \{ e \in \mathbb{R}^n \ : \ Ce = 0 \} \tag{6.51}$$

then from the previous proposition:

Corollary 6.1 *An ideal sliding motion takes place on $\mathcal{S}_o$ defined above.*

Proof

Consider the quadratic form

$$V_s(e_y) = e_y^{\mathrm{T}} P_2 e_y \tag{6.52}$$

Differentiating and substituting from equation (6.44) then arguing as in Proposition 6.1, it follows that

$$\begin{aligned}\dot{V}_s &= -e_y^{\mathrm{T}} Q_2 e_y + 2e_y^{\mathrm{T}} P_2 \mathcal{A}_{21} e_1 + 2e_y^{\mathrm{T}} P_2 (\nu - \mathcal{D}_2 \xi) \\ &\leq 2\|P_2 e_y\| \|\mathcal{A}_{21} e_1\| - 2\gamma_o \|\mathcal{D}_2\| \|P_2 e_y\|\end{aligned}$$

Now

$$\|P_2 e_y\|^2 = (P_2^{1/2} e_y)^{\mathrm{T}} P_2 (P_2^{1/2} e_y) \geq \lambda_{min}(P_2) \|P_2^{1/2} e_y\|^2 = \lambda_{min}(P_2) V_s$$

and hence in the domain $\Omega = \{(e_1, e_y) \ : \ ||\mathcal{A}_{21}e_1|| < ||\mathcal{D}_2||\gamma_o - \eta\}$ where η is a small positive scalar, it follows that

$$\dot{V}_s < -2\eta||P_2e_y|| \leq -2\eta\sqrt{\lambda_{min}(P_2)}\sqrt{V_s} \tag{6.53}$$

From Proposition 6.1 the output error e_y enters Ω in finite time and remains there. Integrating (6.53) and arguing as in Section 3.6.1 it follows that an ideal sliding motion takes place on $\mathcal{S}_o$ in finite time. ■

If $\hat{x}$ represents the state estimate for x and $e = \hat{x} - x$ then the robust observer can conveniently be written as

$$\dot{\hat{x}}(t) = A\hat{x}(t) + Bu(t) - G_lCe(t) + G_n\nu \tag{6.54}$$

where the linear gain

$$G_l = T_o^{-1}\begin{bmatrix} \mathcal{A}_{12} \\ \mathcal{A}_{22} - \mathcal{A}_{22}^s \end{bmatrix} \tag{6.55}$$

and the nonlinear gain

$$G_n = ||\mathcal{D}_2||T_o^{-1}\begin{bmatrix} 0 \\ I_p \end{bmatrix} \tag{6.56}$$

and

$$\nu = \begin{cases} -\rho(t,y,u)\frac{P_2Ce}{||P_2Ce||} & \text{if } Ce \neq 0 \\ 0 & \text{otherwise} \end{cases} \tag{6.57}$$

Even in the special case when $D = B$ the observer formulation (6.54) to (6.57) is different from that of Walcott & Żak (1987) for the case when $p > m$ since their results guarantee sliding will take place on the surface in the error space given by $\{e \in \mathbb{R}^n \ : \ FCe = 0\}$. This does not imply $Ce = 0$ since the null space of F is nonempty and therefore the observer does not necessarily track the system outputs perfectly. In the above formulation this is guaranteed. The usefulness of Proposition 6.1 will depend on being able to identify the class of systems which can be placed in the canonical form of equation (6.39). This will be addressed in the following section.

6.3.2 Existence Conditions

Let (A, D, C) represent the linear part of the uncertain system given in (6.37) which represents the propagation of the uncertainty ξ through to the output. Consider the problem of constructing an observer for the uncertain system of the form

$$\dot{z}(t) = Az(t) + Bu(t) - G_lCe(t) + G_n\nu \tag{6.58}$$

where $e = z - x$, ν is discontinuous about the hyperplane $\mathcal{S}_o = \{e \in \mathbb{R}^n \ : \ Ce = 0\}$ and $G_l, G_n \in \mathbb{R}^{n \times p}$ are appropriate gain matrices. The purpose of this section is to determine the class of systems for which the observer (6.58) provides quadratic stability of the error system despite the presence of bounded matched uncertainty. The canonical form from Section 5.4 will provide an intermediate step for establishing the canonical form in Section 6.3.1 from which the observer was designed.

The presentation below is almost identical to that in Lemma 5.3 except a different partition of the system matrix has been employed commensurate with the partition of the outputs.

Lemma 6.1 *Let the triple (A, D, C) represent a linear system with $p > q$ and suppose rank$(CD) = q$. Then a change of coordinates exists so that the triple with respect to the new coordinates $(\bar{A}, \bar{D}, \bar{C})$ has the following structure:*

(*a*) *The system matrix can be written as*

$$\bar{A} = \left[\begin{array}{c|c} A_{11} & A_{12} \\ \hline A_{211} & \multirow{2}{*}{A_{22}} \\ A_{212} & \end{array}\right] \tag{6.59}$$

where $A_{11} \in \mathbb{R}^{(n-p)\times(n-p)}$, $A_{211} \in \mathbb{R}^{(p-q)\times(n-p)}$ and when partitioned have the structure

$$A_{11} = \begin{bmatrix} A_{11}^o & A_{12}^o \\ 0 & A_{22}^o \end{bmatrix} \quad \text{and} \quad A_{211} = \begin{bmatrix} 0 & A_{21}^o \end{bmatrix} \tag{6.60}$$

where $A_{11}^o \in \mathbb{R}^{r\times r}$ and $A_{21}^o \in \mathbb{R}^{(p-q)\times(n-p-r)}$ for some $r \geq 0$ and the pair (A_{22}^o, A_{21}^o) is completely observable. Furthermore, the eigenvalues of A_{11}^o are the invariant zeros of (A, D, C).

(*b*) *The disturbance distribution matrix has the form*

$$\bar{D} = \begin{bmatrix} 0 \\ D_2 \end{bmatrix} \tag{6.61}$$

where $D_2 \in \mathbb{R}^{q\times q}$ is nonsingular.

(*c*) *The output distribution matrix has the form*

$$\bar{C} = \begin{bmatrix} 0 & T \end{bmatrix} \tag{6.62}$$

where $T \in \mathbb{R}^{p\times p}$ and is orthogonal

Proof
Identical to Lemmas 5.3 and 5.5. ■

In order to ensure compatibility in the partition of the state-space matrices in the statement of Lemma 6.1, let

$$A_{21} = \begin{bmatrix} A_{211} \\ A_{212} \end{bmatrix} \quad \text{and} \quad \bar{D} = \begin{bmatrix} 0 \\ \bar{D}_2 \end{bmatrix} \tag{6.63}$$

where $\bar{D}_2$ is defined as

$$\bar{D}_2 = \begin{bmatrix} 0 \\ D_2 \end{bmatrix} \begin{matrix} \updownarrow p-q \\ \updownarrow q \end{matrix} \tag{6.64}$$

The main result of this section (and the chapter) will now be proved.

Proposition 6.2 *A sliding mode observer of the form (6.58) which rejects the uncertainty class in (6.37) exists if and only if the nominal linear system satisfies*

- *rank* $(CD) = q$
- *any invariant zeros of* (A, D, C) *must lie in* $\mathbb{C}_-$.

Proof
(proof of necessity)
Let G_l and G_n be appropriate gain matrices so that $A_0 = A - G_l C$ is stable, and assume an ideal sliding mode insensitive to the uncertainty exists on the hyperplane in the error space given by $\mathcal{S}_o$. The error system satisfies

$$\dot{e}(t) = A_0 e(t) - D\xi(t, x, u) + G_n \nu \tag{6.65}$$

For a unique equivalent control[3] to exist, $\det(CG_n) \neq 0$ and arguing as in Section 3.4, the sliding motion satisfies

$$\dot{e}(t) = \left(I - G_n(CG_n)^{-1}C\right) A_0 e(t) + \left(I - G_n(CG_n)^{-1}C\right) D\xi(t, x, u) \tag{6.66}$$

To be insensitive to the uncertainty it follows that

$$\left(I - G_n(CG_n)^{-1}C\right) D = 0$$

or equivalently

$$D = G_n(CG_n)^{-1}(CD) \tag{6.67}$$

Since by assumption $\text{rank}(D) = q$, it follows immediately from equation (6.67) that $\text{rank}(CD) = q$. Therefore it can be assumed without loss of generality that the system (A, D, C) is in the canonical form given in Lemma 6.1. If the nonlinear gain matrix is partitioned so that

$$G_n = \begin{bmatrix} G_1 \\ G_2 \end{bmatrix} \begin{matrix} \updownarrow n-p \\ \updownarrow p \end{matrix} \tag{6.68}$$

then $CG_n = TG_2$ and so $\det(G_2) \neq 0$. From equation (6.66) and using arguments similar to those presented in Section 3.4, it follows that the poles of the (linear) reduced-order motion are given by

$$\lambda\left((A_0)_{11} - G_1 G_2^{-1}(A_0)_{21}\right) \tag{6.69}$$

where $(A_0)_{11}$ and $(A_0)_{21}$ represent the top left and bottom left sub-blocks of the closed-loop matrix A_0 partitioned in a compatible way to the canonical form. By definition the matrix $A_0 = A - G_l C$, so

$$(A_0)_{11} = A_{11} - (G_l C)_{11}$$

where $(G_l C)_{11}$ represents the top left sub-block of the square matrix $G_l C$. However, it is easy to check that $(G_l C)_{11} = 0$ for all $G_l \in \mathbb{R}^{n \times p}$ and so $(A_0)_{11} = A_{11}$. Similarly it can be shown that $(A_0)_{21} = A_{21}$ and consequently

$$\lambda\left((A_0)_{11} - G_1 G_2^{-1}(A_0)_{21}\right) = \lambda\left(A_{11} - G_1 G_2^{-1} A_{21}\right) \tag{6.70}$$

[3]This expression is borrowed from control terminology and is defined as the value of ν required to maintain an ideal sliding motion.

From equation (6.67) it follows that

$$G_1 G_2^{-1} \bar{D}_2 = 0$$

which after considering the structure of $\bar{D}_2$ implies

$$G_1 G_2^{-1} = \begin{bmatrix} \bar{G} & 0 \end{bmatrix}$$

where $\bar{G} \in \mathbb{R}^{(n-p)\times(p-q)}$ and therefore from the definition of A_{21} it follows that

$$A_{11} - G_1 G_2^{-1} A_{21} = A_{11} - \bar{G} A_{211}$$

By construction the pair (A_{11}, A_{211}) is such that

$$\{\text{zeros of } (A, D, C)\} = \lambda\left(A_{11}^o\right) \subset \lambda\left(A_{11} - \bar{G}A_{211}\right) \quad \text{for all } \bar{G} \in \mathbb{R}^{(n-p)\times(p-q)}$$

and therefore for a stable sliding motion any invariant zeros must lie in $\mathbb{C}_-$.

(proof of sufficiency)
Conversely, let (A, D, C) represent the system and suppose $\text{rank}(CD) = q$ and any invariant zeros lie in $\mathbb{C}_-$. Without loss of generality it can be assumed that the system is already in the canonical form of Lemma 6.1 where the matrix A_{11}^o is stable. As a consequence there exists a matrix $L \in \mathbb{R}^{(n-p)\times(p-q)}$ such that $A_{11} + LA_{211}$ is stable. Define a nonsingular transformation as

$$T_L = \begin{bmatrix} I_{n-p} & \bar{L} \\ 0 & T \end{bmatrix} \tag{6.71}$$

where

$$\bar{L} = \begin{bmatrix} L & 0_{(n-p)\times q} \end{bmatrix}$$

After changing coordinates with respect to T_L, the new output distribution matrix becomes

$$\mathcal{C} = C T_L^{-1} = \begin{bmatrix} 0 & I_p \end{bmatrix}$$

From the definition of $\bar{L}$ and $\bar{D}_2$

$$\bar{L}\bar{D}_2 = \begin{bmatrix} L & 0 \end{bmatrix} \begin{bmatrix} 0 \\ D_2 \end{bmatrix} = 0$$

and so the uncertainty distribution matrix is given by

$$\mathcal{D} = T_L D = \begin{bmatrix} \bar{L}\bar{D}_2 \\ T\bar{D}_2 \end{bmatrix} = \begin{bmatrix} 0 \\ T\bar{D}_2 \end{bmatrix}$$

Finally, if $\mathcal{A} = T_L A T_L^{-1}$, it can be shown by direct evaluation that

$$\mathcal{A}_{11} = A_{11} + LA_{211}$$

which is stable by choice of L. The system triple $(\mathcal{A}, \mathcal{D}, \mathcal{C})$ is now in the canonical form (6.39) and from Proposition 6.1 a robust observer exists. ∎

This canonical form of Lemma 6.1 can be obtained by using the mfile commands described below.

mfile: generates the canonical form of Lemma 6.1

```
% Canonical form for the observer in section 6.3.1. If the dimension
% of the observable subspace is non zero, p1 is a vector of dimension
% nn-pp-r which specifies the desired stable poles of the subsystem
% where r is the number of invariant zeros of (A,D,C).

% Establish the number of inputs and outputs
[nn,qq]=size(D);
[pp,nn]=size(C);

% Change coordinates so the output distribution matrix is [0 I]
nc = null(C);
Tc=[nc'; C];
Ac=Tc*A*inv(Tc);
Dc=Tc*D;
Cc=C*inv(Tc);

% Partition the input distribution matrix conformably
Dc1=Dc(1:nn-pp,:);
Dc2=Dc(nn-pp+1:nn,:);

% Finds a tranformation to bring about a special structure in
% the input and output distribution matrices
[T,temp]=qr(Dc2);
T=(flipud(T'))';
clear temp
Tb=[eye(nn-pp) -Dc1*inv(Dc2'*Dc2)*Dc2'; zeros(pp,nn-pp) T'];

Aa=Tb*Ac*inv(Tb);           % In this new coordinate system
Da=Tb*Dc;                   % we have C=[0 T] and D=[0 D2']'
Ca=Cc*inv(Tb);              % A has no special structure yet

A11=Aa(1:nn-pp,1:nn-pp);
A211=Aa(nn-pp+1:nn-qq,1:nn-pp);

% The aim is to put (A11,A211) in the observability canonical form
[Ab,Bb,Cb,Tobs,k]=obsvf(A11,zeros(nn-pp,1),A211,1000*eps);

% r is the dimension of the unobservable subspace and
% the number of invariant zeros of the system (A,D,C)
r=nn-pp-sum(k);
fprintf('Dimension of the unobservable subspace is %.0f \n',r);

Ta=[Tobs zeros(nn-pp,pp);zeros(pp,nn-pp) eye(pp)];

Af=Ta*Aa*inv(Ta);
Df=Ta*Da;
Cf=Ca*inv(Ta);

% Calculates a gain matrix L so that A11+L*A211 is stable
if nn-pp-r>0
```

```
  A22o=Af(r+1:nn-pp,r+1:nn-pp);
  A21o=Af(nn-pp+1:nn-qq,r+1:nn-pp);
  Ltemp=place(A22o',A21o',p1)';
  L=-inv(Tobs)*[zeros(pp-qq,1:r),Ltemp];
else
  L=zeros(nn-pp,pp-qq);
end
Lbar=[L zeros(nn-pp,qq)];

Build the final tranformation needed to obtain the canonical form
TL=[eye(nn-pp) Lbar ; zeros(pp,nn-pp) T];

Acal=TL*Af*inv(TL);          % In this state-space representation
Dcal=TL*Df;                  % C=[0 I]; D=[0 D2']' and the A matrix
Ccal=Cf*inv(TL);             % now has a stable top left sub-block

D2=Dcal(nn-pp+1:nn,:);
```

Remarks

- In the special case where $q = p$ an observer of the form (6.58) which is insensitive to the uncertainty in (6.37) exists if and only if

  (1) $\det(CD) \neq 0$

  (2) the invariant zeros of (A, D, C) lie in $\mathbb{C}_-$.

  That is, the triple (A, D, C) is minimum phase and relative degree 1. In this case the restriction that $\det(CD) \neq 0$ guarantees the existence of exactly $n-p$ invariant zeros and therefore the reduced-order sliding motion is totally determined by these zeros.

- The minimum phase, relative degree 1 condition arises totally from the uncertainty class considered.

- For an observer of the form (6.58) to exist, the nominal linear system need not be completely observable. For example, consider the second-order system

$$A = \begin{bmatrix} -1 & a_{12} \\ 0 & a_{22} \end{bmatrix} \qquad D = \begin{bmatrix} 0 \\ d_2 \end{bmatrix} \qquad C = \begin{bmatrix} 0 & 1 \end{bmatrix}$$

  where $d_2 \neq 0$ and any uncertainty is bounded as in Section 6.3. This system is not completely observable but is minimum phase and relative degree 1, so a robust sliding mode observer can be designed.

- Lemma 6.1 and Proposition 6.2 yield an algorithm for the construction of a robust discontinuous observer that is suitable for implementation on any standard numerical matrix computation package such as MATLAB and does not require the use of symbolic manipulation.

The gain matrices associated with the observer are generated by the following commands.

mfile: computes the observer gains

```
% The commands below calculate the gain matrices which define the
% observer. The vector p2 is of dimension pp and specifies the
% desired stable poles of the error system.

% Calculate Luenberger gains in the modified coordinates
a12=Acal(1:nn-pp,nn-pp+1:nn);
a22=Acal(nn-pp+1:nn,nn-pp+1:nn);
a22s=diag(p2,0);

% Compute the Lyapunov matrix to scale the output error
P2=lyap(a22s',eye(p));

% Compute the gain matrices in the original coordinates
Gl=inv(Tc)*inv(Tb)*inv(TL)*[a12; (a22-a22s)];
Gn=norm(D2)*inv(Tc)*inv(Tb)*inv(TL)*[zeros(nn-pp,pp); eye(pp)];
```

6.4 THE WALCOTT–ŻAK OBSERVER REVISITED

Consider the uncertain linear system (6.24) given earlier in Section 6.2.4. It will be shown that if a Walcott–Żak observer represented by the pair (G, P) exists for a given system and, in particular if the Lyapunov matrix (6.28) is known, then this observer can be analysed within the framework developed in Proposition 6.1. It is necessary to introduce first a lemma which considers the effect of a change of basis on the structural constraint.

Lemma 6.2 *Let the system* (A_0, B, C) *be given with* A_0 *stable, and let* $(\tilde{A}_0, \tilde{B}, \tilde{C})$ *be related to* (A_0, B, C) *by a nonsingular similarity transformation* T. *If* P *is a Lyapunov matrix for* A_0 *which satisfies the structural constraint* $C^{\mathrm{T}}F^{\mathrm{T}} = PB$ *then the matrix* $\tilde{P} = (T^{-1})^{\mathrm{T}}PT^{-1}$ *is a Lyapunov matrix for* $\tilde{A}_0$ *which satisfies the structural constraint* $\tilde{C}^{\mathrm{T}}F^{\mathrm{T}} = \tilde{P}\tilde{B}$.

Proof
Since P is a Lyapunov matrix for A_0

$$PA_0 + A_0^{\mathrm{T}}P = -Q \tag{6.72}$$

for some symmetric positive definite Q. Pre- and post-multiplying (6.72) by $(T^{-1})^{\mathrm{T}}$ and T^{-1} respectively and substituting for $\tilde{A}_0$ and $\tilde{P}$ implies

$$\tilde{P}\tilde{A}_0 + \tilde{A}_0^{\mathrm{T}}\tilde{P} = -(T^{-1})^{\mathrm{T}}QT^{-1} \tag{6.73}$$

Since $\tilde{P}$ and $(T^{-1})^{\mathrm{T}}QT^{-1}$ are symmetric positive definite, equation (6.73) implies $\tilde{P}$ is a Lyapunov matrix for $\tilde{A}_0$. It can be verified by direct evaluation that the structural constraint $\tilde{C}^{\mathrm{T}}F^{\mathrm{T}} = \tilde{P}\tilde{B}$ is satisfied. ∎

The main result of this section will now be proved.

Proposition 6.3 *Let (A, B, C) be a system for which there exists a Walcott–Żak observer characterised by the matrix pair (P, F). Then there exists a nonsingular similarity transformation so that the triple with respect to the new coordinates $(\bar{A}, \bar{B}, \bar{C})$ and the observer pair $(\bar{P}, F)$ exhibits the following properties:*

(1) The system matrix can be written as

$$\bar{A} = \begin{bmatrix} \bar{A}_{11} & \bar{A}_{12} \\ \bar{A}_{21} & \bar{A}_{22} \end{bmatrix}$$

where $\bar{A}_{11} \in \mathbb{R}^{(n-p)\times(n-p)}$ and is stable.

(2) The input distribution matrix

$$\bar{B} = \begin{bmatrix} 0 \\ \tilde{P}_{22}F^{\mathrm{T}} \end{bmatrix}$$

where $\tilde{P}_{22} \in \mathbb{R}^{p\times p}$ and is symmetric positive definite.

(3) The output distribution matrix

$$\bar{C} = \begin{bmatrix} 0 & I_p \end{bmatrix}$$

(4) The Lyapunov matrix has the block diagonal structure

$$\bar{P} = \begin{bmatrix} \bar{P}_1 & 0 \\ 0 & \bar{P}_2 \end{bmatrix}$$

where the matrices $\bar{P}_1 \in \mathbb{R}^{(n-p)\times(n-p)}$ and $\bar{P}_2 \in \mathbb{R}^{p\times p}$.

Proof
Let (P, F) represent the nonlinear components of the observer and let G be an appropriate (Luenberger) gain matrix so that $A_0 = A - GC$ is stable. If the output distribution matrix is not in the form $C = [0 \ \ I_p]$, apply transformation (6.2) in Section 6.2 and use Lemma 6.2 to calculate the observer pair in these new coordinates. For convenience assume this is already the case. Partition the Lyapunov matrix as

$$P = \begin{bmatrix} P_{11} & P_{12} \\ P_{12}^{\mathrm{T}} & P_{22} \end{bmatrix}$$

where $P_{11} \in \mathbb{R}^{(n-p)\times(n-p)}$, $P_{12} \in \mathbb{R}^{(n-p)\times p}$ and $P_{22} \in \mathbb{R}^{p\times p}$. Because P is symmetric positive definite, the sub-block P_{11} is also symmetric positive definite and in particular is nonsingular. Change coordinates with respect to the transformation

$$T = \begin{bmatrix} I_{n-p} & P_{11}^{-1}P_{12} \\ 0 & I_p \end{bmatrix}$$

In the new coordinate system

$$\bar{C} = CT^{-1} = \begin{bmatrix} 0 & I_p \end{bmatrix}$$

and so the third property is satisfied.

Let

$$P^{-1} = \begin{bmatrix} \tilde{P}_{11} & \tilde{P}_{12} \\ \tilde{P}_{21} & \tilde{P}_{22} \end{bmatrix}$$

for appropriately defined sub-blocks. By assumption $B = P^{-1}C^{\mathrm{T}}F^{\mathrm{T}}$ and therefore

$$\begin{aligned} \bar{B} = TB &= \begin{bmatrix} I & P_{11}^{-1}P_{12} \\ 0 & I_p \end{bmatrix} \begin{bmatrix} \tilde{P}_{11} & \tilde{P}_{12} \\ \tilde{P}_{21} & \tilde{P}_{22} \end{bmatrix} \begin{bmatrix} 0 \\ I_p \end{bmatrix} F^{\mathrm{T}} \\ &= \begin{bmatrix} 0 \\ \tilde{P}_{22}F^{\mathrm{T}} \end{bmatrix} \qquad \text{(since } P_{11}\tilde{P}_{12} + P_{12}\tilde{P}_{22} = 0\text{)} \end{aligned}$$

and so the second property is proved.

From Lemma 6.2 the matrix $\bar{P} = (T^{-1})^{\mathrm{T}}PT^{-1}$ is a Lyapunov matrix for $\bar{A}_0$ and by direct computation

$$\begin{aligned} \bar{P} &= \begin{bmatrix} I_{n-p} & 0 \\ -P_{12}^{\mathrm{T}}P_{11}^{-1} & I_p \end{bmatrix} \begin{bmatrix} P_{11} & P_{12} \\ P_{12}^{\mathrm{T}} & P_{22} \end{bmatrix} \begin{bmatrix} I_{n-p} & -P_{11}^{-1}P_{12} \\ 0 & I_p \end{bmatrix} \\ &= \begin{bmatrix} P_{11} & 0 \\ 0 & (P_{22} - P_{12}^{\mathrm{T}}P_{11}^{-1}P_{12}) \end{bmatrix} \end{aligned}$$

and so $\bar{P}$ has the block diagonal structure.

By definition $\bar{P}$ is a Lyapunov matrix for $\bar{A}_0$ and therefore

$$\bar{P}\bar{A}_0 + \bar{A}_0^{\mathrm{T}}\bar{P} = -\bar{Q} \tag{6.74}$$

for some $\bar{Q}$ which is symmetric positive definite. If $(\bar{A}_0)_{11}$ and $\bar{Q}_{11}$ are the top left sub-blocks of $\bar{A}_0$ and $\bar{Q}$ respectively then from equation (6.74) it follows that

$$P_{11}(\bar{A}_0)_{11} + (\bar{A}_0)_{11}^{\mathrm{T}}P_{11} = -\bar{Q}_{11} \tag{6.75}$$

where both P_{11} and $\bar{Q}_{11}$ are symmetric positive definite. Equation (6.75) represents a Lyapunov equation for $(\bar{A}_0)_{11}$ which is therefore stable. However, because of the structure of $\bar{C}$, if $\bar{A}$ is partitioned as in the proposition statement, it follows immediately that $\bar{A}_{11} = (\bar{A}_0)_{11}$ and the first property has been demonstrated. ■

The above proposition has shown that given a system (A, B, C) for which there exists a Walcott–Żak observer then there exists a nonsingular similarity transformation so that in these coordinates the system triple $(\bar{A}, \bar{B}, \bar{C})$ can be written as

$$\begin{aligned} \dot{x}_1(t) &= \bar{A}_{11}x_1(t) + \bar{A}_{12}y(t) \\ \dot{y}(t) &= \bar{A}_{21}x_1(t) + \bar{A}_{22}y(t) + \bar{B}_2u(t) + \bar{B}_2\xi \end{aligned} \tag{6.76}$$

where the matrix $\bar{A}_{11}$ is stable. This is identical to the uncertain system given in equation (6.39) and is thus appropriate for the design and analysis method presented in Proposition 6.1. This suggests an explicit method for designing a Walcott–Żak observer:

(1) First establish the canonical form of Lemma 6.1.

(2) Establish the canonical form of Section 6.3.1 using the coordinate transformation (6.71) in the proof of Proposition 6.3.

(3) The linear gain G can be used from the observer design in Section 6.3.1 and the matrix

$$F^{\mathrm{T}} = P_2 B_2 \tag{6.77}$$

It can then be verified that the block diagonal Lyapunov matrix associated with the quadratic form in equation (6.48) satisfies the constraint (6.28).

MATLAB commands to carry out the observer design process are given below under the assumption that $D = B$.

mfile: computes the gains for a Walcott–Żak observer

```
% The commands below design a Walcott-Zak observer for the triple
% (A,B,C) in the canonical form of Lemma 5.6. The vector p2 is of
% dimension pp and specifies stable design poles for the error system.

% Calculate Luenberger gain in the modified coordinates
a12=Acal(1:nn-pp,nn-pp+1:nn);
a22=Acal(nn-pp+1:nn,nn-pp+1:nn);
a22s=diag(p2,0);

% Compute F and the linear gain matrix G in the original coordinates
P2=lyap(a22s',eye(pp));
G=inv(Tc)*inv(Tb)*inv(TL)*[a12; (a22-a22s)];
F=D2'*P2;
```

6.4.1 Example 2: Pendulum

The equations of motion in state-space form for a pendulum can be written as

$$\begin{aligned} \dot{x}_1 &= x_2 \\ \dot{x}_2 &= -\sin(x_1) \end{aligned} \tag{6.78}$$

Therefore

$$A = \begin{bmatrix} 0 & 1 \\ 0 & 0 \end{bmatrix} \quad \text{and} \quad B = \begin{bmatrix} 0 \\ 1 \end{bmatrix}$$

and the matched 'uncertain' bounded function $\xi(t, x_1, x_2) = -\sin(x_1)$. By design the output distribution matrix $C = \begin{bmatrix} 1 & 1 \end{bmatrix}$.

Consider the problem of finding a robust observer. Change coordinates with respect to

$$T_c = \begin{bmatrix} 1 & 0 \\ 1 & 1 \end{bmatrix}$$

so that the output is a state variable. The system triple becomes

$$\tilde{A} = T_c A T_c^{-1} = \begin{bmatrix} -1 & 1 \\ -1 & 1 \end{bmatrix}$$

and

$$\tilde{B} = T_c B = \begin{bmatrix} 0 \\ 1 \end{bmatrix} \quad \text{and} \quad \tilde{C} = C T_c^{-1} = \begin{bmatrix} 0 & 1 \end{bmatrix}$$

which is in the canonical form (6.39). A robust observer exists for this system because $\tilde{A}_{11} = -1$, which is stable. Letting the design matrix $A_{22}^s = -1$ results in $\lambda(A_0) = \{-1, -1\}$. Defining $Q_2 = 2$ and solving the Lyapunov equation for A_{22}^s and Q_2 gives $P_2 = 1$. In the original coordinates the gain matrices become

$$G_l = \begin{bmatrix} 1 \\ 1 \end{bmatrix} \quad \text{and} \quad G_n = \begin{bmatrix} 0 \\ 1 \end{bmatrix}$$

and the observer becomes

$$\frac{d}{dt}\begin{bmatrix} \hat{x}_1 \\ \hat{x}_2 \end{bmatrix} = \begin{bmatrix} -1 & 0 \\ -1 & -1 \end{bmatrix}\begin{bmatrix} \hat{x}_1 \\ \hat{x}_2 \end{bmatrix} + \begin{bmatrix} 1 \\ 1 \end{bmatrix} y + \begin{bmatrix} 0 \\ 1 \end{bmatrix} \nu$$

6.4.2 Pendulum Simulation

Figure 6.4 demonstrates the nonlinear observer tracking the output from the pendulum when the initial conditions of the true states and observer states are deliberately set to different values

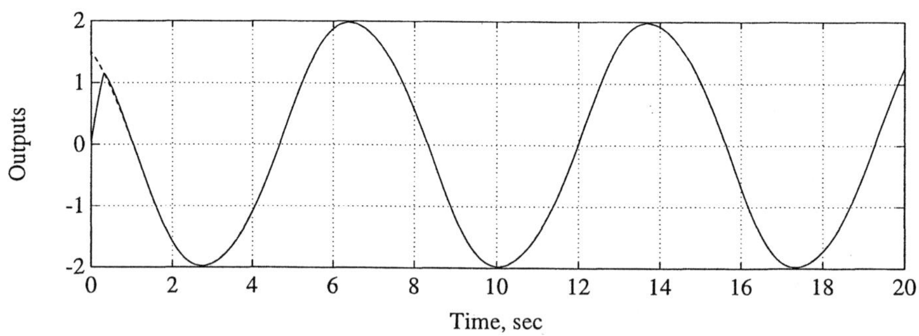

Figure 6.4: Comparison between outputs from the observer and system

It can be seen from Figure 6.4 that after 0.5 second a 'sliding motion' takes place. A comparison of the true and estimated states is given in Figure 6.5. After approximately 4 seconds, visually perfect replication of the states is taking place. If the nonlinear component is removed by setting ρ to zero, the resulting Luenberger Observer behaves as shown in Figure 6.6. There appears to be a distinct phase discrepancy between the outputs of the system and the outputs of the observer; this is due to the presence of the nonlinear sine term.

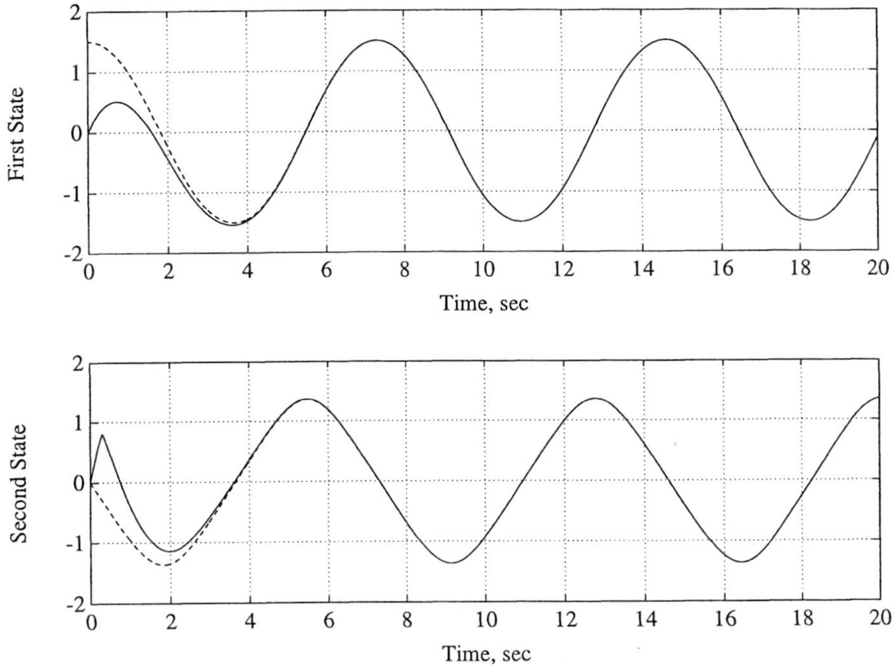

Figure 6.5: Comparison between system and observer states

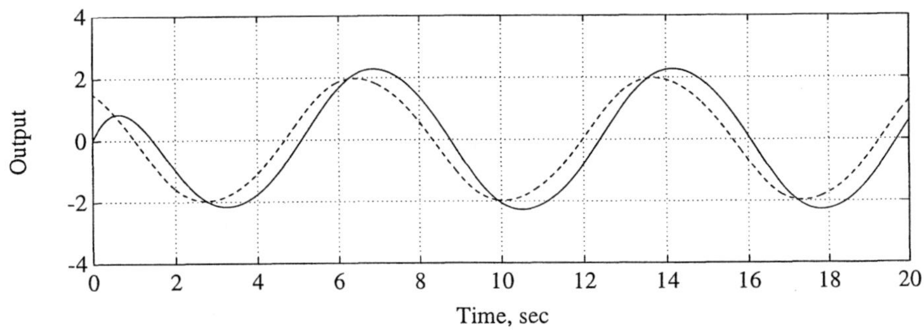

Figure 6.6: Comparison between outputs from the observer and the system

6.5 SLIDING MODE OBSERVERS FOR FAULT DETECTION

The fundamental purpose of a fault detection and isolation (FDI) scheme is to generate an alarm when an abnormal condition, such as a component malfunction or variation in operating condition, develops in the process being monitored and to identify the source or location. An extensively studied FDI methodology is the observer-based approach (Patton *et al.*, 1989) which analyses residuals formed from the difference between the actual outputs and the outputs from an observer. In the absence of faults, the residuals are designed to be small: once a fault occurs, residuals are intended to react by becoming larger (or greater than some predefined threshold), thus signifying the presence of a fault. These schemes usually employ full-order Luenberger observers in which the gain matrix is designed either to make certain residuals sensitive to certain faults and not others, or else to make the residual vector lie in a specific direction in response to a particular fault, thus enabling the source to be identified. This section considers the use of the sliding mode observer defined in Section 6.3 to reconstruct the fault rather than analyse residuals.

Consider a nominal linear system subject to certain faults described by

$$\dot{x}(t) = Ax(t) + Bu(t) + Df_i(t) \tag{6.79}$$

$$y(t) = Cx(t) + f_o(t) \tag{6.80}$$

where $A \in \mathbb{R}^{n\times n}$, $B \in \mathbb{R}^{n\times m}$, $C \in \mathbb{R}^{p\times n}$, $D \in \mathbb{R}^{n\times q}$ with $q \leq p < n$ and the matrices B, C and D are of full rank. The functions $f_i(t)$ and $f_o(t)$ are deemed to represent actuator and sensor faults respectively. It is assumed that the states of the system are unknown and only the signals $u(t)$ and $y(t)$ are available.

The objective is to synthesise an observer to generate a state estimate $\hat{x}(t)$ such that a sliding mode is established in which the output error

$$e_y(t) = \hat{y}(t) - y(t) \tag{6.81}$$

is forced to zero in finite time. The particular observer structure described in Section 6.3 will be considered, namely

$$\dot{\hat{x}}(t) = A\hat{x}(t) + Bu(t) - G_l e_y(t) + G_n\nu \tag{6.82}$$

where G_l is defined in (6.55), the nonlinear gain

$$G_n = T_o^{-1}\begin{bmatrix} 0 \\ I_p \end{bmatrix} \tag{6.83}$$

and

$$\nu = \begin{cases} -\rho(t,y,u)\|\mathcal{D}_2\|\frac{P_2 e_y}{\|P_2 e_y\|} & \text{if } e_y \neq 0 \\ 0 & \text{otherwise} \end{cases} \tag{6.84}$$

Notice how this represents a slight modification to (6.56) and (6.57). It will be shown that, provided a sliding motion can be attained, estimates of $f_i(t)$ and $f_o(t)$ can be computed from approximating the equivalent control.

6.5.1 Reconstruction of the Input Fault Signals

Consider initially the case when $f_o = 0$.

Assume that an observer has been designed as in Section 6.3 and that a sliding motion has been established. During the sliding motion, $e_y = 0$ and $\dot{e}_y = 0$. In the coordinates of the canonical form in Section 6.3.1, equation (6.44) becomes

$$0 = \mathcal{A}_{21}e_1(t) - \mathcal{D}_2 f_i(t) + \nu_{eq} \tag{6.85}$$

where ν_{eq} is the equivalent control. From (6.43) it follows that $e_1(t) \to 0$ and therefore

$$\nu_{eq} \to \mathcal{D}_2 f_i(t) \tag{6.86}$$

One way to recover the equivalent control is by the use of a low pass filter as described in Section 1.2. Here an alternative approach will be employed: suppose that the discontinuous component in (6.41) is replaced by the continuous approximation

$$\nu_\delta = -\rho \|\mathcal{D}_2\| \frac{P_2 e_y}{\|P_2 e_y\| + \delta} \tag{6.87}$$

where δ is a small positive scalar. The equivalent control can be approximated by (6.87) to any required accuracy by a small enough choice of δ. Since $\text{rank}(\mathcal{D}_2) = q$ it follows from (6.86) that

$$f_i(t) \approx -\rho \|\mathcal{D}_2\| (\mathcal{D}_2^{\mathrm{T}} \mathcal{D}_2)^{-1} \mathcal{D}_2^{\mathrm{T}} \frac{P_2 e_y(t)}{\|P_2 e_y(t)\| + \delta} \tag{6.88}$$

The key point is that the signal on the right-hand side of the equation above can be computed on-line and depends only on the output estimation error e_y, thus the fault f_i can be approximated to any degree of accuracy.

6.5.2 Detection of Faults at the Output

Now consider the case when $f_i = 0$ and consider the effect of $f_o(t)$.

In this situation since $y(t) = Cx(t) + f_o(t)$ it follows that

$$e_y(t) = Ce(t) - f_o(t) \tag{6.89}$$

After some manipulation, it follows that in the coordinates of the canonical form in Section 6.3.1 the state estimation error is given by

$$\dot{e}_1(t) = \mathcal{A}_{11} e_1(t) + \mathcal{A}_{12} f_o(t) \tag{6.90}$$
$$\dot{e}_y(t) = \mathcal{A}_{21} e_1(t) + \mathcal{A}_{22}^s e_y(t) - \dot{f}_o(t) + \mathcal{A}_{22} f_o(t) + \nu \tag{6.91}$$

Note that $f_o(t)$ and $\dot{f}_o(t)$ appear as output disturbances and thus ρ in equation (6.84) must be chosen to be sufficiently large to maintain sliding in the presence of these disturbances. Arguing as before, provided a sliding motion can be attained,

$$0 = \mathcal{A}_{21} e_1 - \dot{f}_o(t) + \mathcal{A}_{22} f_o(t) + \nu_{eq} \tag{6.92}$$

Thus, for slowly varying faults, provided the dynamics of the sliding motion are sufficiently fast,

$$\nu_{eq} \approx -(\mathcal{A}_{22} - \mathcal{A}_{21} \mathcal{A}_{11}^{-1} \mathcal{A}_{12}) f_o \tag{6.93}$$

As in the previous section, the equivalent control ν_{eq} can be calculated from (6.87) and consequently, if $(\mathcal{A}_{22} - \mathcal{A}_{21} \mathcal{A}_{11}^{-1} \mathcal{A}_{12})$ is nonsingular, the fault signal can be obtained from equation (6.93).

Note that from the Schur expansion

$$\det(\mathcal{A}) = \det(\mathcal{A}_{11}) \det(\mathcal{A}_{22} - \mathcal{A}_{21} \mathcal{A}_{11}^{-1} \mathcal{A}_{12}) \tag{6.94}$$

and thus $(\mathcal{A}_{22} - \mathcal{A}_{21} \mathcal{A}_{11}^{-1} \mathcal{A}_{12})$ is nonsingular if and only if $\det \mathcal{A} \neq 0$.

Remarks

Note that even if $(\mathcal{A}_{22} - \mathcal{A}_{21}\mathcal{A}_{11}^{-1}\mathcal{A}_{12})$ is singular, it may still be possible to reconstruct some of the sensor faults f_o, depending on the structure of the rank deficiency. An example of this will be considered in the next section.

The fault detection approach adopted here is based on the equivalent control concept and thus does not interact in any way with the use of the sliding mode observer as a state estimator. A set of states compatible with the current plant output will always be produced; this is not the case with those fault detection methods which rely on breaking the sliding motion in order to detect faults.

6.5.3 Example: Inverted Pendulum

Consider the inverted pendulum introduced in Section 5.6.3. Recall that a linearisation of the system about the equilibrium point at the origin generated the triple

$$A = \begin{bmatrix} 0 & 0 & 1 & 0 \\ 0 & 0 & 0 & 1 \\ 0 & -1.9333 & -1.9872 & 0.0091 \\ 0 & 36.9771 & 6.2589 & -0.1738 \end{bmatrix} \qquad B = \begin{bmatrix} 0 \\ 0 \\ 0.3205 \\ -1.0095 \end{bmatrix}$$

$$C = \begin{bmatrix} 1 & 0 & 0 & 0 \\ 0 & 1 & 0 & 0 \\ 0 & 0 & 1 & 0 \end{bmatrix} \tag{6.95}$$

In the analysis which follows, the control law developed in Section 5.6.3 will be used to control the system.

In this particular situation any actuator faults will occur in the input channel, hence in the notation of Section 6.3 the fault distribution matrix $D = B$. It can readily be established that the conditions of Proposition 6.2 are satisfied and thus all the preceding analysis is valid.

In the canonical form of (6.39) the system described in (6.95) becomes

$$\mathcal{A} = \left[\begin{array}{c|ccc} -10.0000 & 0 & -67.6603 & 31.4960 \\ \hline 0 & 0 & 0 & 1.0000 \\ 1.0000 & 0 & 9.8548 & -3.1496 \\ 0.0091 & 0 & -1.8437 & -2.0158 \end{array}\right] \qquad \mathcal{D} = \left[\begin{array}{c} 0 \\ \hline 0 \\ 0 \\ 0.3205 \end{array}\right]$$

$$\mathcal{C} = \left[\begin{array}{c|ccc} 0 & 1 & 0 & 0 \\ 0 & 0 & 1 & 0 \\ 0 & 0 & 0 & 1 \end{array}\right] \tag{6.96}$$

where by design $\mathcal{A}_{11} = -10$. In this particular case $\mathcal{A}_{22}^s = \text{diag}(-11, -12, -13)$ which furnishes the linear component of the observer with poles approximately three times faster than the closed-loop poles of the controlled plant. The symmetric positive definite matrix P_2 has been selected as the unique solution to the Lyapunov equation

$$P_2\mathcal{A}_{22} + \mathcal{A}_{22}^{\mathrm{T}}P_2 = -I \tag{6.97}$$

In this particular design the scalar function $\rho = 75$ and the observer design is complete.

6.5.4 Simulations of Different Fault Conditions

It can be verified that the eigenvalues of A are $\{0,\ 5.8702,\ -6.3965,\ -1.6347\}$, so from the argument in Section 6.5.2 the steady-state gain from f_o to ν_{eq} is singular. In fact

$$(\mathcal{A}_{22} - \mathcal{A}_{21}\mathcal{A}_{11}^{-1}\mathcal{A}_{12}) = \begin{bmatrix} 0 & 0 & -1 \\ 0 & -3.0888 & 0 \\ 0 & 1.9052 & 1.9872 \end{bmatrix} \tag{6.98}$$

which is clearly rank deficient. However, if $\nu_{eq,i}$ and $f_{o,i}$ represent the ith components of the vectors ν_{eq} and f_o, from equation (6.93) and using the particular $\mathcal{A}_{22} - \mathcal{A}_{21}\mathcal{A}_{11}^{-1}\mathcal{A}_{12}$ from (6.98) it is apparent that

$$\nu_{eq,1} \approx f_{o,3} \tag{6.99}$$

$$\nu_{eq,2} \approx 3.0888 f_{o,2} \tag{6.100}$$

It is also clear that any fault in the first output channel has no direct long-term effect on ν_{eq}. Furthermore, because of the structure of $\mathcal{D}_2$ in (6.96) it can be verified that

$$(\mathcal{D}_2^{\mathrm{T}}\mathcal{D}_2)^{-1}\mathcal{D}_2^{\mathrm{T}} = [\ 0\ \ 0\ \ 3.1200\ ]$$

and so from (6.86)

$$\nu_{eq,3} \approx 0.3205 f_i \tag{6.101}$$

Thus the three components of the equivalent control, properly scaled, provide estimates of $f_{o,3}$, $f_{o,2}$ and f_i respectively and may be used as detector signals.

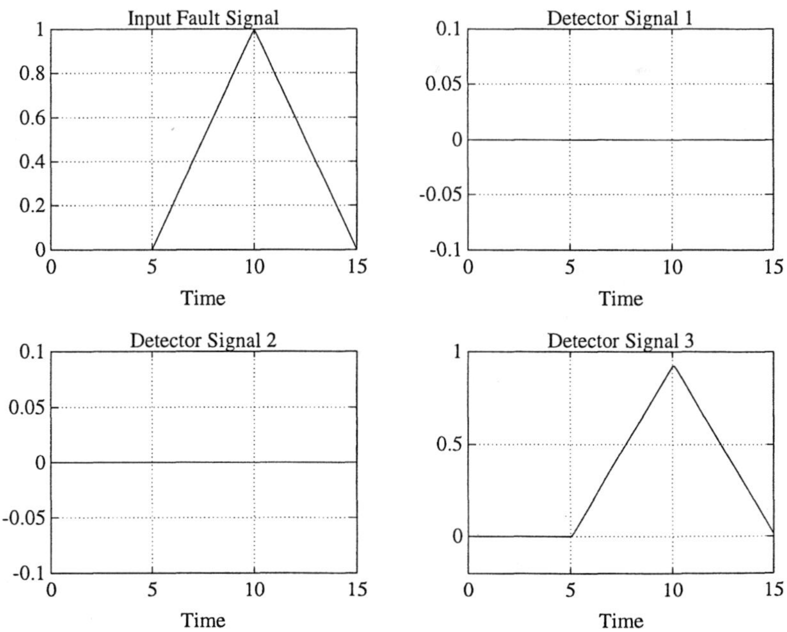

Figure 6.7: Response of the detection signals to a fault in the input channel

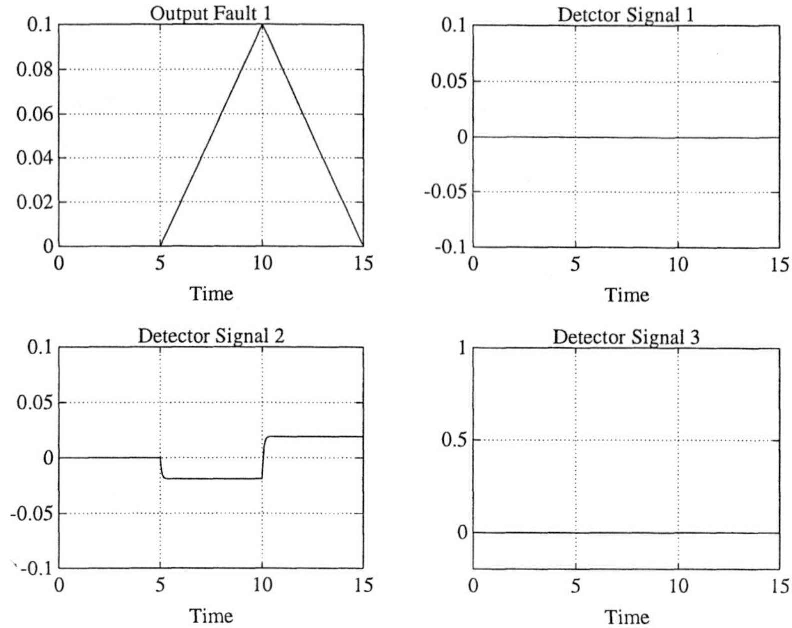

Figure 6.8: Response of detection signals to a fault in the first output channel

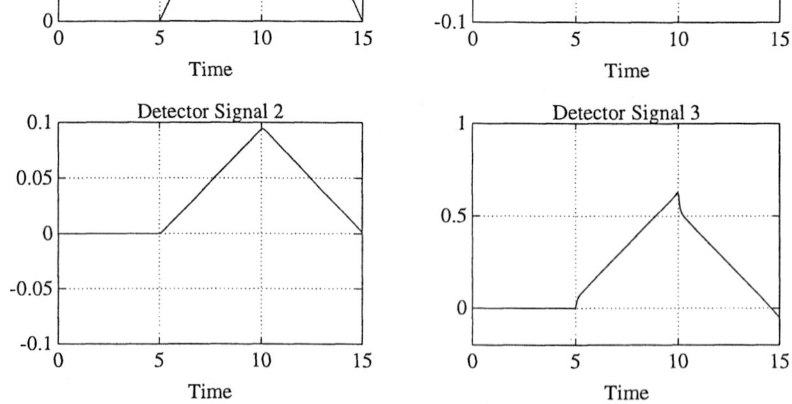

Figure 6.9: Response of detection signals to a fault in the second output channel

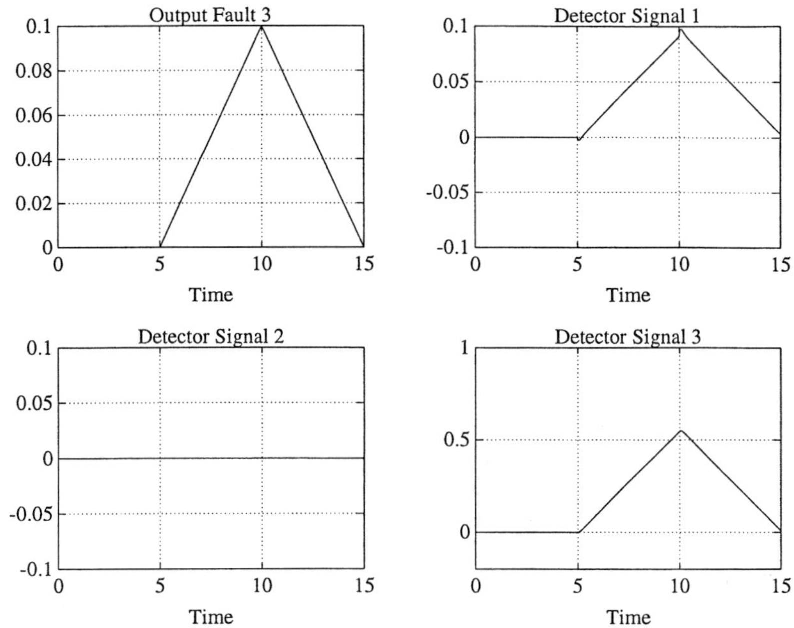

Figure 6.10: Response of detection signals to a fault in the third output channel

The following nonlinear simulation results show the response of these three detector signals to different fault conditions. Figure 6.7 shows the effect of a ramp in the input channel. As predicted by the theory, the third detector signal reproduces the fault signal whilst not affecting the other two signals. Figure 6.8 shows the effect of a ramp in the first output channel. As predicted by the theory, the detector signals do not reproduce this fault signal (although the second detector signal approximates the gradient of the fault signal). Figures 6.9 and 6.10 show that the appropriate detector signal reproduces the ramp fault signals in output channels 2 and 3. In both cases the detector signal 3 is influenced as well.

Remark
This example highlights the difference between the Walcott–Żak approach and the observer described in Section 6.3. In the Walcott–Żak approach, sliding takes place on the surface in the error space given by $\{e \in \mathbb{R}^n \ : \ FCe = 0\}$, so in this particular example the equivalent control would have only one component, making it difficult to distinguish between faults in different channels.

6.6 SUMMARY

Different sliding mode strategies have been employed for the problem of state reconstruction using only input and output information. A framework has been introduced to incorporate the effects of bounded, matched or indeed other structured system uncertainty. This effectively considers the problem first posed by Walcott and Żak. Here, however, the emphasis is placed upon the numerical tractability of the solution method. A different nonlinear structure is used which guarantees sliding on the surface in the error space described by the null space of the input distribution matrix. Also the precise class of systems to which these results apply has explicitly been identified. In the case of matched uncertainty, the class of applicable systems is identical to those for which the direct output feedback designs described in Chapter 5 were applicable. In practice such an observer may form part of a control strategy where the estimated states would be used in a state feedback control law. The next chapter examines the overall closed-loop performance of the observer when used in conjunction with a robust output-tracking state feedback control law.

6.7 NOTES AND REFERENCES

Despite fruitful research and development activity in the area of variable structure control theory, relatively few authors have considered the application of the underlying principles to the problem of observer designs. The earliest work in this field, which was the approach described in Section 6.2.1, appears originally in Utkin (1992).

The approach described in Section 6.2.3 is conceptually similar to that proposed by Slotine *et al.* (1987) in that the output errors are fed back in both a linear and discontinuous way. The problem formulation, however, is quite different since Slotine *et al.* propose an observer for the system $\dot{x} = f(t, x)$ and seek only to extend the region in which sliding takes place, the so-called *sliding patch*, by incorporating a linear term.

The work of Walcott & Żak (1987) seeks global error convergence for a class of uncertain systems. This strategy, although intuitively appealing, necessitates the use of algebraic manipulation tools to effectively solve an associated constrained Lyapunov problem for systems of reasonable order. In addition to the original papers (Walcott & Żak, 1987,1988) the approach is discussed in detail in Żak & Walcott (1990). This collection also describes a *hyperstability* approach to observer design by Balestrino & Innocenti (1990), based on the concept of positive realness.

The pendulum in Section 6.4.1 is the example used by Walcott *et al.* (1987) to demonstrate that their sliding mode observer compares favourably with direct approaches to nonlinear observer designs when the nonlinearities present in the systems are assumed to be perfectly known.

Sreedhar *et al.* (1993) consider a sliding mode observer approach to fault detection, although in their design procedure it is assumed that the states of the system are available. A different approach is adopted by Hermans & Zarrop (1996), who attempt to design an observer of the form used in Section 6.3 in such a way that the sliding motion is destroyed in the presence of a fault.

Chapter 7

Observer-Based Output Tracking Controllers

7.1 INTRODUCTION

This chapter considers the development of schemes to provide output tracking of a constant demand signal under the proviso that only output information is available. The full state information integral action and model-reference schemes described in Chapter 4 will be employed in conjunction with a sliding mode observer to provide estimates of the unknown states. It will be shown that the resulting overall control schemes retain the insensitivity properties of their full state information counterparts. This is an alternative approach to that taken in Chapter 5 and enables previously developed control laws from Chapter 4 to be exploited.

7.2 SYSTEM DESCRIPTION AND OBSERVER FORMULATION

Consider an uncertain dynamical system of the form

$$\begin{aligned} \dot{x}(t) &= Ax(t) + Bu(t) + B\xi(t,x,u) \\ y(t) &= Cx(t) \end{aligned} \tag{7.1}$$

where $x \in \mathbb{R}^n$, $u \in \mathbb{R}^m$ and $y \in \mathbb{R}^p$ with $m \leq p < n$. Assume that the nominal linear system (A, B, C) is known and that the matrices B and C are both of full rank. The function $\xi : \mathbb{R}_+ \times \mathbb{R}^n \times \mathbb{R}^m \rightarrow \mathbb{R}^n$ is unknown and represents any nonlinearities plus any model uncertainties in the system and is assumed to satisfy

$$\|\xi(t,x,u)\| \leq k_1 \|u(t)\| + \alpha(t,y) \tag{7.2}$$

for some known positive scalar $k_1 < 1$ and known function $\alpha : \mathbb{R}_+ \times \mathbb{R}^p \rightarrow \mathbb{R}_+$. In addition, it is assumed that the nominal linear system satisfies

(A1) the pair (A, B) is controllable

(A2) the Markov parameter CB is full rank

(A3) the invariant zeros of (A, B, C) are in $\mathbb{C}_-$

The assumption on the input/output dimensions is required since the sliding mode observer formulations of Chapter 6 require there to be at least as many outputs as inputs.

It can be assumed without loss of generality that the system is already in the regular form, so that

$$A = \begin{bmatrix} A_{11} & A_{12} \\ A_{21} & A_{22} \end{bmatrix} \quad B = \begin{bmatrix} 0 \\ B_2 \end{bmatrix} \quad C = \begin{bmatrix} C_1 & C_2 \end{bmatrix} \tag{7.3}$$

where $A_{11} \in \mathbb{R}^{(n-m)\times(n-m)}$, $B_2 \in \mathbb{R}^{m\times m}$ and the matrix $C_2 \in \mathbb{R}^{p\times m}$. The square matrix B_2 is nonsingular because the input distribution matrix is assumed to be of full rank.

The sliding mode observer that will be considered is essentially that of Walcott & Żak (1988) which was introduced in Section 6.2.4. If $z(t)$ represents an estimate for the true states and

$$e(t) = z(t) - x(t) \tag{7.4}$$

is the state estimation error, then the proposed observer has the form

$$\dot{z}(t) = Az(t) + Bu(t) - GCe(t) + B\nu_o(t) \tag{7.5}$$

The output error feedback gain matrix G is chosen so that the closed-loop matrix $A_0 = A - GC$ is stable and has a Lyapunov matrix P satisfying

$$PA_0 + A_0^{\mathrm{T}}P = -Q \tag{7.6}$$

for some positive definite design matrix Q and the structural constraint

$$PB = C^{\mathrm{T}}F^{\mathrm{T}} \tag{7.7}$$

for some matrix $F \in \mathbb{R}^{m\times p}$. The discontinuous vector ν_o is given by

$$\nu_o = \begin{cases} -\rho_o(u_l, y)\frac{FCe}{\|FCe\|} & \text{if } FCe \neq 0 \\ 0 & \text{otherwise} \end{cases} \tag{7.8}$$

where $\rho_o(u_l, y)$ is a scalar function depending on the linear component of the control law which, loosely speaking, is an upper bound on the norm of the uncertainty.

As shown in Section 6.4, assumptions A2 and A3 are necessary and sufficient conditions for the existence of such an observer which induces a sliding motion on

$$\mathcal{S}_o = \{\, e \in \mathbb{R}^n \;:\; FCe = 0 \,\} \tag{7.9}$$

Suitable modifications of the full-state feedback controller from Chapter 4 will be described in the following sections.

7.3 AN INTEGRAL ACTION CONTROLLER

Assume for simplicity that the system is square, i.e. $p = m$. This additional assumption, although not essential, is convenient since the observer formulation requires there to be at least as many outputs as inputs; conversely, the proposed control law requires at least as many inputs as outputs – a square system is therefore essential.[1] In this situation the square matrix C_2 is also nonsingular because $C_2B_2 = CB$ which is nonsingular by assumption.

The assumption of squareness results in a further convenience – namely an explicit solution for the choice of gain matrices F and G can be obtained in the original coordinates without the apparent need for the coordinate transformation introduced in Section 6.3. This will be demonstrated in the following subsection.

7.3.1 Nonlinear Observer Formulation (For Square Plants)

The approach employed here uses the ideas developed in Chapter 6. For non-square systems, the observer structure in equations (7.5) to (7.8) is different from that described in Section 6.3. However, for square systems they are identical except for a nonsingular scaling of the hyperplane. In the special case of a square plant satisfying A2 and A3, an analytic expression will be obtained for the components G and F. Assume the linear part of the uncertain system (7.1) is in the regular form of (7.3). Since the sub-block C_2 is nonsingular, the linear transformation

$$T_c = \begin{bmatrix} I_{n-p} & 0 \\ C_1 & C_2 \end{bmatrix} \tag{7.10}$$

is also nonsingular. The transformation $x \mapsto T_c x$ induces a set of coordinates in which the system triple $(\mathcal{A}, \mathcal{B}, \mathcal{C})$ is given by

$$\mathcal{A} = \begin{bmatrix} \mathcal{A}_{11} & A_{12}C_2^{-1} \\ C_1\mathcal{A}_{11} + C_2A_{21} - C_2A_{22}C_2^{-1}C_1 & C_1A_{12}C_2^{-1} + C_2A_{22}C_2^{-1} \end{bmatrix} \tag{7.11}$$

with $\mathcal{A}_{11} = A_{11} - A_{12}C_2^{-1}C_1$ and where the input and output distribution matrices are given by

$$\mathcal{B} = \begin{bmatrix} 0 \\ C_2B_2 \end{bmatrix} \qquad \mathcal{C} = \begin{bmatrix} 0 & I_p \end{bmatrix} \tag{7.12}$$

From Lemma 5.1 it follows that the eigenvalues of $\mathcal{A}_{11}$ are the invariant zeros of $(\mathcal{A}, \mathcal{B}, \mathcal{C})$ and consequently by assumption the matrix $\mathcal{A}_{11}$ is stable. The system in (7.11) and (7.12) is thus in the canonical form for sliding mode observer design described in Section 6.3. Using the synthesis procedure described there, for a given stable matrix A_{22}^s, the Lyapunov matrices P_1 and P_2 can be obtained from equations (6.47) and (6.45). Because the transformation to obtain the canonical form can be explicitly determined, it follows that the Lyapunov matrix P and the gain matrix G can be determined explicitly. In the coordinates of the regular form, via the mapping in Lemma 6.2, the gain matrix

$$G = \begin{bmatrix} A_{12}C_2^{-1} \\ A_{22}C_2^{-1} - C_2^{-1}A_{22}^s \end{bmatrix} \tag{7.13}$$

the scaling matrix

$$F = (P_2C_2B_2)^{\mathrm{T}} \tag{7.14}$$

[1]Technically, for a nonsquare system, a subset of m outputs could be identified as the 'controlled outputs'. In this way an alternative output distribution matrix $C_m \in \mathbb{R}^{m\times n}$ can be isolated to be used for the development of the control law; recall Section 5.7.1.

and

$$P = \begin{bmatrix} I & 0 \\ C_1 & C_2 \end{bmatrix}^{\mathrm{T}} \begin{bmatrix} P_1 & 0 \\ 0 & P_2 \end{bmatrix} \begin{bmatrix} I & 0 \\ C_1 & C_2 \end{bmatrix} = \begin{bmatrix} P_1 + C_1^{\mathrm{T}} P_1 C_1 & C_1^{\mathrm{T}} P_2 C_2 \\ C_2^{\mathrm{T}} P_2 C_1 & C_2^{\mathrm{T}} P_2 C_2 \end{bmatrix} \tag{7.15}$$

Remark
It should be noted that from a design point of view, only the Lyapunov matrix P_2 is required for the observer synthesis.

A suitable mfile to accomplish an observer design is given below.

mfile: designs an observer explicitly for a square system

```
% Commands to produce an observer for the special case of square systems
% It is assumed that (A,B,C) is in regular form and the conditions for
% the existence of a robust observer are satisfied.
% The pth order vector po is assumed to contain stable poles
% Q2 is a symmetric positive definite matrix of order p

[nn,mm]=size(B);
[pp,nn]=size(C);

C1=C(:,1:nn-pp);
C2=C(:,nn-pp+1:nn);
T=[eye(nn-pp) zeros(nn-pp,pp); C1 C2];

Astar=T*A*inv(T);          % In this state-space representation
Bstar=T*B;                 % C=[0 Ip]; B=[0 B2']' and the A matrix
Cstar=C*inv(T);            % now has a stable top left sub-block

% Partition the system matrix to use in the observer gain matrix
A11=Astar(1:nn-pp,1:nn-pp);
A21=Astar(nn-pp+1:nn,1:nn-pp);
A12=A(1:nn-pp,nn-pp+1:nn);
A22=A(nn-pp+1:nn,nn-pp+1:nn);
B2=Bstar(nn-pp+1:nn,:);

% Setup the linear null space dynamics
a22s=diag(po,0);
P2=lyap(a22s',Q2);

% Calculate the linear and nonlinear observer gain matrices
Gl=[A12*inv(C2); (A22*inv(C2)-inv(C2)*a22s)];
Gn=B;
F=B2'*P2;
```

7.3.2 State Feedback Integral Action Control Law (Reprise)

Consider initially the development of a tracking control law for the nominal linear system

$$\dot{x}(t) = Ax(t) + Bu(t) \tag{7.16}$$

where the matrix pair (A, B) is assumed to be in regular form as in (7.3). The control law that will be described here is similar to that of Section 4.4.2 except for a modification which makes allowance for the fact that an observer is used to generate state estimates. As in Section 4.4.2 consider the introduction of additional states $x_r \in \mathbb{R}^p$ satisfying

$$\dot{x}_r(t) = r(t) - y(t) \tag{7.17}$$

where the differentiable signal $r(t)$ satisfies

$$\dot{r}(t) = \Gamma\left(r(t) - R\right) \tag{7.18}$$

with $\Gamma \in \mathbb{R}^{p\times p}$ a stable design matrix and R a constant demand vector. Augment the states with the integral action states to obtain

$$\tilde{x} = \begin{bmatrix} x_r \\ x \end{bmatrix} \tag{7.19}$$

and partition the augmented states as

$$\tilde{x} = \begin{bmatrix} \tilde{x}_1 \\ \tilde{x}_2 \end{bmatrix} \tag{7.20}$$

where $\tilde{x}_1 \in \mathbb{R}^n$ and $\tilde{x}_2 \in \mathbb{R}^m$.

The proposed controller seeks to induce a sliding motion on the surface

$$\mathcal{S} = \{\tilde{x} \in \mathbb{R}^{n+p} \ : \ S\tilde{x} = S_r r\} \tag{7.21}$$

where $S \in \mathbb{R}^{m\times(n+p)}$ and $S_r \in \mathbb{R}^{p\times p}$. Assume, as in Section 4.4.2, that the hyperplane system matrix has the form

$$S = S_2 \begin{bmatrix} M & I_m \end{bmatrix} \tag{7.22}$$

where $M \in \mathbb{R}^{m\times n}$ and the scaling matrix $S_2 = \Lambda B_2^{-1}$. The linear component of the control law proposed in Section 4.4.2 is given by

$$u_L(\tilde{x}, r) = L\tilde{x} + L_r r + L_{\dot{r}}\dot{r} \tag{7.23}$$

where

$$L = -\Lambda^{-1}(S\tilde{A} - \Phi S) \tag{7.24}$$

$$L_r = -\Lambda^{-1}(\Phi S_r + S_2 M B_r) \tag{7.25}$$

$$L_{\dot{r}} = \Lambda^{-1} S_r \tag{7.26}$$

Φ is a stable design matrix and B_r is defined in equation (4.103). The discontinuous vector is given by

$$\nu_c = \begin{cases} -\rho_c(u_L, y)\Lambda^{-1}\frac{\bar{P}_2(S\tilde{x}-S_r r)}{\|\bar{P}_2(S\tilde{x}-S_r r)\|} & \text{if } S\tilde{x} \neq S_r r \\ 0 & \text{otherwise} \end{cases} \tag{7.27}$$

where $\bar{P}_2$ is a symmetric positive definite matrix satisfying

$$\bar{P}_2^{-1}\Phi^{\mathrm{T}} + \Phi\bar{P}_2^{-1} = -\hat{Q}_2 \tag{7.28}$$

for some positive definite design matrix $\hat{Q}_2$. The reason for the use of the modified Lyapunov equation (7.28) will be made clear in Section 7.3.3 in the proof of Lemma 7.1. The positive scalar function which multiplies the unit vector component of the controller is given by

$$\rho_c(u_L, y) = \|\Lambda\| \frac{(k_1\|u_L\| + \alpha(t,y) + k_1\gamma_c\|\Lambda^{-1}\| + \gamma_o)}{(1 - k_1\,\kappa(\Lambda))} + \gamma_c \tag{7.29}$$

where γ_o and γ_c are positive design scalars and $\Lambda \in \mathbb{R}^{m\times m}$ is a nonsingular diagonal design matrix satisfying

$$k_1\,\kappa(\Lambda) < 1 \tag{7.30}$$

A corresponding scalar function to premultiply the unit vector component in the observer from equation (7.8) is

$$\rho_o(u_L, y) = \frac{(k_1\|u_L\| + \alpha(t,y) + k_1\gamma_c\|\Lambda^{-1}\| + \gamma_o)}{(1 - k_1\,\kappa(\Lambda))} \tag{7.31}$$

Note that by construction

$$\rho_c(u_L, y) = \|\Lambda\|\rho_o(u_L, y) + \gamma_c \tag{7.32}$$

7.3.3 Closed-Loop Analysis

In this section the effect of using the state estimates in the control law will be explored. In particular, stability of the combined closed-loop system will be demonstrated.

From equations (7.1), (7.5) and (7.17) the system representing the uncertain error and observer dynamics is given by

$$\dot{e}(t) = A_0 e(t) + B\nu_o(t) - B\xi(t, x, \hat{u}) \tag{7.33}$$

$$\dot{x}_r(t) = r(t) - Cz(t) + e_y(t) \tag{7.34}$$

$$\dot{z}(t) = Az(t) - Ge_y(t) + B(\hat{u}(t) + \nu_o(t)) \tag{7.35}$$

where $e_y = Ce$ and $\hat{u}$ is the control action obtained from using the state estimates $z(t)$ instead of $x(t)$ in equations (7.23) and (7.27). It is convenient to repartition equations (7.34) and (7.35) to obtain coordinates

$$\begin{bmatrix} \tilde{z}_1 \\ \tilde{z}_2 \end{bmatrix} = \begin{bmatrix} x_r \\ z \end{bmatrix} \tag{7.36}$$

where $\tilde{z}_1 \in \mathbb{R}^n$ and $\tilde{z}_2 \in \mathbb{R}^m$. Change coordinates using the transformation

$$\bar{T} = \begin{bmatrix} I_n & 0 \\ S_1 & S_2 \end{bmatrix}$$

in Section 4.4.2 to generate a new set of coordinates

$$\begin{bmatrix} \tilde{z}_1 \\ \bar{z}_2 \end{bmatrix} = \bar{T} \begin{bmatrix} \tilde{z}_1 \\ \tilde{z}_2 \end{bmatrix} \tag{7.37}$$

It can be verified that, in terms of $(\tilde{z}_1, \bar{z}_2)$, equations (7.33) to (7.35) become

$$\dot{\tilde{z}}_1(t) = \bar{A}_{11}\tilde{z}_1(t) + \bar{A}_{12}\left(\bar{z}_2(t) - S_r r(t)\right) + \left(B_r + \bar{A}_{12}S_r\right) r(t) - \bar{G}_1 e_y(t) \tag{7.38}$$

$$\dot{\bar{z}}_2(t) = \Phi\left(\bar{z}_2(t) - S_r r(t)\right) + \Lambda\hat{\nu}_c + S_r\dot{r}(t) - (S_1\bar{G}_1 + S_2G_2)e_y(t) + \Lambda\nu_o \tag{7.39}$$

where the output error gain matrix

$$\bar{G}_1 \triangleq \begin{bmatrix} -I_p \\ G_1 \end{bmatrix} \tag{7.40}$$

where $G_1 = A_{12}C_2^{-1}$ and $G_2 = A_{22}C_2^{-1} - C_2^{-1}A_{22}^s$, i.e. a partition of G from equation (7.13), and the sliding motion system matrices $\bar{A}_{11} = \tilde{A}_{11} - \tilde{A}_{12}M$ and $\bar{A}_{12} = \tilde{A}_{12}S_2^{-1}$ where $\tilde{A}_{11}$ and $\tilde{A}_{12}$ are the augmented system matrices in (4.102). Here the linear component of the control action has been included and the equations simplified. For convenience define

$$e_r(t) = r(t) - R \tag{7.41}$$

and therefore from equation (7.18) it follows that the evolution of $e_r(t)$ satisfies

$$\dot{e}_r(t) = \Gamma e_r(t) \tag{7.42}$$

If the affine change of coordinates $(\tilde{z}_1, \bar{z}_2) \mapsto \zeta$ where

$$\zeta \triangleq \begin{bmatrix} \tilde{z}_1 + (\bar{A}_{11})^{-1}(\bar{A}_{12}S_r + B_r)R \\ \bar{z}_2 - S_r r \end{bmatrix} = \begin{bmatrix} \zeta_1 \\ \zeta_2 \end{bmatrix} \tag{7.43}$$

is introduced, then

$$\dot{\zeta}_1(t) = \bar{A}_{11}\zeta_1(t) + \bar{A}_{12}\zeta_2(t) + \left(B_r + \bar{A}_{12}S_r\right) e_r(t) - \bar{G}_1 e_y(t) \tag{7.44}$$

$$\dot{\zeta}_2(t) = \Phi\zeta_2(t) + \Lambda\hat{\nu}_c(t) - \bar{G}_2 e_y(t) + \Lambda\nu_o(t) \tag{7.45}$$

where

$$\bar{G}_2 \triangleq S_1\bar{G}_1 + S_2G_2 \tag{7.46}$$

Equations (7.44) and (7.45) may then be written more conveniently as

$$\dot{\zeta}(t) = A_c\zeta(t) + \bar{G}_r e_r(t) - \bar{G}Ce(t) + \bar{\Lambda}\left(\nu_o(t) + \hat{\nu}_c(t)\right) \tag{7.47}$$

for appropriate choices of gain matrices $\bar{G}$, $\bar{G}_r$ and $\bar{\Lambda}$. Thus the closed-loop system is determined by (7.47) together with the observer error system

$$\dot{e}(t) = A_0 e(t) + B\left(\nu_o(t) - \xi(t, x, \hat{u})\right) \tag{7.48}$$

The linear closed-loop dynamics are defined by the matrix

$$A_c = \begin{bmatrix} \bar{A}_{11} & \bar{A}_{12} \\ 0 & \Phi \end{bmatrix}$$

which is stable, with eigenvalues given by $\lambda(\bar{A}_{11}) \cup \lambda(\Phi)$. For the closed-loop analysis which follows, a block diagonal Lyapunov matrix for A_c will be sought of the form

$$\bar{P} = \begin{bmatrix} \bar{P}_1 & 0 \\ 0 & \bar{P}_2 \end{bmatrix} \tag{7.49}$$

where the lower right matrix sub-block $\bar{P}_2$ is the Lyapunov matrix for Φ used in the unit vector component of the control law. Using the Lyapunov matrix P for A_0 given in (7.6) define

$$P_G \triangleq \begin{bmatrix} P & 0 \\ 0 & \bar{P} \end{bmatrix} \tag{7.50}$$

where the matrix $\bar{P}$ will be chosen so that P_G is a Lyapunov matrix for the overall closed-loop system matrix

$$A_G \triangleq \begin{bmatrix} A_0 & 0 \\ -\bar{G}C & A_c \end{bmatrix} \tag{7.51}$$

It can be easily verified that

$$P_G A_G + A_G^{\mathrm{T}} P_G = \begin{bmatrix} -Q & -C^{\mathrm{T}}\bar{G}^{\mathrm{T}}\bar{P} \\ -\bar{P}\bar{G}C & \bar{P}A_c + A_c^{\mathrm{T}}\bar{P} \end{bmatrix} \tag{7.52}$$

where Q is defined in the Lyapunov equation (7.6). The expression on the right-hand side of equation (7.52) is negative definite if and only if

$$\bar{P}A_c + A_c^{\mathrm{T}}\bar{P} + \bar{P}\bar{G}CQ^{-1}C^{\mathrm{T}}\bar{G}^{\mathrm{T}}\bar{P} < 0 \tag{7.53}$$

Consider the problem of finding a block diagonal $\bar{P}$ as in equation (7.49) so that

$$\bar{P}A_c + A_c^{\mathrm{T}}\bar{P} + \bar{P}\bar{G}CQ^{-1}C^{\mathrm{T}}\bar{G}^{\mathrm{T}}\bar{P} + \bar{P}\bar{Q}\bar{P} = 0 \tag{7.54}$$

for some symmetric positive definite design matrix $\bar{Q}$.

Lemma 7.1 *A block diagonal matrix of the form given in (7.49) exists which satisfies the Riccati equation (7.54) for a family of positive definite matrices $\bar{Q}$. In particular, a solution can be obtained where $\bar{P}_2$ is the unique symmetric positive definite solution to*

$$\bar{P}_2^{-1}\Phi^{\mathrm{T}} + \Phi\bar{P}_2^{-1} = -\hat{Q}_2 \tag{7.55}$$

where $\hat{Q}_2$ is any symmetric positive definite matrix such that $\hat{Q}_2 > \bar{G}_2 Q_{22} \bar{G}_2^{\mathrm{T}}$; and $\bar{P}_1$ is the unique symmetric positive definite solution to

$$\bar{P}_1^{-1}\bar{A}_{11}^{\mathrm{T}} + \bar{A}_{11}\bar{P}_1^{-1} = -\bar{A}_{12}\bar{P}_2^{-1}\hat{Q}_2^{-1}\bar{P}_2^{-1}\bar{A}_{12}^{\mathrm{T}} - \hat{Q}_1 \tag{7.56}$$

where $\hat{Q}_1$ is any symmetric positive definite matrix such that

$$\begin{aligned} \hat{Q}_1 \;>\; & \left(\bar{A}_{12}\bar{P}_2^{-1} + \bar{G}_1 Q_{22}\bar{G}_2^{\mathrm{T}}\right)\left(\hat{Q}_2 - \bar{G}_2 Q_{22}\bar{G}_2^{\mathrm{T}}\right)^{-1}\left(\bar{P}_2^{-1}\bar{A}_{12}^{\mathrm{T}} + \bar{G}_2 Q_{22}\bar{G}_1^{\mathrm{T}}\right) \\ & + \bar{G}_1 Q_{22}\bar{G}_1^{\mathrm{T}} - \bar{A}_{12}\bar{P}_2^{-1}\hat{Q}_2^{-1}\bar{P}_2^{-1}\bar{A}_{12}^{\mathrm{T}} \end{aligned} \tag{7.57}$$

and $Q_{22} \triangleq CQ^{-1}C^{\mathrm{T}}$

Proof

Solving the Riccati equation given in (7.54) is equivalent[2] to finding a block diagonal solution to the Lyapunov equation

$$A_c\bar{P}^{-1} + \bar{P}^{-1}A_c^{\mathrm{T}} = -\hat{Q} \tag{7.58}$$

where

$$\hat{Q} = \bar{G}CQ^{-1}C^{\mathrm{T}}\bar{G}^{\mathrm{T}} + \bar{Q} \tag{7.59}$$

From the definition of A_c it follows that

$$A_c\bar{P}^{-1} + \bar{P}^{-1}A_c^{\mathrm{T}} = \begin{bmatrix} \bar{P}_1^{-1}\bar{A}_{11}^{\mathrm{T}} + \bar{A}_{11}\bar{P}_1^{-1} & \bar{A}_{12}\bar{P}_2^{-1} \\ \bar{P}_2^{-1}\bar{A}_{12}^{\mathrm{T}} & \bar{P}_2^{-1}\Phi^{\mathrm{T}} + \Phi\bar{P}_2^{-1} \end{bmatrix}$$

and therefore a parametrisation of $\hat{Q}$ is given by

$$\hat{Q} = \begin{bmatrix} \hat{Q}_1 + \bar{A}_{12}\bar{P}_2^{-1}\hat{Q}_2^{-1}\bar{P}_2^{-1}\bar{A}_{12}^{\mathrm{T}} & -\bar{A}_{12}\bar{P}_2^{-1} \\ -\bar{P}_2^{-1}\bar{A}_{12}^{\mathrm{T}} & \hat{Q}_2 \end{bmatrix}$$

where the matrices $\bar{P}_1$ and $\bar{P}_2$ solve the pair of Lyapunov equations (7.55) and (7.56) where $\hat{Q}_1$ and $\hat{Q}_2$ are arbitrary symmetric positive definite matrices of appropriate dimension. If

$$Q_{22} \triangleq CQ^{-1}C^{\mathrm{T}} \tag{7.60}$$

then it follows that

$$\bar{G}CQ^{-1}C^{\mathrm{T}}\bar{G}^{\mathrm{T}} = \begin{bmatrix} \bar{G}_1Q_{22}\bar{G}_1^{\mathrm{T}} & \bar{G}_1Q_{22}\bar{G}_2^{\mathrm{T}} \\ \bar{G}_2Q_{22}\bar{G}_1^{\mathrm{T}} & \bar{G}_2Q_{22}\bar{G}_2^{\mathrm{T}} \end{bmatrix}$$

Rearranging equation (7.59) it follows that

$$\bar{Q} = \begin{bmatrix} \hat{Q}_1 + \bar{A}_{12}\bar{P}_2^{-1}\hat{Q}_2^{-1}\bar{P}_2^{-1}\bar{A}_{12}^{\mathrm{T}} - \bar{G}_1Q_{22}\bar{G}_1^{\mathrm{T}} & -\bar{A}_{12}\bar{P}_2^{-1} - \bar{G}_1Q_{22}\bar{G}_2^{\mathrm{T}} \\ -\bar{P}_2^{-1}\bar{A}_{12}^{\mathrm{T}} - \bar{G}_2Q_{22}\bar{G}_1^{\mathrm{T}} & \hat{Q}_2 - \bar{G}_2Q_{22}\bar{G}_2^{\mathrm{T}} \end{bmatrix} \tag{7.61}$$

which by careful choice of $\hat{Q}_1$ and $\hat{Q}_2$ can be made positive definite. This statement will be justified as follows. A necessary condition for $\bar{Q}$ to be positive definite is that

$$\hat{Q}_2 > \bar{G}_2Q_{22}\bar{G}_2^{\mathrm{T}} \tag{7.62}$$

Let $\hat{Q}_2$ be any positive definite matrix satisfying the matrix inequality above. Solving the Lyapunov equation (7.55) provides the matrix $\bar{P}_2$. Consequently, each element in the matrix $\bar{Q}$ given in equation (7.61) is specified except for $\hat{Q}_1$. A necessary and sufficient condition from (7.61) for $\bar{Q}$ to be positive definite is that $\hat{Q}_1$ satisfies (7.57). $\bar{P}_1$ can then be obtained from (7.56) and a block diagonal solution to the Riccati equation (7.54) is given by

$$\bar{P} = \begin{bmatrix} \bar{P}_1 & 0 \\ 0 & \bar{P}_2 \end{bmatrix}$$

■

[2]Since the intention is to find a positive definite solution, the matrix $\bar{P}$ will be invertible.

Let P_r be a Lyapunov matrix for Γ satisfying

$$P_r\Gamma + \Gamma^{\mathrm{T}} P_r = -\bar{G}_r^{\mathrm{T}} \bar{Q}^{-1} \bar{G}_r - Q_r \tag{7.63}$$

for some symmetric positive definite matrix Q_r.

The main result relating to the expected closed-loop performance of the controller/observer pair will now be proved.

Proposition 7.1 *The quadratic form* $V(e,\zeta,e_r) \triangleq e^{\mathrm{T}}Pe + \zeta^{\mathrm{T}}\bar{P}\zeta + e_r^{\mathrm{T}}P_r e_r$ *is a Lyapunov function for the closed-loop system given in (7.47) and (7.48).*

Proof
Taking derivatives along the trajectory gives

$$\begin{aligned}\dot{V} \;=\; & -e^{\mathrm{T}}Qe - 2e^{\mathrm{T}}PB\xi + 2e^{\mathrm{T}}PB\nu_o - \zeta^{\mathrm{T}}\bar{P}\bar{Q}\bar{P}\zeta - \zeta^{\mathrm{T}}\bar{P}\bar{G}CQ^{-1}C^{\mathrm{T}}\bar{G}^{\mathrm{T}}\bar{P}\zeta \\ & - 2\zeta^{\mathrm{T}}\bar{P}\bar{G}Ce + 2\zeta^{\mathrm{T}}\bar{P}\bar{\Lambda}\nu_o + 2\zeta^{\mathrm{T}}\bar{P}\bar{\Lambda}\hat{\nu}_c + 2\zeta^{\mathrm{T}}\bar{P}\bar{G}_r e_r \\ & - e_r^{\mathrm{T}}Q_r e_r - e_r^{\mathrm{T}}\bar{G}_r^{\mathrm{T}}\bar{Q}^{-1}\bar{G}_r e_r \end{aligned} \tag{7.64}$$

Utilising the structural constraint (7.7) and the uncertainty structure given in equation (7.2) it follows that

$$\begin{aligned} 2e^{\mathrm{T}}PB\,(\nu_o - \xi) &= 2e^{\mathrm{T}}C^{\mathrm{T}}F^{\mathrm{T}}\nu_o - 2e^{\mathrm{T}}C^{\mathrm{T}}F^{\mathrm{T}}\xi \\ &\le 2\|FCe\|\,(k_1\|\hat{u}\| + \alpha(t,y) - \rho_o(\hat{u}_L,y)) \end{aligned} \tag{7.65}$$

Recall from equation (7.31) that

$$\rho_o(\hat{u}_L,y) = \left(k_1\|\hat{u}_L\| + \alpha(t,y) + k_1\gamma_c\|\Lambda^{-1}\| + \gamma_o\right) / \left(1 - k_1\,\kappa(\Lambda)\right)$$

which on rearranging yields

$$\begin{aligned} \rho_o(\hat{u}_L,y) &= k_1\left(\|\hat{u}_L\| + \|\Lambda^{-1}\|(\rho_o(\hat{u}_L,y)\|\Lambda\| + \gamma_c)\right) + \alpha(t,y) + \gamma_o \\ &= k_1\left(\|\hat{u}_L\| + \|\Lambda^{-1}\|\rho_c(\hat{u}_L,y)\right) + \alpha(t,y) + \gamma_o \\ &> k_1\left(\|\hat{u}_L\| + \|\hat{\nu}_c\|\right) + \alpha(t,y) + \gamma_o \\ &\ge k_1\|\hat{u}_L + \hat{\nu}_c\| + \alpha(t,y) + \gamma_o \\ &= k_1\|\hat{u}\| + \alpha(t,y) + \gamma_o \end{aligned} \tag{7.66}$$

Note the second equality uses the fact that $\rho_c = \|\Lambda\|\rho_o + \gamma_c$ from (7.32). From inequalities (7.65) and (7.66) it follows that

$$2e^{\mathrm{T}}C^{\mathrm{T}}F^{\mathrm{T}}\nu_o - 2e^{\mathrm{T}}PB\xi \;\le\; -2\gamma_o\|FCe\| \tag{7.67}$$

Also

$$\begin{aligned} 2\zeta^{\mathrm{T}}\bar{P}\bar{\Lambda}(\nu_o + \hat{\nu}_c) &= 2\zeta_2^{\mathrm{T}}\bar{P}_2\Lambda\nu_o + 2\zeta_2^{\mathrm{T}}\bar{P}_2\Lambda\hat{\nu}_c \\ &= 2\zeta_2^{\mathrm{T}}\bar{P}_2\Lambda\nu_o - 2\rho_c(\hat{u}_L,y)\|\bar{P}_2\zeta_2\| \\ &\le 2\|\bar{P}_2\zeta_2\|\,(\rho_o(\hat{u}_L,y)\|\Lambda\| - \rho_c(\hat{u}_L,y)) \\ &= -2\gamma_c\|\bar{P}_2\zeta_2\| \end{aligned} \tag{7.68}$$

since $\|\Lambda\|\rho_o = \rho_c - \gamma_c$. Using inequalities (7.67) and (7.68) in equation (7.64) it follows that

$$
\begin{aligned}
\dot{V} &\leq -e^{\mathrm{T}}Qe - 2\gamma_o\|FCe\| - \zeta^{\mathrm{T}}\bar{P}\bar{Q}\bar{P}\zeta - \zeta^{\mathrm{T}}\bar{P}\bar{G}CQ^{-1}C^{\mathrm{T}}\bar{G}^{\mathrm{T}}\bar{P}\zeta - 2\zeta^{\mathrm{T}}\bar{P}\bar{G}Ce \\
&\quad - 2\gamma_c\|\bar{P}_2\zeta_2\| + 2\zeta^{\mathrm{T}}\bar{P}\bar{G}_re_r - e_r^{\mathrm{T}}Q_re_r - e_r^{\mathrm{T}}\bar{G}_r^{\mathrm{T}}\bar{Q}^{-1}\bar{G}_re_r \\
&\equiv -(e + Q^{-1}C^{\mathrm{T}}\bar{G}^{\mathrm{T}}\bar{P}\zeta)^{\mathrm{T}}Q(e + Q^{-1}C^{\mathrm{T}}\bar{G}^{\mathrm{T}}\bar{P}\zeta) - \zeta^{\mathrm{T}}\bar{P}\bar{Q}\bar{P}\zeta - 2\gamma_o\|FCe\| \\
&\quad - 2\gamma_c\|\bar{P}_2\zeta_2\| + 2\zeta^{\mathrm{T}}\bar{P}\bar{G}_re_r - e_r^{\mathrm{T}}Q_re_r - e_r^{\mathrm{T}}\bar{G}_r^{\mathrm{T}}\bar{Q}^{-1}\bar{G}_re_r \\
&\equiv -(e + Q^{-1}C^{\mathrm{T}}\bar{G}^{\mathrm{T}}\bar{P}\zeta)^{\mathrm{T}}Q(e + Q^{-1}C^{\mathrm{T}}\bar{G}^{\mathrm{T}}\bar{P}\zeta) - e_r^{\mathrm{T}}Q_re_r - 2\gamma_o\|FCe\| \\
&\quad - 2\gamma_c\|\bar{P}_2\zeta_2\| - (\zeta - \bar{P}^{-1}\bar{Q}^{-1}\bar{G}_re_r)^{\mathrm{T}}\bar{P}\bar{Q}\bar{P}(\zeta - \bar{P}^{-1}\bar{Q}^{-1}\bar{G}_re_r) \\
&< 0 \qquad \text{if } (e, \zeta, e_r) \neq 0
\end{aligned}
$$

■

Corollary 7.1 *Sliding motions are induced on the surfaces $\mathcal{S}_o$ defined in (7.9) and*

$$\hat{\mathcal{S}} = \{(\zeta_1, \zeta_2) \in \mathbb{R}^{n+p} \ : \ \zeta_2 = 0\} \tag{7.69}$$

Proof
It is sufficient to demonstrate that a sliding motion occurs on the surface in the combined state space given by

$$\mathcal{S}_c = \{(e, \zeta_1, \zeta_2) \in \mathbb{R}^{2n+p} \ : \ Ce = 0 \text{ and } \zeta_2 = 0\}$$

To this end, let η be a small positive scalar such that $\eta < \gamma_c$ and $\eta < \gamma_0$, and define a domain

$$\Omega_\eta \triangleq \{(e, \zeta_1, \zeta_2) \in \mathbb{R}^{2n+p} \ : \ \|(CB)^{-1}CA_0e\| < \gamma_o - \eta \text{ and } \|\bar{G}_2Ce\| < \gamma_c - \eta\}$$

From Proposition 7.1 it follows that in finite time the states (e, ζ) enter Ω_η and remain there. It will be shown that a sliding motion takes place in this domain.

Consider the quadratic form

$$V_c = e^{\mathrm{T}}C^{\mathrm{T}}P_2Ce + \zeta_2^{\mathrm{T}}\bar{P}_2\zeta_2 \tag{7.70}$$

This is clearly positive definite with respect to the surface $\mathcal{S}_c$ defined above. From (7.48) it follows that

$$C\dot{e}(t) = CA_0e(t) + C_2B_2\left(\nu_o(t) - \xi(t, x, \hat{u})\right)$$

and from the definition of F in equation (7.14) it follows that $P_2 = F^{\mathrm{T}}\left(CB\right)^{-1}$. Therefore

$$P_2C\dot{e}(t) = F^{\mathrm{T}}\left(CB\right)^{-1}CA_0e(t) + F^{\mathrm{T}}\left(\nu_o(t) - \xi(t, x, \hat{u})\right)$$

and using inequalities (7.67) it can be verified that

$$e^{\mathrm{T}}C^{\mathrm{T}}P_2C\dot{e} \leq \|FCe\|\|\left(CB\right)^{-1}CA_0e\| - \gamma_o\|FCe\|$$

Taking derivatives of the quadratic form (7.70) along the trajectories, and using equation (7.45) in conjunction with inequality (7.68) it follows that

$$
\begin{aligned}
\dot{V}_c &= 2e^{\mathrm{T}}C^{\mathrm{T}}P_2C\dot{e} + \zeta_2^{\mathrm{T}}\bar{P}_2\dot{\zeta}_2 + \dot{\zeta}_2^{\mathrm{T}}\bar{P}_2\zeta_2 \\
&\leq 2e^{\mathrm{T}}C^{\mathrm{T}}P_2C\dot{e} - 2\zeta_2^{\mathrm{T}}\bar{P}_2\bar{G}_2Ce - 2\gamma_c\|\bar{P}_2\zeta_2\| \\
&\leq -2\|FCe\|\left(\gamma_o - \|(CB)^{-1}CA_0e\|\right) - 2\|\bar{P}_2\zeta_2\|\left(\gamma_c - \|\bar{G}_2Ce\|\right)
\end{aligned}
$$

Therefore in the domain Ω_η

$$\begin{aligned}\dot{V}_c &\leq -2\eta\left(\|FCe\| + \|\bar{P}_2\zeta_2\|\right)\\ &\leq -2\eta r_1 r_2^{-1/2}\sqrt{V_c}\end{aligned}$$

where the strictly positive scalars r_1 and r_2 are defined as

$$r_1 = \min\{\underline{\sigma}(F), \underline{\sigma}(\bar{P}_2)\} \quad \text{and} \quad r_2 = \max\{\bar{\sigma}(\bar{P}_2), \bar{\sigma}(P_2)\}$$

respectively. This represents a sufficient condition for sliding on the surface $\mathcal{S}_c$. ∎

Corollary 7.2 *The output of the uncertain system $y(t)$ tracks the constant demand vector R asymptotically.*

Proof
From Corollary 7.1 a sliding motion is achieved, which implies from equation (7.44) that

$$\dot{\zeta}_1(t) = \bar{A}_{11}\zeta_1(t) + \left(\bar{A}_{12}S_r + B_r\right)e_r(t)$$

where

$$\dot{e}_r(t) = \Gamma e_r(t)$$

and both $\bar{A}_{11}$ and Γ are stable. For this *linear* system, stability of ζ_1 implies $\dot{\zeta}_1 \to 0$ which implies $\dot{\tilde{z}}_1 \to 0$. From the definition of $\tilde{z}_1$, the first p states are the integral action states. Consequently, $\dot{x}_r \to 0$ and it follows that $y \to r$ asymptotically. ∎

Remark
An intuitive argument supporting this analysis runs as follows. Consider equation (7.33); although it is related to the uncertainty and hence to the estimated states, the arguments presented in Section 6.3.1 indicate that a sliding motion is attained on $\mathcal{S}_o$ in finite time. It follows therefore that in finite time equation (7.35) becomes

$$\dot{z}(t) = Az(t) + B\hat{u}(t) - B\nu_{eq} \tag{7.71}$$

where ν_{eq} is the 'equivalent control' necessary to maintain the motion on $\mathcal{S}_o$ and is a function of the uncertainty. The control law $\hat{u}$ can be thought of as 'controlling' the dynamical system (7.71) given above. The term $B\nu_{eq}$ can be considered as matched uncertainty and, by an appropriate choice of sliding mode control law, it can therefore be rejected. The output $\hat{y} = Cz$ therefore behaves in an ideal way. However, since sliding occurs on the hyperplane $\mathcal{S}_o$, it follows that the output of the uncertain system y is identically equal to $\hat{y}$. Therefore the output of the uncertain system behaves in an ideal nominal way.

7.3.4 Design and Implementation Issues

The observer described in Section 7.3.1 has only the stable matrix A_{22}^s and its associated Lyapunov matrix P_2 in the way of design freedom. As a result, this section concentrates on the design of the nonlinear control law. The hyperplane defined in Section 7.3.2 is given by

$$\mathcal{S} = \{\tilde{x} \in \mathbb{R}^{n+p} \ : \ S\tilde{x} = S_r r\} \tag{7.72}$$

where r represents the reference signal, $S_r \in \mathbb{R}^{m\times m}$ is a free design parameter and

$$S = S_2 \begin{bmatrix} M & I_m \end{bmatrix}$$

where $M \in \mathbb{R}^{m\times n}$. The square design matrix S_2 has no effect on the dynamics of the reduced-order motion and serves only as a scaling of the hyperplane. The sliding motion is given by

$$\dot{\tilde{x}}_1(t) = (\tilde{A}_{11} - \tilde{A}_{12}M)\tilde{x}_1(t) + \left(\tilde{A}_{12}S_2^{-1}S_r + B_r\right) r(t) \tag{7.73}$$

where the pair $(\tilde{A}_{11}, \tilde{A}_{12})$ is defined in (4.102) and B_r is defined in equation (4.103). It was shown in Lemma 4.1 that assumptions A1 and A3 guarantee that $(\tilde{A}_{11}, \tilde{A}_{12})$ is completely controllable and consequently an appropriate matrix M may be chosen by any of the design techniques described in Chapter 4.

The choice of S_r may be viewed as affecting the values of the integral states since, from equation (7.73), at steady state

$$\bar{A}_{11}\tilde{x}_1 + \left(\tilde{A}_{12}S_2^{-1}S_r + B_r\right) r = 0 \tag{7.74}$$

One possibility is to choose a value of S_r so that, for the nominal system at steady-state, the integral action states are zero. Equation (7.74) can be rewritten as

$$\tilde{x}_1 = -\bar{A}_{11}^{-1}\left(\tilde{A}_{12}S_2^{-1}S_r + B_r\right) r \tag{7.75}$$

because $\bar{A}_{11}$ is stable by design and therefore invertible. The steady-state values of the integral action states are given by

$$x_r = -B_r^{\mathrm{T}}\bar{A}_{11}^{-1}\left(\tilde{A}_{12}S_2^{-1}S_r + B_r\right) r \tag{7.76}$$

because of the special form of B_r in (4.103). Define K_s to be the static gain of the square system $(\bar{A}_{11}, \tilde{A}_{12}, B_r^{\mathrm{T}})$, then provided K_s is nonsingular, choosing

$$S_r = S_2 K_s^{-1} B_r^{\mathrm{T}} \bar{A}_{11}^{-1} B_r \tag{7.77}$$

implies $x_r = 0$. The following lemma demonstrates that the assumptions A1 and A3 in Section 7.2 guarantee that K_s is nonsingular.

Lemma 7.2 *The static gain of the square system $(\bar{A}_{11}, \tilde{A}_{12}, B_r^{\mathrm{T}})$ is nonsingular.*

Proof
Let K_s represent the static gain of the system $(\bar{A}_{11}, \tilde{A}_{12}, B_r^{\mathrm{T}})$, then by direct evaluation it follows that

$$\begin{bmatrix} I & 0 \\ B_r^{\mathrm{T}}\bar{A}_{11}^{-1} & K_s \end{bmatrix} \equiv \begin{bmatrix} \tilde{A}_{11} & \tilde{A}_{12} \\ B_r^{\mathrm{T}} & 0 \end{bmatrix} \begin{bmatrix} I & 0 \\ -M & I \end{bmatrix} \begin{bmatrix} \bar{A}_{11}^{-1} & -\bar{A}_{11}^{-1}\tilde{A}_{12} \\ 0 & I \end{bmatrix}$$

Therefore K_s is of full rank if and only if

$$\det \begin{bmatrix} \tilde{A}_{11} & \tilde{A}_{12} \\ B_r^{\mathrm{T}} & 0 \end{bmatrix} \neq 0$$

Substituting for $\tilde{A}_{11}$ and $\tilde{A}_{12}$ from equation (4.102) and for B_r from (4.103) it follows that

$$\det\begin{bmatrix} \tilde{A}_{11} & \tilde{A}_{12} \\ B_r^{\mathrm{T}} & 0 \end{bmatrix} \neq 0 \;\Leftrightarrow\; \det\begin{bmatrix} 0 & -C_1 & -C_2 \\ 0 & A_{11} & A_{12} \\ I & 0 & 0 \end{bmatrix} \neq 0$$
$$\Leftrightarrow\; \det\begin{bmatrix} -C_1 & -C_2 \\ A_{11} & A_{12} \end{bmatrix} \neq 0$$

In the proof of Lemma 4.1 it was shown that

$$\det\begin{bmatrix} -C_1 & -C_2 \\ A_{11} & A_{12} \end{bmatrix} = 0 \;\Leftrightarrow\; (A, B, C) \text{ has invariant zeros at the origin}$$

This is guaranteed by assumption A3 in Section 7.2 and so the gain K_s is nonsingular.

■

An mfile to design a time-varying hyperplane driven by the reference signal is given below.

mfile: to design a time-varying hyperplane

```
% Commands to make a unit vector controller incorporating integral action.
% It assumes the triple (A,B,C) is already in regular form
% For simplicity robust pole placement is used for the hyperplane design
% (with suitable modification any of the methods in Chap 4 could be used)
% the nth order vector p1 specifies the reduced order motion poles
% the mth order vector p2 specifies the range space dynamics

[nn,mm]=size(B);
[pp,nn]=size(C);

% Augment the original (A,B) with the integrator states
bigA=[zeros(pp,pp) -C; zeros(nn,pp) A];
bigB=[zeros(pp,mm); B];

% Since it is assumed that (A,B) is in regular form the augmented
% pair (bigA,bigB) is also automatically in regular form
A11=bigA(1:pp+nn-mm,1:pp+nn-mm);
A12=bigA(1:pp+nn-mm,nn+pp-mm+1:nn+pp);
B2=bigB(nn+pp-mm+1:nn+pp,:);

% Hyperplane design using robust pole placement
M=vplace(A11,A12,p1);
A11s=A11-A12*M;
S2=eye(mm);            % For simplicity
S=S2*[M eye(mm)];

% Form the range space dynamics system matrix
Phi=diag(p2,0);
P2hat=lyap(Phi,eye(mm));
P2=inv(P2hat);
```

```
% Calculate the time varying component of the switching function
Br=[eye(pp) ; zeros(nn-mm,pp)];
Sr=-S2*inv(Br'*inv(A11s)*A12)*Br'*inv(A11s)*Br;

% Computes the gains for the feedback and feed-forward components
L= -inv(S*bigB)*S*bigA + inv(S*bigB)*Phi*S;
Lr= -inv(S*bigB)*(Phi*Sr+S(:,1:pp));
Lrdot=inv(S*bigB)*Sr;
```

From the design perspective, apart from the scalar functions which premultiply the unit vector components, the only dependence between the controller and observer is through the choice of the controller Lyapunov matrix $\bar{P}_2$. This matrix defines the unit vector component of the control law

$$\nu_c = \begin{cases} -\rho_c(u_L, y)\Lambda^{-1}\frac{\bar{P}_2(S\tilde{x}-S_r r)}{\|\bar{P}_2(S\tilde{x}-S_r r)\|} & \text{if } S\tilde{x} \neq S_r r \\ 0 & \text{otherwise} \end{cases} \tag{7.78}$$

and satisfies

$$\bar{P}_2^{-1}\Phi^{\mathrm{T}} + \Phi\bar{P}_2^{-1} = -\hat{Q}_2 \tag{7.79}$$

where Φ is the stable design matrix defining the linear range space dynamics. From Lemma 7.1 the matrix $\hat{Q}_2$ is a symmetric positive definite design matrix which must satisfy equation (7.62), namely

$$\hat{Q}_2 > \bar{G}_2 Q_{22}\bar{G}_2^{\mathrm{T}}$$

For a given observer design, the right-hand side of the above equation is completely determined and therefore appears to restrict the choice of $\bar{P}_2$. However, this is not so. Let $\hat{Q}_2$ be *any symmetric positive definite matrix* and let $\bar{P}_2$ be the unique solution to the Lyapunov equation (7.79). If λ is any positive scalar then the matrix pair $(\bar{P}_2^\lambda, \hat{Q}_2^\lambda) \triangleq (\frac{1}{\lambda}\bar{P}_2, \lambda\hat{Q}_2)$ satisfies

$$(\bar{P}_2^\lambda)^{-1}\Phi^{\mathrm{T}} + \Phi(\bar{P}_2^\lambda)^{-1} = -\hat{Q}_2^\lambda \tag{7.80}$$

It is easy to verify that

$$\hat{Q}_2^\lambda > \bar{G}_2 Q_{22}\bar{G}_2^{\mathrm{T}} \quad \Leftrightarrow \quad \lambda > \lambda_{max}\left(\hat{Q}_2^{-1/2}\bar{G}_2 Q_{22}\bar{G}_2^{\mathrm{T}}\hat{Q}_2^{-1/2}\right) \tag{7.81}$$

and consequently for any given $\hat{Q}_2$, for a large enough value of λ, the matrix $\bar{P}_2^\lambda$ is an appropriate choice for the unit vector control component ν_c. However, from (7.78) it can be seen that ν_c is *independent* of the value of λ. As a result, the controller and observer can be designed independently in the sense that any observer designed as in Section 7.3.1 and any control law of the form given in Section 7.3.2 will provide closed-loop stability for the uncertain system, provided the (uncertainty dependent) scalar functions multiplying the unit vector components are chosen so that equation (7.32) holds.

7.4 EXAMPLE: A TEMPERATURE CONTROL SCHEME

A linear model relating temperature change in a furnace to changes in the valve position regulating the fuel flow is given by

$$A = \begin{bmatrix} 0 & 1 & 0 \\ 0 & 0 & 1 \\ -0.0001 & -0.0082 & -0.1029 \end{bmatrix} \quad B = \begin{bmatrix} 0 \\ 0 \\ 1 \end{bmatrix}$$

$$C = \begin{bmatrix} 0.0001 & 0.0022 & 0.0053 \end{bmatrix} \tag{7.82}$$

This linear model has been obtained by system identification from a nonlinear furnace simulation based on the *zone method* (Hottel & Sarofim, 1967). In this approach the surfaces and volumes that comprise the furnace are divided into subsurfaces and subvolumes which are small enough to be considered isothermal. The integro-differential equations governing radiation exchange become algebraic and finite difference equations which can be solved numerically. The solution of these equations is numerically intense and such models do not lend themselves easily to the problem of designing control laws. As such the linear model given in (7.82) will be used for this purpose; the inherent nonlinearities present will be treated as bounded uncertainty and will be incorporated into the uncertainties associated with the valve dynamics and any external disturbances. The states of the linear system representation have no physical meaning and are therefore not available to the control law; an observer-based approach is therefore necessary. The control law will then be tested on a nonlinear simulation model.

Conveniently the realisation given in (7.82) is already in regular form. The invariant zeros of this linear system are given by $\{-0.3749, -0.0358\}$ and therefore all the preceding theory is valid. The rest of the section looks at synthesising and analysing sliding mode temperature controllers based on this nominal linear model.

Before discussing the design procedure adopted, it should be noted that because of the structure of the realisation, any variation in the elements in the last row of the system matrix (which are the coefficients of the characteristic polynomial) occurs in $\mathcal{R}(B)$, and so can be considered as matched uncertainty. The design will therefore be insensitive to changes in the poles of the system.

7.4.1 Observer Design

Using the notation of Section 7.3.1 it follows that

$$\begin{bmatrix} A_{11} & A_{12} \\ A_{21} & A_{22} \end{bmatrix} = \left[\begin{array}{cc|c} 0 & 1 & 0 \\ 0 & 0 & 1 \\ \hline -0.0001 & -0.0082 & -0.1029 \end{array}\right]$$

and

$$\begin{bmatrix} C_1 & C_2 \end{bmatrix} = \left[\begin{array}{cc|c} 0.0001 & 0.0022 & 0.0053 \end{array}\right]$$

If the stable design matrix $A_{22}^s = -0.2$, it follows immediately from equation (7.13) that

$$G = \begin{bmatrix} A_{12}C_2^{-1} \\ A_{22}C_2^{-1} - C_2^{-1}A_{22}^s \end{bmatrix} = \begin{bmatrix} 0 \\ 188.8498 \\ 18.3328 \end{bmatrix}$$

Because the system is single-input single-output there is no need to compute the sliding surface matrix F. All that remains is to compute the nonlinear scalar gain function given in (7.31). In this case, the approach that has been adopted is to let

$$\rho_o(u_L, y) = r_1|y| + r_2|u_L(\cdot)| + \gamma_o \tag{7.83}$$

where $u_L(\cdot)$ represents the linear component of the control action and the positive scalars r_1, r_2 and γ_o are to be chosen empirically. This will be discussed in more detail in a later subsection.

7.4.2 Controller Design

Because the system is already in regular form, the augmented system from equation (4.102) can easily be identified as

$$\begin{bmatrix} \tilde{A}_{11} & \tilde{A}_{12} \\ \tilde{A}_{21} & A_{22} \end{bmatrix} = \left[\begin{array}{cccc|c} 0 & -0.0001 & -0.0022 & -0.0053 \\ 0 & 0 & 1 & 0 \\ 0 & 0 & 0 & 1 \\ \hline 0 & -0.0001 & -0.0082 & -0.1029 \end{array}\right]$$

As predicted from the theory, the pair $(\tilde{A}_{11}, \tilde{A}_{12})$ is completely controllable. The poles of the ideal sliding motion have been chosen to be $\{-0.025,\ -0.03 \pm 0.025j\}$, which represent dynamics marginally faster than the dominant pole of the open-loop plant. The unique M such that $\lambda(\tilde{A}_{11} - \tilde{A}_{12}M) = \{-0.025,\ -0.03 \pm 0.025j\}$ is given by

$$M = \begin{bmatrix} -0.5372 & 0.0019 & 0.0822 \end{bmatrix}$$

For single-input single-output systems, no additional design freedom is provided by the scalar Λ. For simplicity it has been chosen so that $S_2 = 1$. The remaining pole, associated with the range space dynamics, has been assigned the value -0.1 by selecting the design matrix $\Phi = -0.1$. Again, because of the single-input single-output nature of the system, there is no need to calculate the Lyapunov matrix $\bar{P}_2$ associated with the unit vector controller. The nonlinear component of the controller is given by

$$\nu_c = -\rho_c(u_L, y)\ \mathrm{sgn}(S\tilde{x} - S_r r) \tag{7.84}$$

where the gain function $\rho_c(\cdot) = \rho_o(\cdot) + \gamma_c$ for some positive scalar γ_c; and the hyperplane matrix is given by

$$S = \begin{bmatrix} -0.5372 & 0.0019 & 0.0822 & 1.0000 \end{bmatrix}$$

As argued in Section 7.3.4, because of the minimum phase relative degree 1 assumption, it is always possible to choose S_r so that $x_r = 0$ for the nominal system at steady state. For the system under consideration, equation (7.77) gives

$S_r = 26.1642$. From equations (7.24) to (7.26) the gains are given by

$$\begin{array}{rcl} L & = & \left[\ 0.0537 \quad -0.0002 \quad -0.0030 \quad -0.0821\ \right] \\ L_r & = & 3.1536 \\ L_{\dot{r}} & = & 26.1642 \end{array} \qquad (7.85)$$

The control law is given by

$$u(t) = L\tilde{x}(t) + L_r r(t) + L_{\dot{r}}\dot{r}(t) + \nu_c \qquad (7.86)$$

where $\tilde{x}$ represents the augmented state and ν_c is defined as in (7.84).

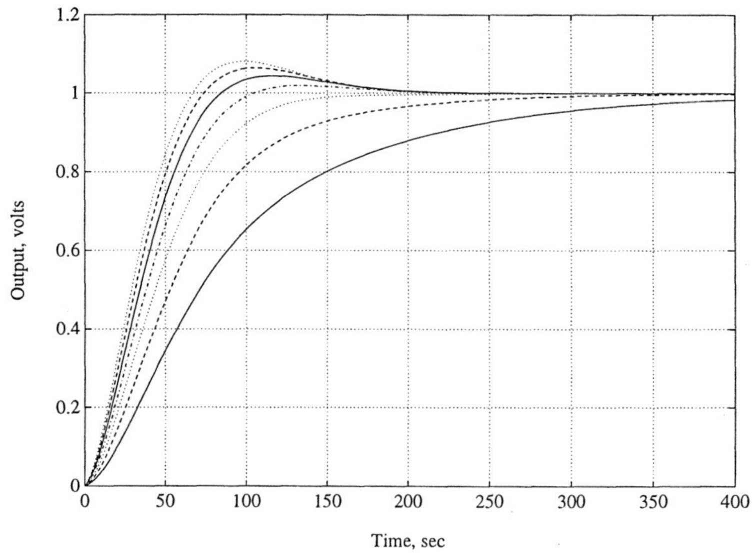

Figure 7.1: Selection of the stable design matrix Γ

The stable matrix Γ from equation (7.18) affects the closed-loop performance. Here it has been chosen to tailor the step response of the nominal closed-loop system. Figure 7.1 shows the step response of the nominal closed-loop system for different values of Γ in the interval $[-0.04, -0.015]$. In practice, overshoot in a thermal process is very undesirable. The heating and cooling characteristics are obviously very different, and overshoot will essentially produce unnecessary time delays as the process is left to cool down. As a consequence, the value of -0.025 has been chosen as a compromise between the conflicting objectives of overshoot and rise time. All that remains is to calculate the scalars that comprise the gain functions in equations (7.83). This is discussed in detail in the following subsection.

7.4.3 Design of the Nonlinear Gain Function

Formally, the nonlinear gains are related to the magnitudes of the uncertainty bounds, which in this situation are not available. An estimate of the design constants in the nonlinear gain function can be obtained by considering the range of allowable inputs. For the system under consideration, the input is restricted to the interval $[0, 4]$ volts. Consequently, it is reasonable to require that the inequality $r_1|y| < 4$ is satisfied. In practice, as a result of the chosen temperature profile to be tracked, it was found that y is of the order 1. In this way, a sensible upper bound on r_1 can be obtained.

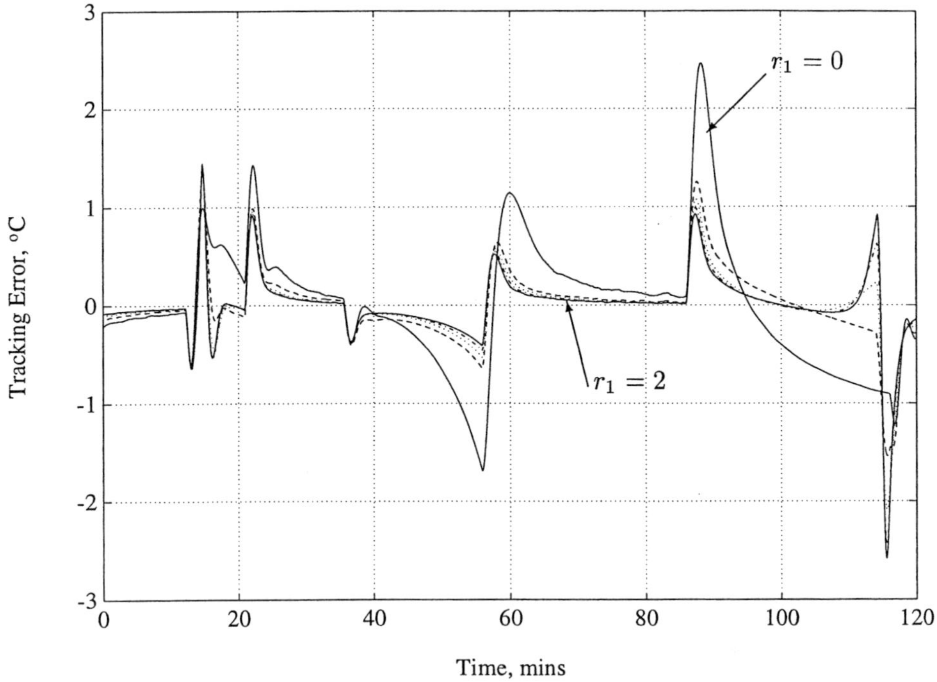

Figure 7.2: Simulations with different nonlinear components

Figure 7.2 represents nonlinear simulation tests, using the observer designed in the previous subsection, with the nonlinear gain

$$\rho_o(u_L, y) = r_1|y| \qquad \text{for } r_1 \in \{0, \tfrac{1}{2}, 1, \tfrac{3}{2}, 2\}$$

This demonstrates the increase in performance obtained as a result of increasing r_1 and hence the nonlinear component of the control action. The final value of r_1 was chosen as a trade-off between the tracking error and the control effort required. In the simulations shown in the next section, a scaling function of the form shown in (7.83) has been employed.

7.4.4 Furnace Simulations

Under typical operating conditions, the controller is required to take the furnace from one operating temperature to another, along a specified trajectory. Figure 7.3 shows a typical temperature demand signal, which will subsequently be used for the nonlinear simulations. The demand comprises a period of low fire, a ramp up to a high temperature, a period of soak then a return to low fire.

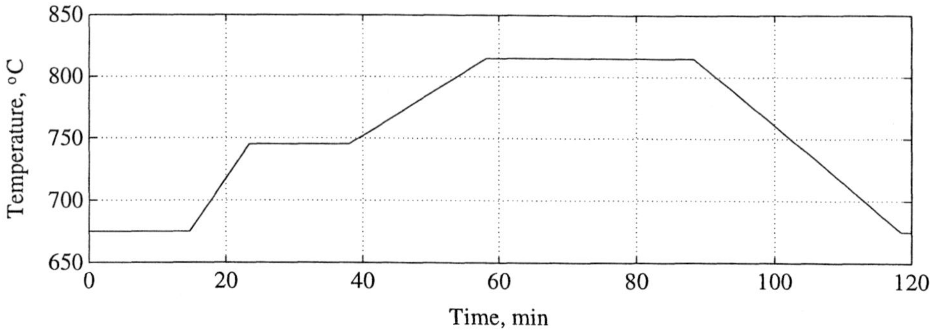

Figure 7.3: Typical temperature reference signal

Consider the differential equation

$$\dot{r}(t) = \Gamma\left(r(t) - R(t - 1/\Gamma)\right) \tag{7.87}$$

which effectively replaces (7.18) in Section 7.3.2. It is assumed that the intervals which define the piecewise linear components of the reference are large enough compared to the time constant of Γ so that 'steady state' occurs. It can be seen that on each interval $\ddot{R}(t) = 0$ and therefore $\dot{R}(t) = \alpha$ for some scalar. If $e_r = r(t) - R(t)$ then from equation (7.87) it follows immediately that

$$\begin{aligned} \dot{e}_r(t) &= \Gamma r(t) - \Gamma R(t - 1/\Gamma) - \alpha \\ &= \Gamma e_r(t) + \Gamma R(t) - \Gamma R(t - 1/\Gamma) - \alpha \\ &= \Gamma e_r(t) \end{aligned}$$

Therefore $e_r(t) \to 0$ and the solution $r(t)$ to equation (7.87) follows the profile in Figure 7.3 asymptotically.

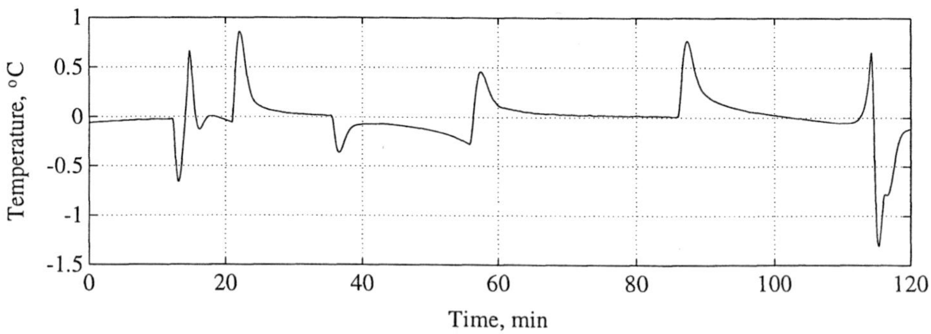

Figure 7.4: Tracking error from the nonlinear simulation

Figure 7.4 shows the closed-loop response of the final controller design on the nonlinear simulation. The tracking error is very good even on those parts of the demand profile that comprise transients between steady-state operating points and for which asymptotic tracking is not guaranteed theoretically.

7.5 A MODEL-REFERENCE APPROACH

The model-following design objective, as described in Section 4.4.1, is to develop a control scheme which forces the plant dynamics to follow the dynamics of an ideal model. The controller should thus force the error between the plant and the model states to zero as time tends to infinity. Consider the multivariable plant

$$\dot{x}(t) = Ax(t) + Bu(t) \tag{7.88}$$

with the corresponding ideal model

$$\dot{w}(t) = A_m w(t) + B_m r(t) \tag{7.89}$$

where $w \in \mathbb{R}^n$ is the state vector of the model, $r \in \mathbb{R}^r$ is an input vector and A_m and B_m are compatibly dimensioned matrices. It is assumed that the ideal model is stable. Define an error state

$$e_x(t) = x(t) - w(t) \tag{7.90}$$

as the difference between the plant and model state responses. This error is required to tend asymptotically to zero. Substituting from equations (7.88) and (7.89), the dynamics of the model-following error system can be determined as

$$\dot{e}_x(t) = A_m e_x(t) + (A - A_m)x(t) + B(u(t) + \xi(t, x, u)) - B_m r(t) \tag{7.91}$$

As in Section 4.4.1, if the conditions for perfect model-following hold, there exist compatibly dimensioned matrices L_x and L_r such that

$$BL_x = A_m - A \tag{7.92}$$

$$BL_r = B_m \tag{7.93}$$

In this case, all forcing terms to the error system (7.91) occur in the range space of the plant input distribution matrix. Equation (7.91) can then be written as

$$\dot{e}_x(t) = A_m e_x(t) + B(u(t) - L_x x(t) - L_r r(t) + \xi(t, x, u)) \tag{7.94}$$

The required system performance is usually specified via the model, and the controller is synthesised to minimise the error between the model states and the controlled plant. If the states x (and hence the error e_x) are available, then a control law of the form described in Section 4.4.1 can be used to ensure $e_x \to 0$. Here, however, it will be assumed that only the output from (7.1) is available and hence an alternative approach is necessary.

To this end, consider the observer structure from Section 7.2 and define

$$e_z(t) = z(t) - w(t) \tag{7.95}$$

i.e. the difference between the ideal model and observer states. Let the control law be of the form

$$u(t) = u_f(t) + \bar{u}(t) \tag{7.96}$$

where $u_f(t) = L_x z(t) + L_r r(t)$ and the component $\bar{u}(t)$, which includes a discontinuous element, will be described later. Arguing as before, and substituting from (7.5) and (7.89), it can easily be shown that

$$\dot{e}_z(t) = A_m e_z(t) + B\bar{u}(t) - GCe(t) + B\nu_o \tag{7.97}$$

where $e(t)$ represents the state observation error. The objective is to establish a control law $\bar{u}$ to induce a sliding motion on the surface

$$\mathcal{S} = \{e_z \in \mathbb{R}^n \quad : \quad s(e_z) = 0\} \tag{7.98}$$

where the switching function $s(e_z) = Se_z$ and the matrix $S \in \mathbb{R}^{m \times n}$. For convenience, partition the hyperplane matrix so that

$$S = \begin{bmatrix} S_1 & S_2 \end{bmatrix} \tag{7.99}$$

with $S_1 \in \mathbb{R}^{m \times (n-m)}$ and $S_2 \in \mathbb{R}^{m \times m}$. By design, assume that $\det(S_2) \neq 0$ and

$$S_2 B_2 = \Lambda \tag{7.100}$$

where Λ is a nonsingular diagonal matrix. Since it has been assumed that (A, B) is in regular form, it follows that (A_m, B) is also in regular form. Also, since $A_m = A + BL_x$ it follows that the sliding motion on $\mathcal{S}$ is governed by

$$\bar{A}_{11} = A_{11} - A_{12}M \tag{7.101}$$

where $M = S_2^{-1}S_1$. A minimum requirement is that M is chosen to ensure that $\bar{A}_{11}$ is stable, and again any of the methods described in Chapter 4 may be utilised. For a given Λ the matrix S_2 is specified, hence the switching function matrix is fully determined by the choice of M and Λ.

Remark

The notation used in this section replicates that of Section 7.3 in the sense that, for example, $\bar{A}_{11}$ was used to denote the system matrix governing sliding in (7.73), although the orders of the matrices are different. This abuse of notation is deliberate and will allow some results to be conveniently quoted since the disparity in order makes no difference to the fundamental results.

The control law component $\bar{u}(t)$ comprises a linear element to stabilise the nominal linear system, and a discontinuous term to induce an ideal sliding motion. Specifically

$$\bar{u}(t) = u_l(t) + \nu_c \tag{7.102}$$

where the linear component is given by

$$u_l(t) = -(SB)^{-1}(SA_m - \Phi S)\, e_z(t) \tag{7.103}$$

where $\Phi \in \mathbb{R}^{m \times m}$ is a stable design matrix. The nonlinear component is defined to be

$$\nu_c = -\rho_c(u_l, y)\Lambda^{-1}\frac{\bar{P}_2 s(t)}{\|\bar{P}_2 s(t)\|} \qquad \text{for } s(t) \neq 0 \tag{7.104}$$

where $\bar{P}_2 \in \mathbb{R}^{m \times m}$ is a symmetric positive definite Lyapunov matrix for Φ. The scalar function $\rho_c(u_l, y)$, which depends only on the magnitude of the uncertainty, is chosen to ensure that

$$\rho_c(u_l, y) = \|\Lambda\|\rho_o(u_l, y) + \gamma_c \tag{7.105}$$

as in Section 7.3.2. Define a linear change of coordinates by

$$\bar{T} = \begin{bmatrix} I & 0 \\ S_1 & S_2 \end{bmatrix} \tag{7.106}$$

which is nonsingular since S_2 is nonsingular. Define new coordinates as $\bar{e}_z = \bar{T}e_z$, which induces a new triple $(\bar{A}, \bar{B}, \bar{C})$. Equation (7.97) becomes

$$\dot{\bar{e}}_z(t) = \bar{A}_c\bar{e}_z(t) + \bar{B}\,(\nu_c + \nu_o) - \bar{G}Ce(t) \tag{7.107}$$

where $\bar{G} = \bar{T}G$ and $\bar{A}_c$ is the closed-loop system matrix obtained by substituting for the linear component. In particular, it can be shown that

$$\bar{A}_c = \begin{bmatrix} \bar{A}_{11} & \bar{A}_{12} \\ 0 & \Phi \end{bmatrix}$$

where $\bar{A}_{11}$ is defined in (7.101) and $\bar{A}_{12} = A_{12}S_2^{-1}$. Further, for the analysis which follows, partition the gain matrix so that

$$G = \begin{bmatrix} \bar{G}_1 \\ \bar{G}_2 \end{bmatrix} \begin{matrix} \updownarrow n-m \\ \updownarrow m \end{matrix} \tag{7.108}$$

The closed-loop configuration is determined by these equations:

$$\dot{w}(t) = A_m w(t) + B_m r(t) \tag{7.109}$$

$$\dot{e}(t) = A_0 e(t) + B\nu_o - B\xi(t, x, u) \tag{7.110}$$

$$\dot{\bar{e}}_z(t) = \bar{A}_c\bar{e}_z(t) - \bar{G}Ce(t) + \bar{B}(\nu_c + \nu_o) \tag{7.111}$$

It should first be noted that equation (7.109) is known, stable and independent of e and $\bar{e}_z$. It follows that it is sufficient to prove the stability of subsystems (7.110) and (7.111).

The stability analysis which follows parallels the development in Section 7.3.3. A block diagonal Lyapunov matrix for $\bar{A}_c$ of the form

$$\bar{P} = \begin{bmatrix} \bar{P}_1 & 0 \\ 0 & \bar{P}_2 \end{bmatrix} \tag{7.112}$$

will be shown to exist where the lower right matrix sub-block $\bar{P}_2$ is the Lyapunov matrix satisfying

$$\bar{P}_2^{-1}\Phi^{\mathrm{T}} + \Phi\bar{P}_2^{-1} = -\hat{Q}_2 \tag{7.113}$$

where $\hat{Q}_2$ is any symmetric positive definite matrix satisfying

$$\hat{Q}_2 > \bar{G}_2 C Q^{-1} C^{\mathrm{T}} \bar{G}_2^{\mathrm{T}} \tag{7.114}$$

Using the Lyapunov matrix P for A_0 given in (7.6), define

$$P_G \triangleq \begin{bmatrix} P & 0 \\ 0 & \bar{P} \end{bmatrix} \tag{7.115}$$

where the matrix $\bar{P}$ will be chosen so that P_G is a Lyapunov matrix for the overall closed-loop system matrix

$$A_G \triangleq \begin{bmatrix} A_0 & 0 \\ -\bar{G}C & \bar{A}_c \end{bmatrix} \tag{7.116}$$

It can be easily verified that

$$P_G A_G + A_G^{\mathrm{T}} P_G = \begin{bmatrix} -Q & -C^{\mathrm{T}}\bar{G}^{\mathrm{T}}\bar{P} \\ -\bar{P}\bar{G}C & \bar{P}\bar{A}_c + \bar{A}_c^{\mathrm{T}}\bar{P} \end{bmatrix} \tag{7.117}$$

where Q is defined in the Lyapunov equation (7.6). The expression on the right-hand side of equation (7.117) is negative definite if and only if

$$\bar{P}\bar{A}_c + \bar{A}_c^{\mathrm{T}}\bar{P} + \bar{P}\bar{G}CQ^{-1}C^{\mathrm{T}}\bar{G}^{\mathrm{T}}\bar{P} < 0 \tag{7.118}$$

The existence of a block diagonal $\bar{P}$ satisfying

$$\bar{P}\bar{A}_c + \bar{A}_c^{\mathrm{T}}\bar{P} + \bar{P}\bar{G}CQ^{-1}C^{\mathrm{T}}\bar{G}^{\mathrm{T}}\bar{P} + \bar{P}\bar{Q}\bar{P} = 0 \tag{7.119}$$

for some symmetric positive definite design matrix $\bar{Q}$ was proved in Lemma 7.1. A similar result to that in the previous section will now be established.

Proposition 7.2 *The quadratic form*

$$V(e, \bar{e}_z) = e^{\mathrm{T}}Pe + \bar{e}_z^{\mathrm{T}}\bar{P}\bar{e}_z \tag{7.120}$$

is a Lyapunov function for the system given in equations (7.110) and (7.111) and an ideal model-matching takes place.

Proof

Taking derivatives along the trajectories implies

$$\begin{aligned} \dot{V} &= -e^{\mathrm{T}}Qe + 2e^{\mathrm{T}}PB\nu_o - 2e^{\mathrm{T}}PB\xi + \bar{e}_z^{\mathrm{T}}(\bar{P}\bar{A}_c + \bar{A}_c^{\mathrm{T}}\bar{P})\bar{e}_z \\ &\quad - 2\bar{e}_z^{\mathrm{T}}\bar{P}\bar{G}Ce + 2\bar{e}_z^{\mathrm{T}}\bar{P}\bar{B}(\nu_c + \nu_o) \end{aligned} \tag{7.121}$$

Substituting from equation (7.119), the expression

$$\begin{aligned} \dot{V} &= -\bar{e}_z^{\mathrm{T}}\bar{P}\bar{Q}\bar{P}\bar{e}_z - (e + Q^{-1}C^{\mathrm{T}}\bar{G}^{\mathrm{T}}\bar{P}\bar{e}_z)^{\mathrm{T}}Q(e + Q^{-1}C^{\mathrm{T}}\bar{G}^{\mathrm{T}}\bar{P}\bar{e}_z) \\ &\quad + 2e^{\mathrm{T}}PB\nu_o - 2e^{\mathrm{T}}PB\xi + 2\bar{e}_z^{\mathrm{T}}\bar{P}\bar{B}(\nu_c + \nu_o) \end{aligned} \tag{7.122}$$

can be established. It follows from the definition of $\rho_o(u, y)$ in (7.31) and manipulations as in the proof of Proposition 7.1 that

$$2e^{\mathrm{T}}PB\nu_o - 2e^{\mathrm{T}}PB\xi \le -2\|FCe\|\gamma_o \tag{7.123}$$

By an argument similar to the proof of Proposition 7.1, it follows that

$$2\bar{e}_z^{\mathrm{T}}\bar{P}\bar{B}(\nu_c + \nu_o) \le -2\gamma_c\|\bar{P}_2 s\| \tag{7.124}$$

Using inequalities (7.123) and (7.124) in conjunction with (7.122), the derivative $\dot{V}$ satisfies

$$\dot{V} \le -\bar{e}_z^{\mathrm{T}}\bar{P}\bar{Q}\bar{P}\bar{e}_z - (e + Q^{-1}C^{\mathrm{T}}\bar{G}^{\mathrm{T}}\bar{P}\bar{e}_z)^{\mathrm{T}}Q(e + Q^{-1}C^{\mathrm{T}}\bar{G}^{\mathrm{T}}\bar{P}\bar{e}_z) - 2\|\bar{P}_2 s\|\gamma_c - 2\|FCe\|\gamma_o$$

and therefore $\dot{V} < 0$ for $e, \bar{e}_z \neq 0$ and quadratic stability is proved.

Furthermore because $e_x = e_z - e$, since $e_z \to 0$ and $e \to 0$, it follows that $e_x \to 0$ and perfect model-following is attained. ∎

Corollary 7.3 *An ideal sliding motion takes place on $\mathcal{S}_o$ and $\mathcal{S}$ in finite time.*

Proof

Similar to Corollary 7.1. ∎

7.5.1 Example: L-1011 Fixed-Wing Aircraft

Consider the lateral axis model of an L-1011 at cruise flight conditions which was discussed in Section 5.7.1. The system triple (A, B, C) is given by

$$A = \begin{bmatrix} 0 & 0 & 1.0000 & 0 & 0 \\ 0 & -0.1540 & -0.0042 & 1.5400 & 0 \\ 0 & 0.2490 & -1.0000 & -5.2000 & 0 \\ 0.0386 & -0.9960 & -0.0003 & -0.1170 & 0 \\ 0 & 0.5000 & 0 & 0 & -0.5000 \end{bmatrix}$$

$$B = \begin{bmatrix} 0 & 0 \\ -0.7440 & -0.0320 \\ 0.3370 & -1.1200 \\ 0.0200 & 0 \\ 0 & 0 \end{bmatrix} \qquad C = \begin{bmatrix} 0 & 1 & 0 & 0 & -1 \\ 0 & 0 & 1 & 0 & 0 \\ 0 & 0 & 0 & 1 & 0 \\ 1 & 0 & 0 & 0 & 0 \end{bmatrix}$$

The ideal model that will be used here is the model of Section 5.7.1. (Recall that a gain matrix L_x was designed using eigenstructure assignment to yield a stable matrix $A_m = A + BL_x$ with $\lambda(A_m) = \{-0.05, -2 \pm 1.5j, -1.5 \pm 1.5j\}$). The feed-forward gain matrix L_r was chosen so that the steady-state gain of (A_m, BL_r, C_m) was unity where the distribution matrix C_m defined in (5.120) identified the bank angle ϕ and sideslip angle β as the 'controlled outputs'.

In the design which follows, the switching function matrix S from (7.99) has been obtained using the quadratic minimisation approach described in Section 4.2.2. Here, as in Section 5.7.1, the symmetric positive definite matrix which defines the cost function has been chosen as

$$Q = \text{diag}(5, 1, 1, 5, 5)$$

which gives sliding mode error poles at

$$\{-0.4470, \ -2.2372, -2.5061\}$$

and a switching function matrix

$$S = \begin{bmatrix} -0.9368 & 0.8593 & -0.4110 & -1.9596 & 0.4444 \\ -2.0310 & -0.3901 & -0.9116 & 0.8491 & -0.2336 \end{bmatrix}$$

All that remains is to design the sliding mode observer. The canonical form of Section 6.3 can be obtained via the output feedback canonical form in equation (5.121) by suitable repartitioning. From (5.121) it follows that

$$A_{11} = \begin{bmatrix} -0.0133 \end{bmatrix} \quad \text{and} \quad A_{211} = \begin{bmatrix} 0 \\ -0.7071 \end{bmatrix}$$

and thus pair (A_{11}, A_{211}) is trivially observable. Furthermore, since the poles of $A_{11} + LA_{211}$ can be placed arbitrarily, there is no loss of design freedom in the placement of the poles of A_0. Here L has been chosen so that $A_{11} + LA_{211} = -3$. In the canonical form of Section 6.3, the system matrix is

$$\mathcal{A} = \begin{bmatrix} -3.0000 & 3.2541 & 0.0082 & -12.4182 & -0.1637 \\ 0.1089 & -0.6417 & -0.0046 & 2.0019 & -0.0000 \\ -0.1761 & 0.2292 & -0.9994 & -5.9468 & -0.0000 \\ 0.7043 & -0.9167 & -0.0026 & 2.8702 & 0.0386 \\ 0 & 0 & 1 & 0 & 0 \end{bmatrix} \qquad (7.125)$$

and the input and output distribution matrices are

$$\mathcal{B} = \begin{bmatrix} 0 & 0 \\ -0.7440 & -0.0320 \\ 0.3370 & -1.1200 \\ 0.0200 & 0 \\ 0 & 0 \end{bmatrix} \qquad \mathcal{C} = \begin{bmatrix} 0 & 1 & 0 & 0 & 0 \\ 0 & 0 & 1 & 0 & 0 \\ 0 & 0 & 0 & 1 & 0 \\ 0 & 0 & 0 & 0 & 1 \end{bmatrix} \qquad (7.126)$$

In the design which follows, $\mathcal{A}_{22}^s = \text{diag}(-4, -4.425, -4.5, -5)$ which results in an observer defined by the matrices

$$F = \begin{bmatrix} -0.0930 & 0.0381 & 0.0022 & 0 \\ -0.0040 & -0.1266 & 0 & 0 \end{bmatrix}$$

and

$$G = \begin{bmatrix} 0 & 1 & 0 & 5.0000 \\ 3.5399 & 0.0055 & -11.4944 & 0 \\ 0.2292 & 3.4256 & -5.9468 & 0 \\ -0.9167 & -0.0026 & 7.3702 & 0.0386 \\ 0.1816 & 0.0101 & -13.4963 & 0 \end{bmatrix}$$

Figure 5.12 shows the response of the closed-loop system to a step change on bank angle. The dotted line represents the ideal model response and the solid line the system response. As in Section 5.7.1, in order to make the error system apparent, the initial state of the observer has been deliberately chosen injudiciously. As can be seen from Figure 7.5, the mismatch between the plant model and the ideal model disappears after a few seconds and the responses are indistinguishable.

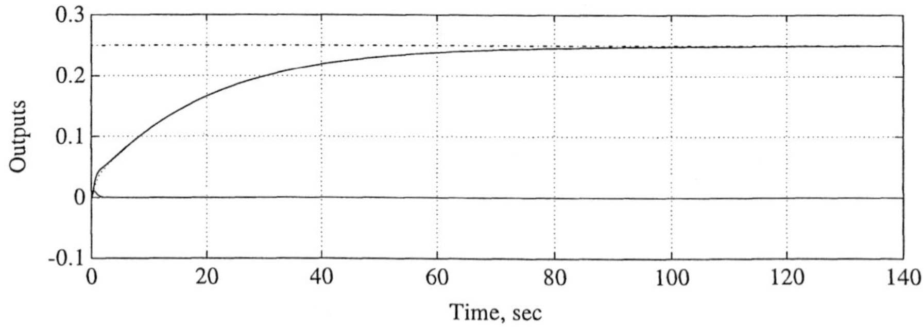

Figure 7.5: System and ideal model response

Figure 7.6 shows the first 10 seconds of the simulation so that the plant and observer discrepancies can be seen. Figure 7.7 shows that a sliding motion takes place on $\mathcal{S}$ after approximately 1 second.

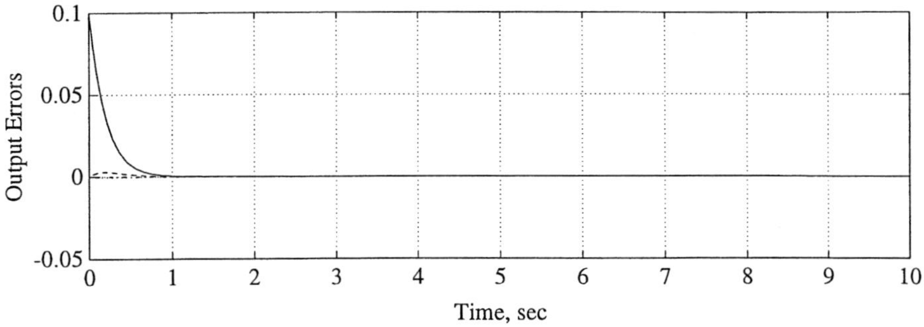

Figure 7.6: Evolution of the errors between plant and observer outputs

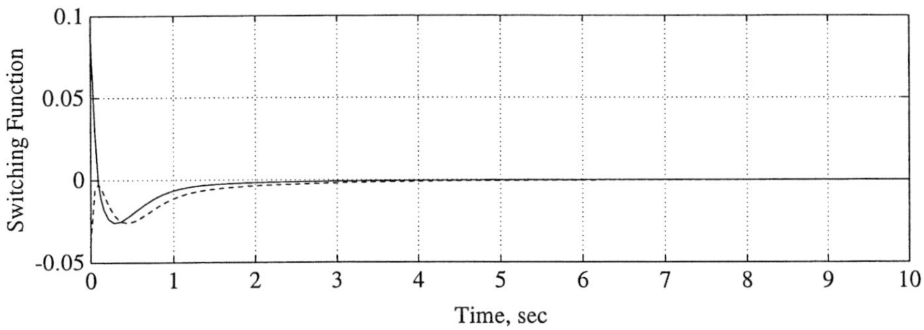

Figure 7.7: Evolution of the error-based switching function

7.6 SUMMARY

Practical nonlinear control strategies have been presented which provide robust output tracking despite the presence of uncertainty. The strategies use only measured output information and construct estimates of the internal states, required for the control law, using a nonlinear observer. It has been demonstrated for the class of controllers and observers considered that a form of separation principle holds. A rigorous analysis of the closed-loop performance of the controller/observer pair has been undertaken; it indicates that in finite time a sliding motion can be attained which rejects any bounded matched uncertainty.

7.7 NOTES AND REFERENCES

The problem of analysing the properties of sliding mode controllers used in conjunction with sliding mode observers has not been studied extensively. Work analysing the closed-loop stability when using a sliding mode controller and an asymptotic observer appears in Utkin (1992), Young & Kwatny (1982) and Breinl & Leitmann (1987). In the latter the well-known 'separation principle' for linear systems is shown to be valid. Walcott & Żak (1988) consider a sliding mode controller/observer pair; they appeal to the results of Breinl & Leitmann (1987) and claim that the separation principle is valid in this situation, but they do not attempt any formal analysis.

The nonlinear furnace simulation in Section 7.4 is based on work by Palmer (1989) which has subsequently been adapted for use as a test bed for controller design.

Chapter 8

Automotive Case Studies

8.1 INTRODUCTION

The automotive industry is interested in manufacturing cars to improved specifications whilst maintaining fixed costs. This has led the industry to show significant interest in robust multivariable control techniques. Inner feedback control loops give the designer the option of improving the performance of a low cost component which may exhibit undesirable open-loop characteristics. Robust control strategies can ensure that components maintain good performance over longer life spans in what is a hostile environment. Multivariable control enables interactions between inputs and outputs to be considered at the initial control design stage rather than considering suites of single-input single-output systems at the design stage and later reconciling interacting behaviour.

This chapter considers two automotive case studies where the sliding mode control techniques developed previously are employed to achieve the above objectives. The first study considers the development of a sliding mode controller for a highly nonlinear actuator system which incorporates both matched and unmatched uncertainty. It employs a very straightforward sliding mode control approach, similar to that outlined in Chapter 4. Implementation issues are discussed and a robustness assessment is made using bench test results. A comparison is made between the performance obtained using classical tools, which are traditionally applied in this application area, and the proposed basic sliding mode control strategy. The second study uses the robust nonlinear controller/observer scheme developed in Chapter 7 for idle speed control of an automotive engine. First a nonlinear controller design based on sliding mode theory is developed. A nonlinear observer is used for state estimation and the resulting nonlinear controller/observer pair is implemented on an engine test facility. The results of rig trials are presented and discussed. The robustness of the control strategy is confirmed.

8.2 AUTOMOTIVE ACTUATOR WITH STICTION

The actuation system under consideration is severely nonlinear and contains significant unmatched as well as matched uncertainty. A critical component is the large stiction torque. The stiction contribution to the unmatched uncertainty poses a particular problem for control system design. Stiction is very difficult to model

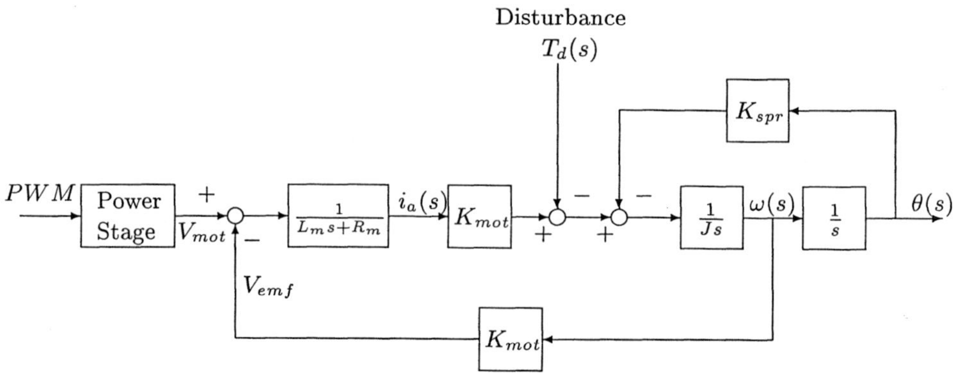

Figure 8.1: Block diagram of actuator configuration

and/or measure and worst case estimates of stiction effects are often the only information that is available to the designer. If tight performance specifications are required, as is the case for this automotive application area, robust performance across the operation range can be difficult to achieve.

The basic behaviour of the actuator is best discussed in terms of the simplified schematic diagram given in Figure 8.1. The actuator comprises a DC motor which directly drives a spring-loaded inertia. The model is developed from the standard linear representation of a DC motor. The winding resistance R_m is assumed linear and constant. This is a considerable simplification as the resistance will change significantly with temperature. However, it will be shown that the effect of R_m is wholly matched, so the proposed control configuration can be made completely insensitive to its effects without further modelling investigations. The inductance L_m varies with both the position output θ and the current flowing in the motor coil. The motor gain K_{mot} also shows significant dependence upon the motor current. The disturbance $T_d(s)$ incorporates stiction effects, the offset torque of the spring, the cogging torque and endstop effects. The magnitude of the stiction effects depends upon output position and the direction depends on angular velocity. The two torques are functions of output position. Endstop effects are significant when the output position nears its limits of movement. In Figure 8.1, V_{emf} denotes the back emf. The controller must be tolerant of changes in rotor inertia due to differences in the type of motor which can be used to build the system. The automotive environment is harsh due to large temperature variations and large levels of vibration and shock. It is expected that many of the system parameters will vary significantly during the life of the actuator. The dynamic behaviour of the actuator is determined by the following states:

- θ angular position of the rotor (rad)
- ω angular velocity of the rotor (rad/sec)
- i_a current flowing in the motor coil (A)

These will be used for controller development.

The design specification requires both large and small step responses to be accomplished within a settling time of 0.1 second, with a 0.1° steady-state error and ideally no overshoot. In addition, the controller must be able to cope with a slow ramp input; practical experience indicates that it is particularly hard to overcome stiction effects swamping the performance under this condition. For reasons of safety, the controller must be able to tolerate a sudden drop to zero spring offset torque corresponding to the spring breaking during continuous operation of the actuator.

Having selected one nominal operating condition, three state variables are defined in terms of the deviations $(.)^d$ from this nominal condition; $x_1 = \theta - \theta^d$, $x_2 = \omega$, $x_3 = i_a - (i_a)^d$, $u = V_{mot} - (V_{mot})^d$. Note that the power stage is simply modelled by an ideal voltage source. The control requirement is to provide close tracking of a reference demand. A further state variable is therefore introduced:

$$x_r = \frac{1}{T_i}\int (\theta_{ref} - x_1)\, dt \tag{8.1}$$

Here θ_{ref} is the difference between the actual position demand and θ^d, and T_i is a design parameter. If, as in Section 4.4.2, an augmented state vector

$$\tilde{x} = \begin{bmatrix} x_r \\ x \end{bmatrix} \tag{8.2}$$

is introduced, then a state-space description is

$$\dot{\tilde{x}}(t) = \tilde{A}\tilde{x}(t) + \tilde{B}u(t) + \tilde{B}_w\theta_{ref} \tag{8.3}$$

with system matrices

$$\tilde{A} = \begin{bmatrix} 0 & -\frac{1}{T_i}e_1^{\mathrm{T}} \\ 0 & A \end{bmatrix} \qquad \tilde{B} = \begin{bmatrix} 0 \\ B \end{bmatrix} \qquad \tilde{B}_w = \begin{bmatrix} \frac{1}{T_i} \\ 0 \end{bmatrix} \tag{8.4}$$

where $e_1^{\mathrm{T}} = [1 \;\; 0 \;\; 0]$ and the matrix (A, B) is the state-space model of a DC motor given in equation (3.112). As observed in Section 3.6.4, the nonlinearities associated with the electrical dynamics and motor gain occur as matched uncertainty, i.e. in the channels implicit in the input distribution matrix, and therefore belong to the class of uncertainty to which any appropriately designed sliding mode control strategy is inherently insensitive. However, the disturbance $T_d(s)$, which is dominated by the friction torque, appears as unmatched uncertainty and its effect must be minimised by an appropriate choice of switching surface. Here a time invariant switching function of the form

$$s(t) = Sx(t) \tag{8.5}$$

will be employed where

$$S = [\; m^{\mathrm{T}} \quad 1 \;] \tag{8.6}$$

and the vector $m \in \mathbb{R}^3$. The performance specifications needed for the actuator translate to a requirement for a fast well-damped sliding mode dynamic behaviour which must be achieved for the range of possible torque disturbance effects. As shown in Lemma 4.1, since the triple (A, B, e_1^{T}) does not have any invariant zeros at the origin, the augmented pair $(\tilde{A}, \tilde{B})$ is completely controllable. Here a design was performed using the robust pole placement technique described in Section 4.2.1, although any preferred robust design philosophy could be used.

A control strategy to ensure that this dynamic performance is attained and maintained must ensure satisfaction of a reachability condition which ensures the sliding function is at least locally attractive. Subject to this constraint, many forms of control are possible. One appropriate choice is given by

$$u = -(S\tilde{B})^{-1}S\tilde{B}_w\theta_{ref} + k^{\mathrm{T}}|x|\,\mathrm{sgn}(s) \tag{8.7}$$

with

$$k_i < -|((S\tilde{B})^{-1}S\tilde{A})_i| \tag{8.8}$$

where $(.)_i$ indicates the ith element of a vector and $|x|$ represents the vector containing the moduli of the state vector components x_i. In practice, it was found that good performance could be obtained using a scalar upper bound for $k^T|x|$, which is very attractive from the implementation point of view. The possibly discontinuous sign function is replaced with a smooth boundary layer approximation as detailed in Section 3.7. It should be noted that the sliding mode control strategy (8.7) requires full-state feedback. However, it is not a practical proposition to measure the angular velocity of the rotor. Some form of estimation or observation is required for practical implementation of the controller. Investigation showed that straightforward numerical differentiation of measurements of output position produced adequate results.

A preliminary performance analysis was first carried out on the controller using a nonlinear simulation model of the actuator. This investigation considered the robustness of the controller to changes in the assumed stiction model and also the effects of noisy measurement devices on the performance of the controller. In addition, power stage nonlinearities were incorporated in the tests. The simulation results maintained the required specification and the controller proceeded to bench test assessment. Practical implementation was carried out on a microprocessor using integer arithmetic.

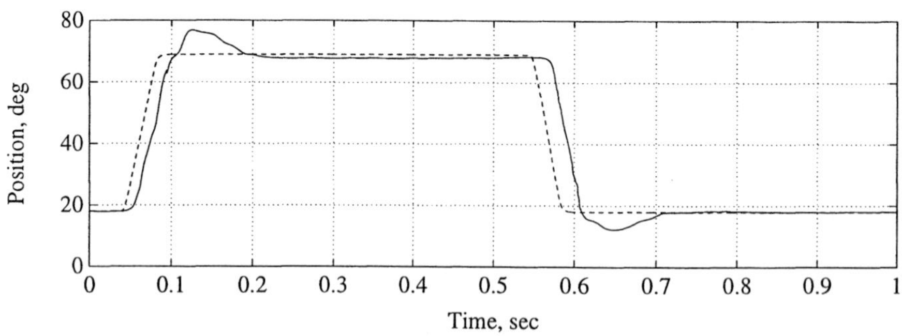

Figure 8.2: Response of the sliding mode controller to a large step demand

Figure 8.2 shows the closed-loop response of the sliding mode controller to a large step demand. For comparison, the response of a third-order linear controller with an integral element is shown in Figure 8.3.

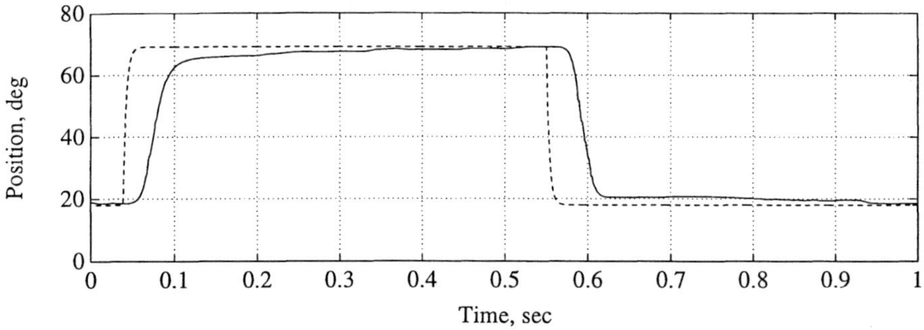

Figure 8.3: Response of the linear controller to a large step demand

This linear controller was designed using a large magnitude frequency response and consequently the requirement of ideally zero overshoot is achieved. Consider now the response of both schemes to a small step demand (Figures 8.4 and 8.5).

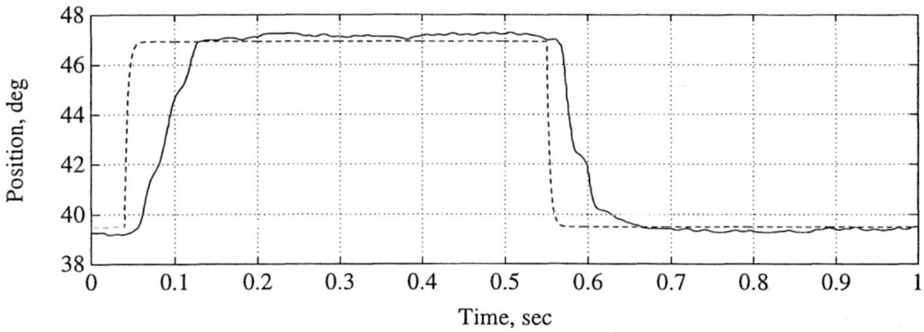

Figure 8.4: Response of the sliding mode controller to a small step demand

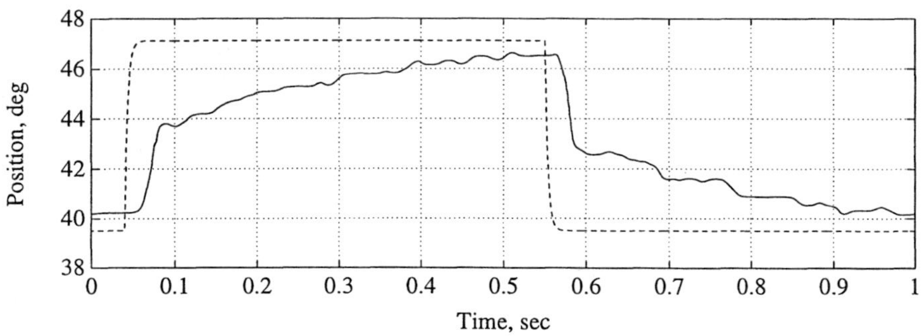

Figure 8.5: Response of the linear controller to a small step demand

The performance of the linear controller has deteriorated significantly and is well outside the desired performance specifications. The sliding mode scheme which was not tuned after implementation to any particular mode of operation provides superior performance that is very close to the required specifications. Figure 8.6 shows a typical sliding mode response to a slow ramp input. Stick–slip behaviour is evident, but stiction effects have not swamped the behaviour of the actuator.

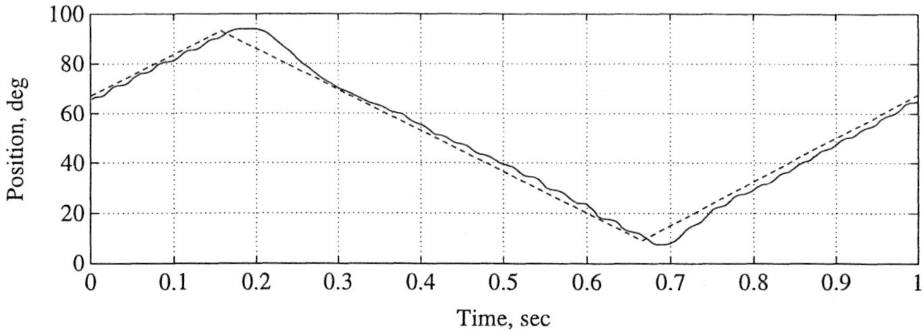

Figure 8.6: Sliding mode controller response to a slow ramp output trajectory

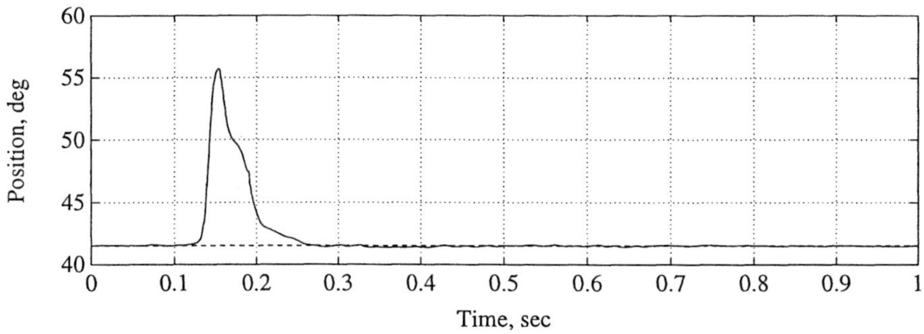

Figure 8.7: Sliding mode controller response to spring failure during operation

Figure 8.7 shows a sliding mode response corresponding to the spring being disengaged at approximately $t = 0.14$ second. After a small transient, the controller regains control with no loss in performance. This is particularly important since robustness to spring failure is an essential safety requirement. It should be noted that the linear controller shows some stick–slip behaviour on the slow ramp and takes approximately three times as long to recover from a broken spring.

8.3 ROBUST CONTROL OF AN AUTOMOTIVE ENGINE

When an automotive engine is idling and an accessory such as headlights, heater, air conditioner, etc, is turned on, an additional torsional load is placed on the engine. If there were no compensation arrangements, the engine speed would fall drastically and the engine would shut down. This effect is not observed when the vehicle is moving since the driver compensates for the additional load by using the throttle. Traditionally, in the automotive industry, this problem of idle speed regulation has been solved by increasing the airflow into the cylinder. This increase is different for different loads or accessories, so this feed-forward control is accomplished via look-up tables. These demand extensive time and effort to formulate. The alternative is to treat the on and off switching of the accessories as a disturbance to be rejected via feedback control.

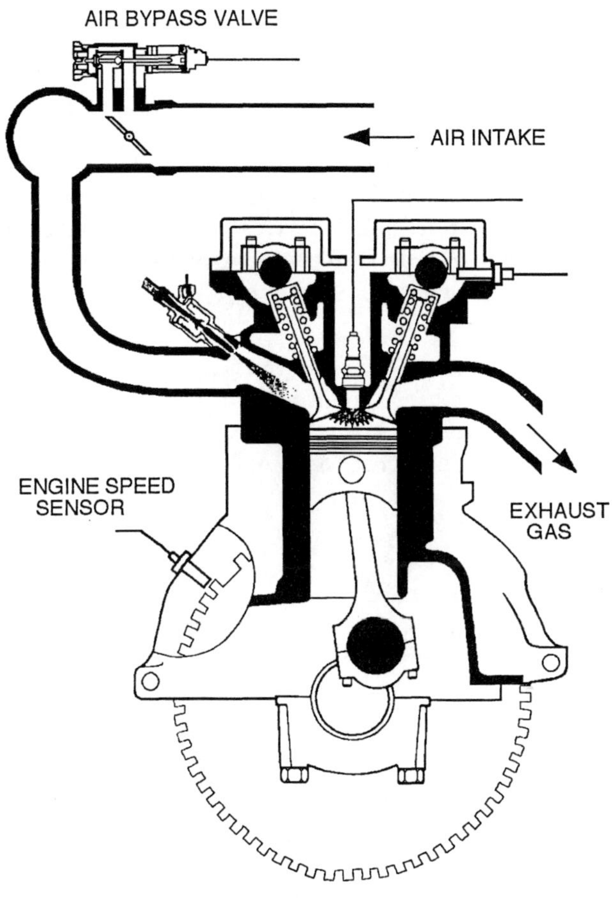

Figure 8.8: Schematic of an engine-block

In spark ignition engines the speed is regulated by controlling the amount of air consumed (Figure 8.8). This is done by throttling the air mass flow rate entering the cylinder through the inlet manifold. An additional valve is placed in parallel to the throttle to act as a bypass path for the incoming air; it is called the *air bypass valve* (ABV). The air intake is controlled using this valve during idling, since the throttle is not used. The input voltage is sent to the valve as a pulse width modulated signal (PWM) and the designed controller varies the duty cycle of the PWM signal.

In an engine cylinder the air–fuel mixture is ignited by a spark plug just before the piston reaches the top end of the cylinder. The position of the piston with respect to the top end of the cylinder affects the output torque and in turn the engine speed. Thus, the timing of the spark ignition or *spark advance* (SPA) can be used to regulate the idle speed. However, this input has only a transient effect. It is worth noting that SPA should not be used during steady-state conditions as it generates emissions. The limits on this input are largely dependent on the engine speed, but a rough measure on the limit is between 5 and 25°. The unit of this input is the angle of the crankshaft with respect to the vertical when the spark ignition takes place.

Obviously the output of the system being considered is the engine speed in rotations per minute or rpm. The speed of the engine is sensed twice during each revolution, which means that the sampling time is dependent on speed; the higher the speed, the shorter the time constant and vice versa. Thus the effective sampling time of the implemented controller is variable, resulting in time delays which alter the phase of the system. The controller should be robust enough to cope with these time delays.

The controller/observer scheme presented in Section 7.3 will be employed. A Ford 1.6 litre Zitec engine is used for the design study; it is a four-cylinder, four-stroke engine. Identified linear models from the two inputs (the ABV and the SPA) to the output (engine speed) were used for controller design. These linear identified models are minimum phase and the product CB is also nonsingular.

8.3.1 Controller Design Issues

The design issues which need to be considered can be summarised in the following steps:

1. The dynamics of the demand $r(t)$ are specified by the spectrum of Γ. The eigenvalues of Γ used in equation (7.18) should not be much faster than the open-loop plant spectrum to avoid unrealistically large controller effort.

2. The sliding surface dynamics are specified by the design matrix M from equation (7.22), which acts as a state feedback matrix for the null space dynamics. Previous knowledge of the plant can be used to define the required closed-loop characteristics; this information can be translated into a set of required eigenvalues for the sliding mode system and the eigenvalues can be assigned via M. This translation can be done using LQR or minimum entropy methods.

3. The scalar constants used in the nonlinear part of the controller and the observer are determined from the error bounds of the identified model. Later on, these constants can be tuned during the nonlinear simulation to improve performance.

Having discussed the general concepts of the controller design, particular attention will now be focused on the controller design issues pertaining to the automotive engine control problem.

8.3.2 Engine Controller Design

The following observations were made during engine controller design:

1. The engine inputs ABV and SPA used for speed control are severely restricted in magnitude. In addition, the engine process is a discrete time process with variable time interval, so it is more prone to input saturation if the controller dynamics are fast. Hence the choice of Γ was made realistically, so as not to saturate the inputs.

2. As discussed earlier, the hyperplane design procedure can be accomplished via traditional linear methods such as LQR or minimum entropy design. Both these methods rely on appropriate weighting of the plant states and plant inputs. Usually these methods cannot be used efficiently with an identified linear model, as the linear model states do not have any physical meaning. However, in the present sliding mode controller formulation, the first state of the augmented plant is the integral error state and it should be as small as possible. Hence this state is heavily weighted. In order to avoid input saturation the inputs are also grossly penalised. (This is particularly true for the SPA input because it has much less authority than the ABV input.) The hyperplane was designed using the minimum entropy method as it gives good disturbance rejection properties.

3. The linear part of the observer was designed by specifying the matrix A_{22}^o in equation (7.13). This observer specification should be as fast as possible. However, as the sample rate for implementation had to be the same as the sample rate of the controller, there was an upper limit on the speed of observer.

4. The nonlinear parts of the controller and observer are fully determined by specifying the scalars k_1, γ_o in equation (7.31), and γ_c in equation (7.29). These scalars were initially set to some small values and tuned on the nonlinear simulation model of the engine. Later on, they were tuned on the actual rig. The parameter γ_c in particular provides a very effective knob for improving the disturbance rejection properties. The function $\alpha(.)$ in equation (7.31) was taken to be zero for the sake of simplicity.

5. In Chapter 7 it is assumed that the plant is square (i.e. the number of inputs equals the number of outputs). However, in this ISC problem there are two inputs and one output. The square plant assumption is needed for observer design only. The first input channel, ABV, has most of the authority as compared to the SPA input channel, which is only available during transient motion, so the observer was designed for the first input only.

The controller was implemented on the test rig using dSPACE,[1] which is an efficient tool for rapid prototyping. It can convert SIMULINK block diagrams into optimised executable C code and then download it onto a microprocessor running as controller host. This greatly facilitates controller tuning during testing. The sliding mode controller/observer pair was first implemented in SIMULINK and subsequently the

[1] dSPACE is the registered trademark of dSPACE GmbH.

observer was translated into a MEX file[2] written in C. The controller structure in SIMULINK is shown in Figure 8.9.

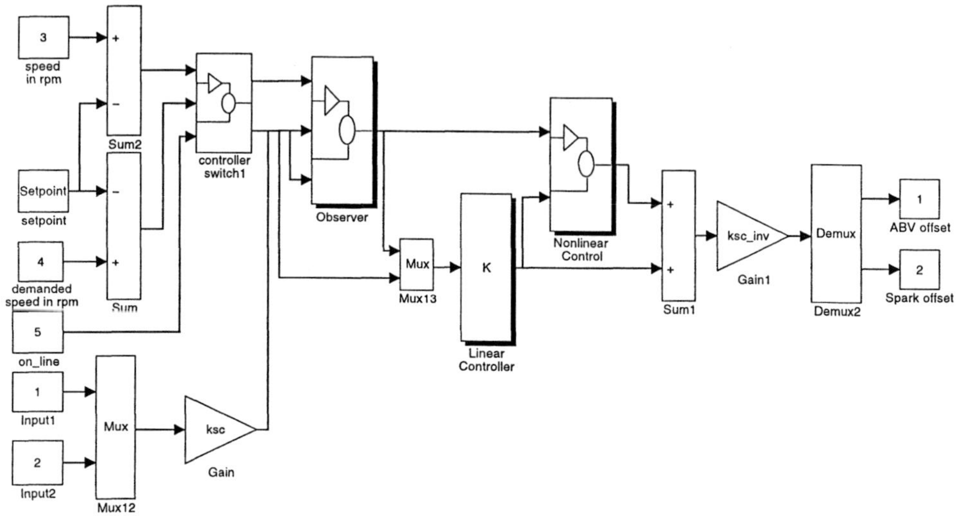

Figure 8.9: Simulink block diagram of controller structure

8.3.3 Implementation Results

As described previously, the controller should exhibit good tracking, robustness and disturbance rejection properties. The set-point speed was chosen to be 880 rpm and the tracking and disturbance rejection properties were verified by applying different loads to the engine and observing the controller performance. Figure 8.10 shows the closed-loop performance on application of the following sequence of loads: first the electric drive fan and headlights were turned on (point A); after the controller had returned the engine to the set-point, the heater and rear window defroster were turned on (point B). The two dips (points A and B) and flares (points C and D) represent the application and release of these loads respectively. From Figure 8.10 it is observed that the controller keeps the speed within the tolerable limits. For the dips, the controller returns the engine to set-point speed at a rate similar to that of the descent and the overshoots are less than 4%. In the case of flares, the magnitude of the speed does not exceed the 10% tolerable limit and the maximum dip is around 17%. This dip can be reduced, but at the cost of increased oscillations, high overshoots and lower robustness. The load presented in Figure 8.10 is a scaled version of measured load (obtained from a hot-wire anemometer located in the air intake).

---

[2]MEX file is the registered trademark of Mathworks, Inc.

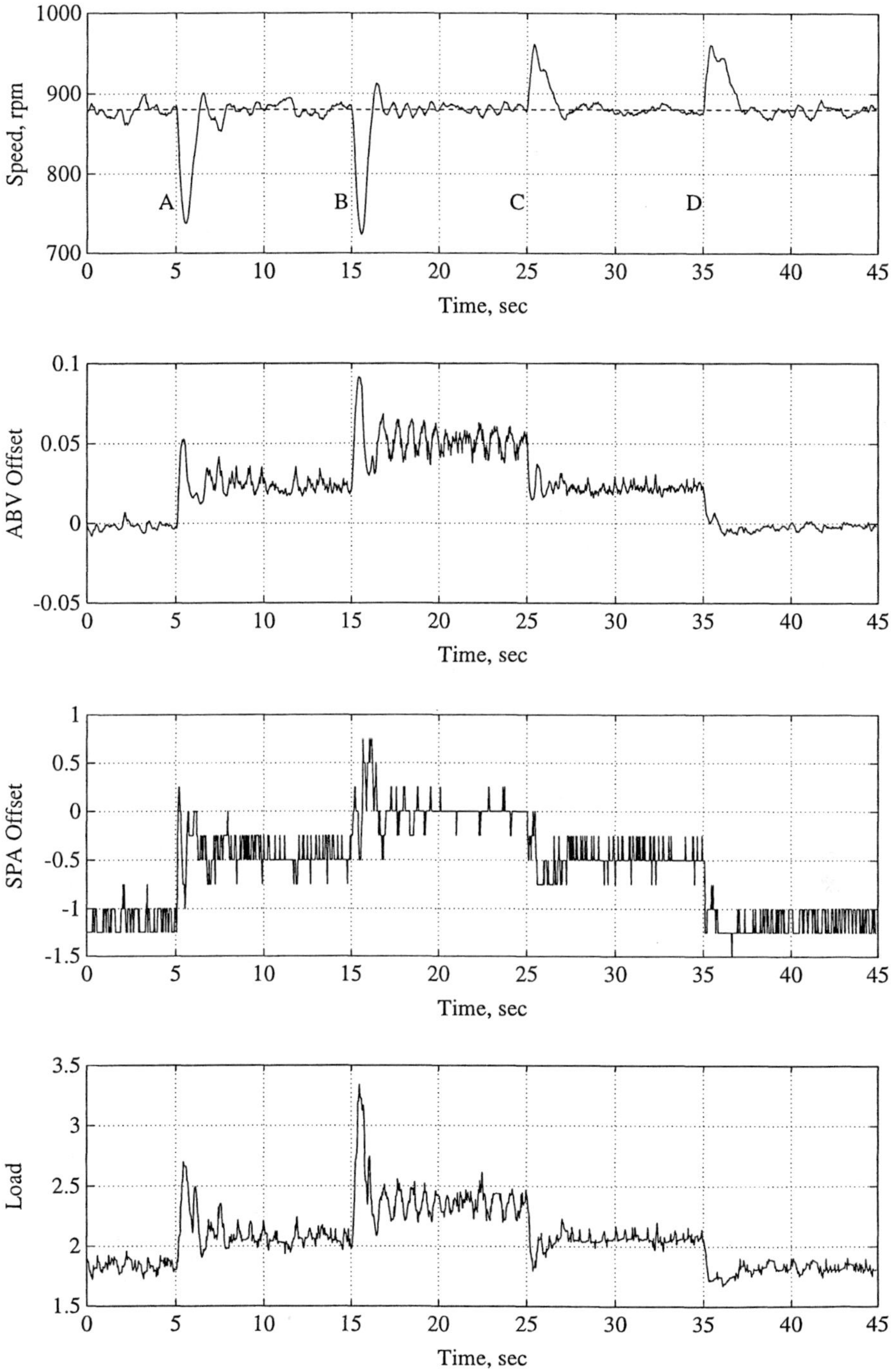

Figure 8.10: Rig results for test 1

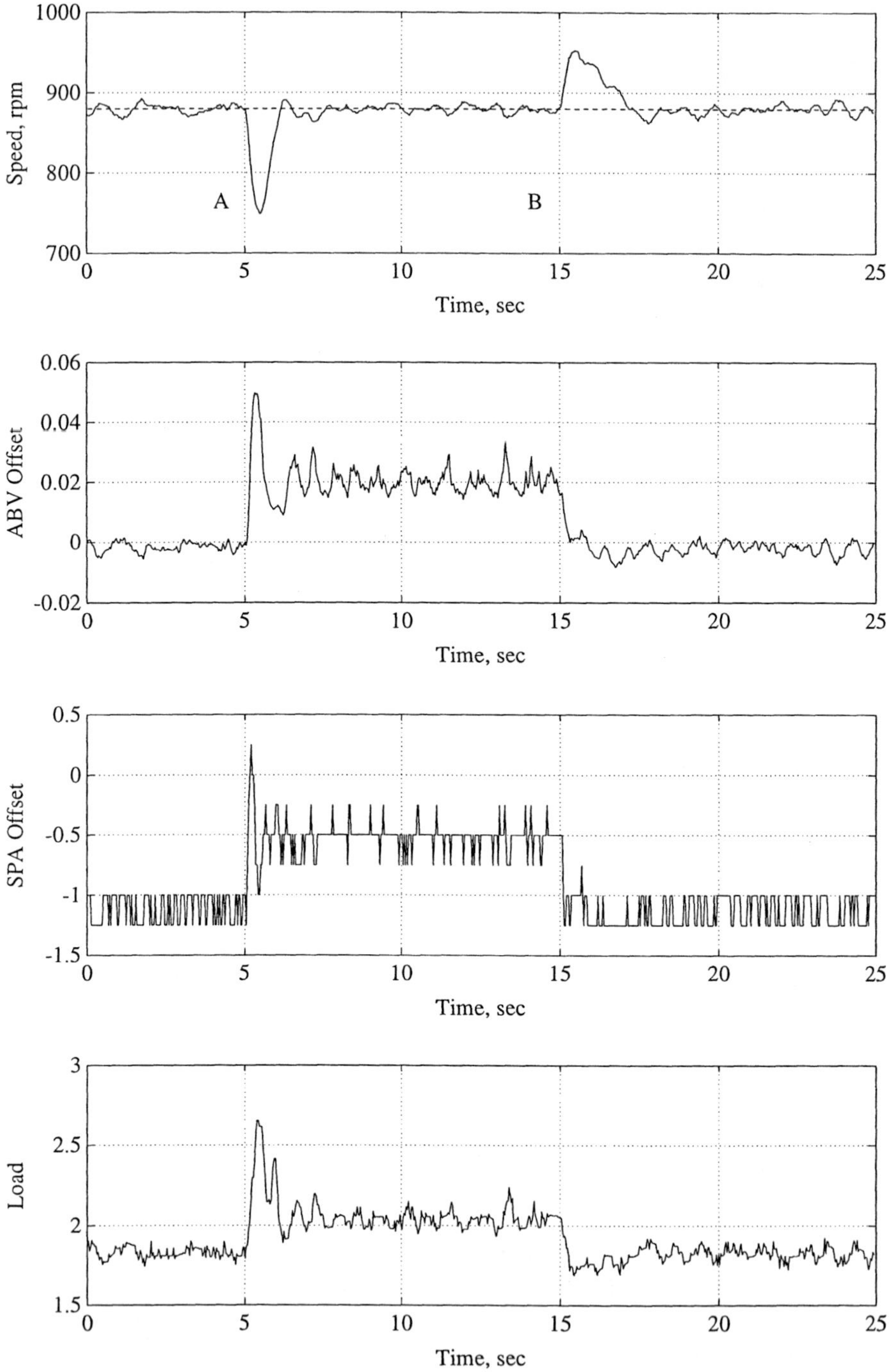

Figure 8.11: Rig results for test 2

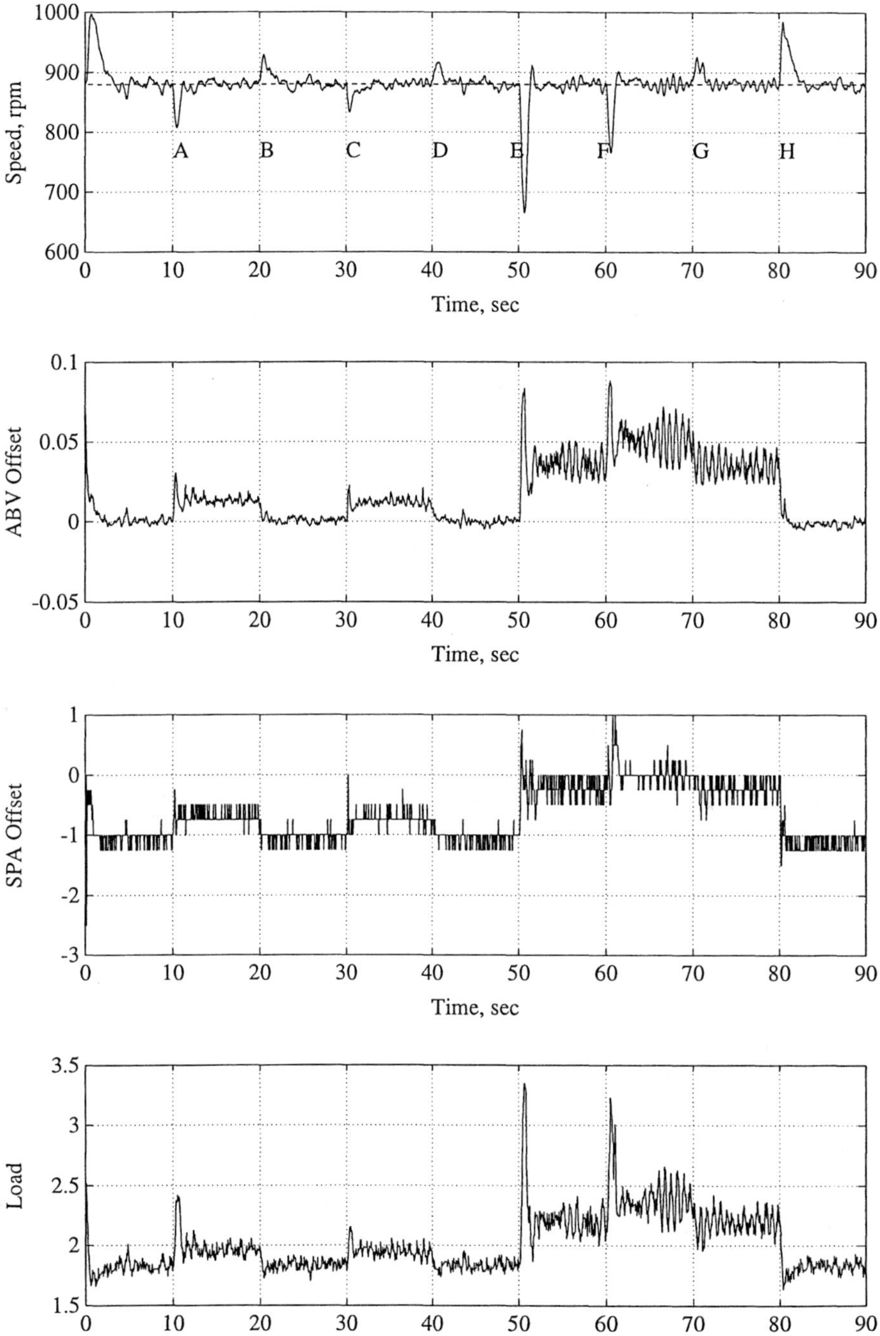

Figure 8.12: Rig results for test 3

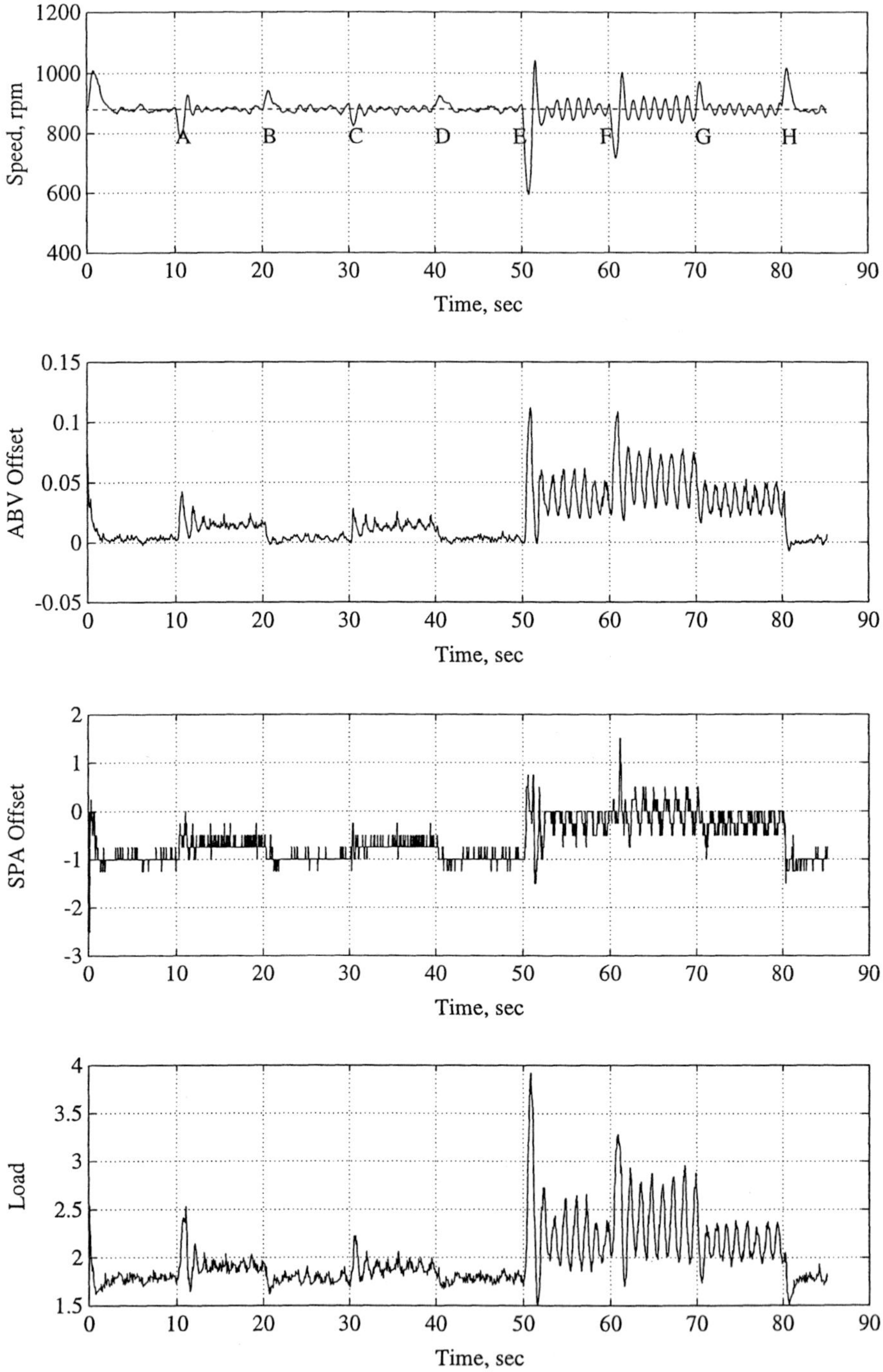

Figure 8.13: Rig results for five sample delays

The next test involves the simultaneous application and release of the air conditioning fan and rear window defroster (Figure 8.11). The dip (point A) corresponds to the application of the accessory load, and the flare (point B) is the consequence of turning the accessory off. The flare has a magnitude of 8% and the dip is around 15%. It should be noted that the controller is rejecting these flares and disturbances within the maximum time limit of 3 seconds. Hence the controller is exhibiting good tracking and disturbance rejection properties. In both these tests, the controller effort is quite low, guaranteeing that the input valve saturations are avoided. The second input channel, SPA, seems to exhibit some chattering, but this is the consequence of noise generated during the combustion process.

The robustness of the controller is demonstrated by the fact that the controller design was performed at a set-point speed of 720 rpm whereas the tests were performed at a speed of 880 rpm.

A further disturbance rejection test was performed with different gains and time delays included. The controller easily stabilised these load tests with a delay of five sample intervals and the gain increased by a factor of 1.5. Figure 8.12 shows a performance test for various loads and accessories, applied and withdrawn at the marked points A to H. The same test was performed with a delay of 5 sample periods on the measured output (Figure 8.13). By comparing Figures 8.12 and 8.13 it is seen that, despite the heavy disturbance due to the delay, the controller is still maintaining reasonable performance.

8.4 SUMMARY

This chapter has considered the application of sliding mode control techniques to automotive problems. The first case study has demonstrated that sliding mode control is an appropriate design technique for nonlinear systems which include significant unmatched uncertainty as well as the usual matched components. The design approach is systematic, involving classification of matched and unmatched uncertainty and switching surface selection to minimise unmatched effects. The methodology provides good performance across a wider range of operation than may be possible with a classical design. A classical design can be tuned well at one operating point but the performance can degrade severely away from this condition. Bench test results support these assertions. The second case study considers the application of a sliding mode controller/observer pair to the problem of idle speed control. The controller shows good performance and robustness, hence demonstrating that the previously described theoretical developments are suitable for practical application. Since the controller/observer pair is designed in a multivariable framework, and assumes that only a subset of state information is measurable, it is applicable to other automotive control problems such as air-to-fuel ratio control.

8.5 NOTES AND REFERENCES

Further details pertaining to the idle speed control problem and the controller design are given in the thesis of Bhatti (1998). This problem has been addressed in the past by various other controller design techniques. Puskorius & Feldkamp (1993,1994) used recurrent neural networks to design a controller for the ISC problem. They used throttle demand and spark advance as control inputs, demonstrating good simulation results. Shim *et al.* (1995) derived a nonlinear automotive engine model and used it for removing engine speed ripple fluctuations. The l_1 optimisation is used in Butts *et al.* (1995) for feedback and feed-forward controller design. Fuzzy logic (Vachtsevanos *et al.*, 1992) and $\mathcal{H}_\infty$ (Williams *et al.*, 1989) have also been used for this purpose.

The details of the linear identification methods used in the second case study can be found in Åström & Wittenmark (1989) and Soderstrom & Stoica (1989).

Boyd & Barratt (1991) give a detailed treatment of the LQR and minimum entropy methods which were used to determine the desired sliding mode transient performance.

Chapter 9

Furnace Control Case Study

9.1 INTRODUCTION

This chapter considers the development of a controller for a high temperature furnace of the type shown in Figure 9.1.

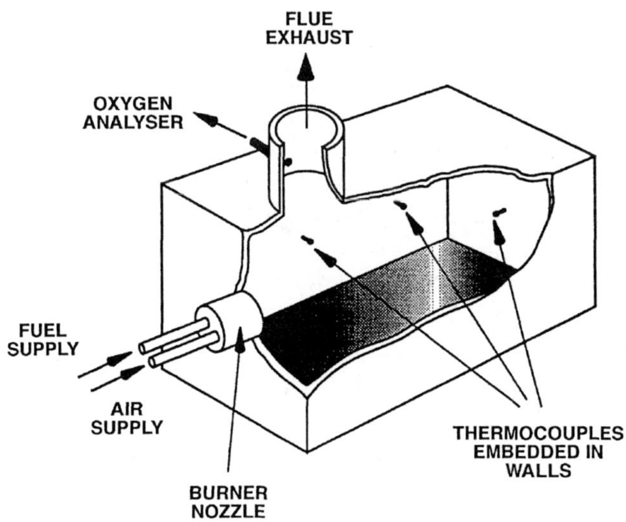

Figure 9.1: High temperature gas-fired furnace

The furnace can be thought of as a gas-filled enclosure bounded by insulating surfaces and containing a heat sink. The heat input is achieved by a burner located in one of the end walls, and the combustion products are evacuated via a flue in the roof. This is perhaps the simplest design possible – a single burner and a single flue – but such a plant could legitimately represent an industrial kiln for the firing of pottery or ceramics. The burner is fed by a fuel and air supply whose flow rates can be independently modulated via motorised butterfly valves present in the feed lines.

Until recently control of temperature was the primary concern. However, in view of future environmental legislation with regard to harmful emissions resulting from combustion, attention is now being focused on the problem of combustion efficiency and its effect on the formation of noxious by-products (Goodhart, 1994). One way of monitoring combustion efficiency is to measure the concentration of excess oxygen present in the flue gases. Oxygen analysers, adapted from automotive technology, situated in the exhaust flue are able to provide on-line measurements of oxygen concentration for use by the control system. Currently, control of both temperature and excess oxygen is achieved by independent single-input single-output control loops. This is clearly not necessarily the optimal solution. This chapter describes the development of a nonlinear multivariable control scheme from design through to implementation for a test facility at the Gas Research Centre at Loughborough, UK.

From a control systems perspective, the outputs of the 'system' are the furnace temperature (as measured by the thermocouple) and the percentage of oxygen present in the combustion products (as measured by the oxygen analyser in the flue). In an industrial situation, the internal furnace temperature would be required to exhibit a specific time/temperature profile comprising, say, a period of low fire, a ramp to a higher temperature, a period of soak and finally a return to ambient temperature. During normal furnace operation, efficient fuel combustion is desirable. For a given mass of fuel, a theoretical mass of oxygen is required to completely oxidise the hydrocarbons – so-called *stoichiometric combustion*. An inadequate air supply will result in incomplete combustion with a corresponding loss in thermal energy release. Conversely, excess air will guarantee complete combustion but will give rise to unnecessary enthalpy losses through the flue due to the increased flue flow rate. Therefore, efficient combustion is ensured by minimising the amount of excess oxygen present in the combustion products in the flue. In addition to the potential energy savings, it is argued by Disdell *et al.* (1994) that accurate control of both temperature and excess oxygen is of paramount importance from the viewpoint of reducing pollutant emissions.

For safety reasons, such furnaces usually operate at a fixed fuel/air ratio as a result of an electronic ratio controller (ERC). The fuel and air flows are measured using heated thermistors placed in bypass lines around orifice plates in the flow paths of the air and gas. The relative flows are compared and fedback to the ERC, which makes appropriate adjustments to the air valve position. The device also provides an additional safety feature in the form of a 'shutdown' alarm which isolates the fuel supply when the furnace persistently operates 'off ratio', i.e. away from the required fuel/air ratio set-point. This framework is usually described as 'gas led' since the airflow is modulated as a result of changes in the fuel flow to maintain the appropriate fuel/air ratio necessary for efficient combustion. An additional input to the ERC is the 'trim signal'. This allows the fuel/air ratio set-point to be dynamically adjusted. In this way, as far as this chapter is concerned, the control inputs to the system are the fuel valve positioner signal and the ERC trim signal. These inputs will be manipulated to ensure the temperature and excess oxygen levels follow specified reference profiles. This multivariable approach is distinct from current practices which control the excess oxygen and temperature via independent single-input single-output control loops. The proposed control scheme is illustrated in Figure 9.2.

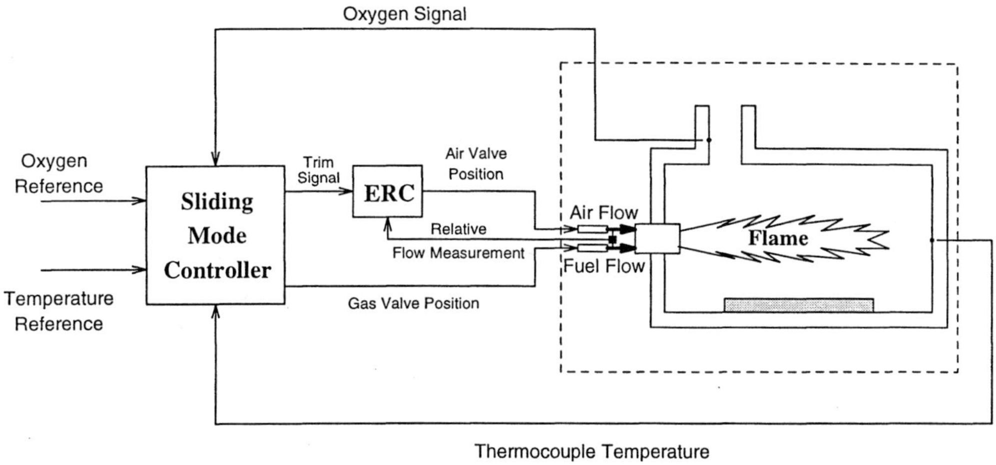

Figure 9.2: Schematic of proposed multivariable control scheme

Physical modelling of such a system, by the zone method[1] for instance, is clearly nontrivial and does not lead to a system of equations conducive to controller synthesis; instead, an identification approach has been adopted. In order to obtain a linear model around which to base the controller designs, identification of the real plant was chosen as the most viable option. A pseudo-random binary signal (PRBS), taking the discrete values 0 and 1.5 volts, was applied to the valve positioner signal in the fuel line. Unfortunately, such a signal could not be used to excite the oxygen trim channel. For reasons of safety, the burner is only permitted to operate 'off ratio' for a short period of time, after which a relay 'alarm' shuts down the flow of fuel to the burner. Using a PRBS signal in the oxygen trim channel with its large, sudden amplitude changes would undoubtedly provoke a shutdown, hence it is not suitable for identification purposes. Instead, an excitation signal comprising the sum of several sinusoids of different frequency and amplitude was used. Such a signal is not ideal for identification purposes but represented the best that could be done under the circumstances.

A low-order, relative degree 1 minimum phase model, which provides good agreement with the input/output data, is given in regular form by

$$A = \begin{bmatrix} -0.0186 & -0.0065 & 0.0190 & 0.0129 \\ 0.0026 & -0.1354 & 0.0310 & 0.0040 \\ -0.0972 & 0.0695 & -0.1273 & 0.0530 \\ -0.0193 & -0.0155 & -0.1121 & -0.4934 \end{bmatrix} \quad B = \begin{bmatrix} 0 & 0 \\ 0 & 0 \\ 0 & -0.0960 \\ 0.4969 & 0.0453 \end{bmatrix}$$

$$C = \begin{bmatrix} 0.6707 & -0.1085 & -0.0286 & 0.0086 \\ -0.2750 & -0.1933 & -0.2175 & 0.0060 \end{bmatrix} \tag{9.1}$$

This system is stable and has invariant zeros at $\{-0.1633, \ -1.2059\}$. The first input channel represents the fuel valve position signal and the second represents the oxygen trim signal. The first output is the furnace temperature as measured by the thermocouple; and the second output is the signal from the zirconia probe

[1]See Section 7.4 for further details and references.

which monitors the percentage of oxygen in the flue. It should be remembered that, because of the limitations imposed on the identification signal, this model may exhibit significant mismatches with the true behaviour of the system.

The following section describes the design procedures adopted to synthesise a controller/observer pair. The design of an observer is discussed first.

9.2 OBSERVER DESIGN

Since the system is square, the reduced-order dynamics of the estimation error are completely determined by the invariant zeros of the system. Following the design procedure in Section 7.3.1, by choosing *a priori* the stable matrix A_{22}^s from (7.13) to be diagonal, any diagonal positive definite matrix P_2 will always be a Lyapunov matrix for A_{22}^s. The diagonal matrix P_2 can then be considered independently as a design matrix. Here it has been chosen in an effort to ensure the diagonal entries of the matrix

$$F = (P_2 C_2 B_2)^{\mathrm{T}}$$

from equation (7.14) are of compatible order. It was reasoned that the diagonal elements of F act as weighting parameters which govern the distribution of the nonlinear action between the individual input channels. By choosing

$$P_2 = \begin{bmatrix} 10 & 0 \\ 0 & 2 \end{bmatrix}$$

it can be shown that

$$F = \begin{bmatrix} 0.0426 & 0.0060 \\ 0.0313 & 0.0423 \end{bmatrix}$$

The diagonal elements of A_{22}^s were chosen to make the condition number of the observer closed-loop matrix $A_0 = A - GC$ small. By inspection, it was found that choosing

$$A_{22}^s = \begin{bmatrix} -0.5 & 0 \\ 0 & -0.5 \end{bmatrix}$$

fulfils this requirement, giving a condition number for A_0 of approximately 38.26. From equation (7.13) the linear gain matrix is given by

$$G = \begin{bmatrix} 1.7201 & -0.3132 \\ 0.6269 & -0.2251 \\ 8.1321 & -2.7820 \\ 0.4512 & 0.4562 \end{bmatrix}$$

As in the single-input single-output design in Section 7.4, the nonlinear gain function will need to be determined empirically.

9.3 CONTROLLER DESIGN

Forming the augmented plant from (4.102) it follows that in the notation of Section 4.4.2

$$\left[\begin{array}{cc} \tilde{A}_{11} & \tilde{A}_{12} \\ \tilde{A}_{21} & A_{22} \end{array}\right] = \left[\begin{array}{cccc|cc} 0 & 0 & -0.6707 & 0.1085 & 0.0286 & -0.0086 \\ 0 & 0 & 0.2750 & 0.1933 & 0.2175 & -0.0060 \\ 0 & 0 & -0.0186 & -0.0065 & 0.0190 & 0.0129 \\ 0 & 0 & 0.0026 & -0.1354 & 0.0310 & 0.0040 \\ \hline 0 & 0 & -0.0972 & 0.0695 & -0.1273 & 0.0530 \\ 0 & 0 & -0.0193 & -0.0155 & -0.1121 & -0.4934 \end{array}\right]$$

Early attempts to select the poles of the reduced-order sliding motion by inspection resulted in controllers which exhibited high levels of control activity. To circumvent this, a preliminary linear LQR design was made for the augmented system, with the cost function biased towards penalising the use of control effort. The four slowest poles of the resulting closed-loop matrix were used as initial values for the poles of the reduced-order motion. These were subsequently manually adjusted in an effort to improve the conditioning of the closed-loop matrix. Ultimately the sliding mode eigenvalues were taken to be $\{-0.0849, -0.0118, -0.0357 \pm 0.0220j\}$. The matrix M, from equation (4.106), which acts as a feedback matrix for the pair $(\tilde{A}_{11}, \tilde{A}_{12})$, was obtained using the robust eigenstructure assignment method described in Section 4.2.1. The resulting matrix which defines the hyperplane is given by

$$M = \left[\begin{array}{cccc} 0.0389 & 0.0414 & -1.5359 & -3.0467 \\ -0.1659 & -0.0407 & 8.4771 & 4.2487 \end{array}\right]$$

The design parameter Λ was chosen as

$$\Lambda = \left[\begin{array}{cc} 0.1 & 0 \\ 0 & 0.1 \end{array}\right]$$

which completes the hyperplane design. If the stable design matrix Φ, which assigns the poles of the range space dynamics, is block diagonal, a diagonal solution for $\bar{P}_2$ can be attained from the modified Lyapunov equation (7.28) by a suitable choice of $\hat{Q}_2$. A diagonal structure for $\bar{P}_2$ is preferable since this 'decouples' the nonlinear components of the control action. The eigenvalues for the range space dynamics were take to be the unused poles from the initial LQR design, namely $\{-0.1800 - 0.2509\}$. The diagonal elements of Φ were arranged so that the slowest pole was associated with the oxygen trim channel. Consciously, every effort was made not to apply an aggressive control signal into this channel to guard against violating the fuel/air ratio error shutdown alarm. The linear components of the control law from equations (7.24) to (7.26) can be shown to be

$$\begin{aligned} L &= \left[\begin{array}{cccccc} 0.0745 & 0.0107 & -3.6948 & -0.5455 & -0.3455 & 0.2117 \\ 0.0729 & 0.0777 & -3.8278 & -0.4594 & -0.6343 & 0.2124 \end{array}\right] \\ L_r &= \left[\begin{array}{cc} 6.9795 & -0.7794 \\ 5.9178 & 2.2003 \end{array}\right] \\ L_{\dot{r}} &= \left[\begin{array}{cc} 26.6345 & -3.2763 \\ 30.6273 & 9.8265 \end{array}\right] \end{aligned}$$

The sliding surface design parameter S_r was chosen so that, in the nominal case, the steady-state values of the integrator states are zero using the procedure described in Section 7.3.4. In this particular case study

$$S = \begin{bmatrix} -0.0297 & -0.0043 & 1.5601 & 0.5655 & 0.0950 & 0.2012 \\ -0.0405 & -0.0432 & 1.5996 & 3.1732 & -1.0415 & 0.0000 \end{bmatrix}$$

$$S_r = \begin{bmatrix} 2.6635 & -0.3276 \\ 3.0627 & 0.9827 \end{bmatrix}$$

The design matrix Γ from equation (7.18) has been chosen to tailor the step response of the closed-loop system in the nominal case. The stable design matrix has been chosen to be diagonal, since it makes no sense to introduce coupling between the reference signals. The multivariable equation (7.18) can be represented as the pair of scalar equations

$$\dot{r}_i(t) = \Gamma_{ii}\left(r_i(t) - R_i(t - 1/\Gamma_{ii}\right) \qquad i = 1, 2 \tag{9.2}$$

where Γ_{11} and Γ_{22} are the diagonal elements of Γ. If $\ddot{R}_i(t) = 0$ on some interval then $r_i(t) \to R_i(t)$ asymptotically (for further details see Section 7.4.4). In this particular design $\Gamma_{11} = -0.02$ and $\Gamma_{22} = -0.05$. This reflects the different speeds of response required in the temperature and oxygen channels respectively. Because of the identification procedure adopted, no information was available to compute the nonlinear gain functions. Furthermore, no linear model was available to test the design before implementation. Instead, a gain function of the form

$$\rho_c(t, y) = r_1\|u_L(\cdot)\| + r_2\|y\| + r_3 \tag{9.3}$$

was used and the scalars were chosen experimentally during the trials at the Gas Research Centre, Loughborough.

9.4 IMPLEMENTATION RESULTS

The control scheme was implemented on a portable PC containing appropriate interface cards to perform the required analogue $\leftrightarrow$ digital conversions. The plant output signals were sampled and the control outputs updated 10 times a second, which represents a high sampling rate compared with the dominant time constant of the process. However, this is usual with high temperature gas-fed plant for reasons of monitoring and safety. The temperature demand profile shown in Figure 9.3 was used, representative of what is required by typical industrial processes. It can be seen that the condition $\ddot{R}_i(t) = 0$ is satisfied almost everywhere since the reference profile is piecewise linear. It is observed by Goodhart (1994) that the excess oxygen reference signal cannot be chosen without regard to the temperature set-points. Care has therefore been taken to provide an excess oxygen profile that is both realistic and attainable.

The scalars comprising the nonlinear gain functions were chosen initially very conservatively under the restriction that the gain $\rho_c(\cdot)$ be bounded by unity. The rationale for this conservatism was that, generally speaking, increasing the nonlinear control component increases the 'aggressiveness' of the control signal, which

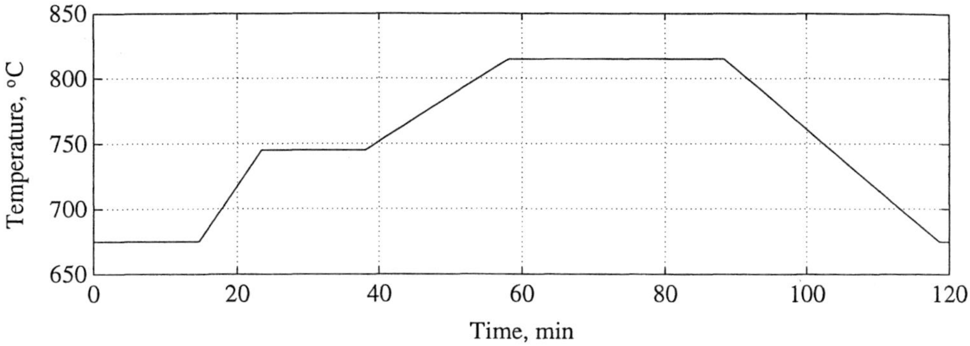

Figure 9.3: Typical temperature reference signal

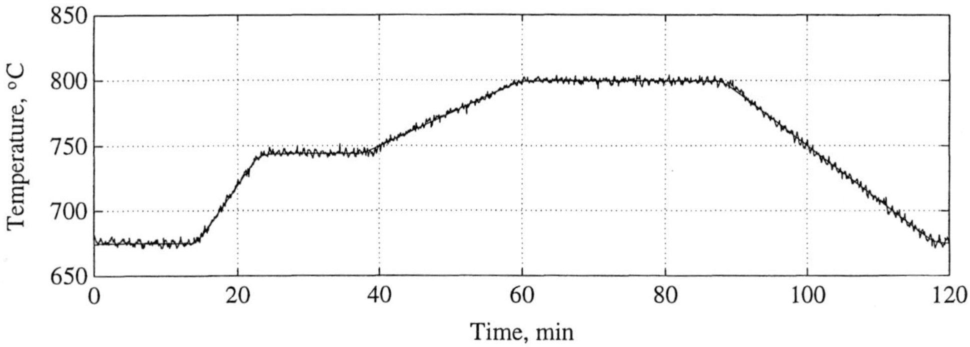

Figure 9.4: Furnace temperature compared to the reference

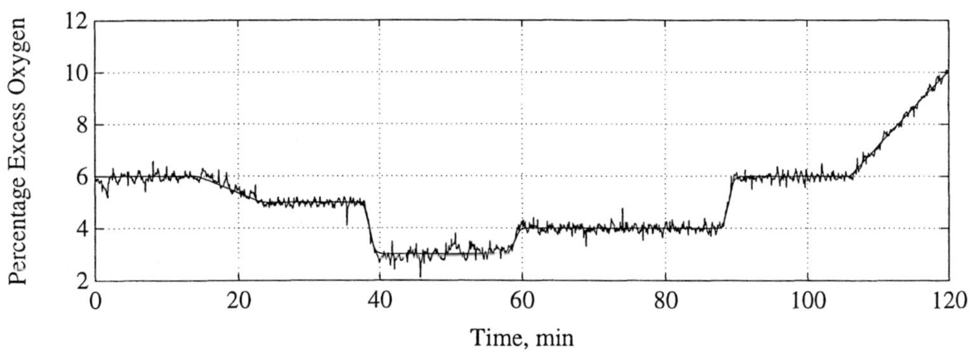

Figure 9.5: Excess oxygen in the flue compared to the reference

was considered to be undesirable in the oxygen trim channel because of the safety shutdown mechanism which exists in this input channel.

Figures 9.4 and 9.5 present results obtained from using the scalar gains

$$\rho_o(u_L, y) = 0.2\ ||u_L(\cdot)|| + 0.2\ ||y|| + 0.25 \tag{9.4}$$

and

$$\rho_c(u_L, y) = 0.1\rho_o(u_L, y) + 0.05 \tag{9.5}$$

Tracking accuracy of approximately 2° was obtained, equivalent to the precision of the measuring device.

9.5 SUMMARY

This chapter has considered the problem of controlling both the temperature and the oxygen content in the combustion products. The oxygen trim signal to the ERC, which alters the fuel/air ratio set-point, has been used as an additional input. The theoretical results of Chapter 7 were once again successfully used to design a controller/observer pair. Because a model was not available to test the controller before implementation, the design scalars that define the nonlinear gain functions were selected *in situ* during the trials.

9.6 NOTES AND REFERENCES

The electronic ratio controller (ERC) and its operation are discussed in Hammond (1992). A comparison of different control methodologies which have been applied to this type of high temperature furnace is given in Goodhart (1994).

Appendix A

Mathematical Preliminaries

A.1 MATHEMATICAL NOTATION

$\mathbb{N}$	the natural numbers
$\mathbb{R}$	the field of real numbers
$\mathbb{C}$	the field of complex numbers
$\text{Re}[z]$	the real part of the complex number z
$\text{Im}[z]$	the imaginary part of the complex number z
z^*	the complex conjugate
$\mathbb{R}_+$	the set of strictly positive real numbers
$\mathbb{C}_-$	the open left half of the complex plane, i.e. $\{z \in \mathbb{C} : \text{Re}[z] < 0\}$
$\mathbb{R}^{n\times m}$	the set of real matrices with n rows and m columns
$\|a\|$	the absolute value of the real number a
$\text{sgn}(\cdot)$	the signum function
A^{T}	the transpose of the matrix A
A^{H}	the complex conjugate transpose of the matrix A
$\det(A)$	the determinant of the square matrix A
A^{-1}	the inverse of the square matrix A
$A^{\dagger}$	the (left) pseudo-inverse of the matrix A
$\text{rank}(A)$	the rank of the matrix A
$\lambda(A)$	the spectrum of the square matrix A, i.e. the set of eigenvalues
$\lambda_{max}(A)$	the largest eigenvalue of the square matrix A
$\lambda_{min}(A)$	the smallest eigenvalue of the square matrix A
$\kappa(A)$	the spectral condition number
$\mathcal{R}(A)$	the range space of the matrix A (viewed as a linear operator)
$\mathcal{N}(A)$	the null space of the matrix A (viewed as a linear operator)
I_n	the $n \times n$ identity matrix
$\underline{\sigma}(A)$	the minimum singular value of A
$\bar{\sigma}(A)$	the maximum singular value of A
$A > 0$	implies the square matrix A is symmetric positive definite
$A > B$	implies the square matrix $A - B$ is symmetric positive definite
$\|\cdot\|$	the Euclidian norm for vectors and the spectral norm for matrices
$\dot{y}$	the derivative of y with respect to time
$\ddot{y}$	the second derivative of y with respect to time
$\equiv$	equivalent to
$\triangleq$	equal to by definition
$\times$	Cartesian product
$\perp$	orthogonal complement
$\langle\, ,\, \rangle$	inner product
$\oplus$	direct sum

A.2 LINEAR ALGEBRA

A.2.1 Vector Spaces and Linear Maps

In general, the definition of a vector space involves the notion of an *arbitrary* field whose elements are called scalars.

Definition A.1 *Let K be a given field and let V be a nonempty set with rules of addition and scalar multiplication which assigns to any $u, v \in V$ a sum $u + v \in V$ and to any $u \in V$, $k \in K$ a product $ku \in V$. Then V is called a vector space over K, and the elements of V are called vectors, if the following axioms hold:*

1. *For any vectors $u, v, w \in V$, $(u + v) + w = u + (v + w)$.*
2. *There is a vector in V, denoted by 0 and called the zero vector, for which $u + 0 = u$ for any vector $u \in V$.*
3. *For each vector $u \in V$ there is a vector in V, denoted by $-u$, for which $u + (-u) = 0$.*
4. *For any vectors $u, v \in V$, $u + v = v + u$.*
5. *For any scalar $k \in K$ and any vectors $u, v \in V$, $k(u + v) = ku + kv$.*
6. *For any scalars $a, b \in K$ and any vector $u \in V$, $(a + b)u = au + bu$.*
7. *For any scalars $a, b \in K$ and any vector $u \in V$, $(ab)u = a(bu)$.*
8. *For the unit scalar, $1 \in K$, $1u = u$ for any vector $u \in V$.*

In this context the particular field K will be either the field of real numbers $\mathbb{R}$ or the field of complex numbers $\mathbb{C}$.

Having defined the vector space, it is now possible to develop the notion of a vector subspace.

Definition A.2 *Let W be a subset of a vector space over a field K. W is called a subspace of V if W itself is a vector space over K with respect to the operations of vector addition and scalar multiplication on V.*

The fact that W is a subspace of V is written $W \subseteq V$.

Definition A.3 *Let V be a vector space over a field K. The vectors $v_1, v_2, \ldots, v_m \in V$ are said to be linearly dependent if there exist scalars $a_1, a_2, \ldots, a_m \in K$, not all of them zero, such that*

$$a_1 v_1 + a_2 v_2 + \ldots + a_m v_m = 0 \tag{A.1}$$

Otherwise the vectors are said to be linearly independent.

It should be noted that equation (A.1) will always hold if $a_1 = a_2 = \ldots = a_m = 0$. If the relation holds only in this case then the vectors are linearly independent.

Let V be a vector space over a field K and let $v_1, v_2, \ldots, v_m \in V$. Any vector in V

of the form

$$a_1v_1 + a_2v_2 + \ldots + a_mv_m \tag{A.2}$$

where the $a_i \in K$ is called a *linear combination* of $v_1, v_2, \ldots, v_m$. The following theorem applies.

Theorem A.1 *Let S be a nonempty subset of V. The set of all linear combinations of the vectors in S, denoted by $\mathcal{L}(S)$, is a subset of V containing S. Furthermore, if W is any other subspace of V containing S, then $\mathcal{L}(S) \subset W$.*

The above theorem states that $\mathcal{L}(S)$ is the smallest subspace of V containing S; it is called the subspace *spanned* by S.

Definition A.4 *A vector space V is said to be of finite dimension n or to be n-dimensional, written $\dim V = n$, if there exist linearly independent vectors $e_1, e_2, \ldots, e_n$ which span V. The set $\{e_1, \ldots, e_n\}$ is then called a basis for V.*

Let A and B be arbitrary sets. Suppose to each $a \in A$ there is assigned a unique element of B; the collection f of such assignments is called a *function* or *mapping* from A into B and is written

$$f : A \rightarrow B \tag{A.3}$$

Let K be an appropriate field. A rectangular array of the form

$$\begin{bmatrix} a_{11} & a_{12} & \ldots & a_{1n} \\ a_{21} & a_{22} & \ldots & a_{2n} \\ \vdots & & & \vdots \\ a_{m1} & a_{m2} & \ldots & a_{mn} \end{bmatrix} \tag{A.4}$$

where the a_{ij} are scalars in K defines a *matrix*. Every such $m \times n$ matrix A determines the mapping $T : K^n \rightarrow K^m$ defined by

$$v \rightarrow Av \tag{A.5}$$

where the product of the matrix and vector on the right-hand side is defined componentwise as

$$(Av)_i = \sum_{j=1}^{n} a_{ij}v_j \tag{A.6}$$

Let V and U be vector spaces over the same field K. A mapping $F : V \rightarrow U$ is called a *linear mapping* if it satisfies the following two conditions:

1. For any $v, w \in V$, $F(v + w) = F(v) + F(w)$.

2. For any $k \in K$ and any $v \in V$, $F(kv) = kF(v)$.

It follows that $F : V \rightarrow U$ is linear if it preserves the two basic operations of a vector space, i.e. vector addition and scalar multiplication. The matrix A in equation (A.4) may be viewed as such a linear mapping.

A.2.2 Properties of Linear Maps (Matrices)

Let T be a linear operator on a vector space V over a field K and suppose $\{e_1, \ldots, e_n\}$ is a basis for V. It follows that $T(e_1), \ldots, T(e_n)$ are vectors in V and so each is a linear combination of the elements of the basis $\{e_i\}$:

$$\begin{aligned} T(e_1) &= a_{11}e_1 + a_{12}e_2 + \ldots + a_{1n}e_n \\ T(e_2) &= a_{21}e_1 + a_{22}e_2 + \ldots + a_{2n}e_n \\ \vdots \quad & \qquad \vdots \qquad\qquad\qquad\qquad \vdots \\ T(e_n) &= a_{n1}e_1 + a_{n2}e_2 + \ldots + a_{nn}e_n \end{aligned} \tag{A.7}$$

The transpose of the above matrix of coefficients is the matrix representation of T relative to the basis $\{e_i\}$. Now define $\{f_1, \ldots, f_n\}$ as another basis of V and suppose in terms of the original basis

$$\begin{aligned} f_1 &= a_{11}e_1 + a_{12}e_2 + \ldots + a_{1n}e_n \\ f_2 &= a_{21}e_1 + a_{22}e_2 + \ldots + a_{2n}e_n \\ \vdots \quad & \qquad \vdots \qquad\qquad\qquad\qquad \vdots \\ f_n &= a_{n1}e_1 + a_{n2}e_2 + \ldots + a_{nn}e_n \end{aligned} \tag{A.8}$$

In this case, the transpose of the above matrix of coefficients, defined by

$$P = \begin{bmatrix} a_{11} & a_{21} & \ldots & a_{n1} \\ a_{12} & a_{22} & \ldots & a_{n2} \\ \vdots & & & \vdots \\ a_{1n} & a_{2n} & \ldots & a_{nn} \end{bmatrix} \tag{A.9}$$

is the transformation matrix from the original basis $\{e_1, \ldots, e_n\}$ to the new basis $\{f_1, \ldots, f_n\}$. As the vectors $f_1, \ldots, f_n$ are linearly independent, the matrix P is invertible. It follows that the inverse P^{-1} is the transformation matrix from the basis $\{f_i\}$ back to the basis $\{e_i\}$.

For matrices $A \in \mathbb{R}^{n\times m}$ and $B \in \mathbb{R}^{n\times m}$, the sum of the two matrices is given by the matrix $C \in \mathbb{R}^{n\times m}$ where each element

$$c_{ij} = a_{ij} + b_{ij} \qquad \text{for } i = 1, \ldots, n, \quad j = 1, \ldots, m$$

The product of the two matrices $A \in \mathbb{R}^{n\times r}$ and $B \in \mathbb{R}^{r\times m}$ is given by the matrix $C \in \mathbb{R}^{n\times m}$ where

$$c_{ij} = \sum_{k=1}^{r} a_{ik}b_{kj} \qquad \text{for } i = 1, \ \ldots, n, \quad j = 1, \ldots, m$$

For a matrix $A \in \mathbb{R}^{n\times m}$ it is sometimes convenient to consider the partitioned matrix representation

$$A = \begin{bmatrix} A_{11} & A_{12} \\ A_{21} & A_{22} \end{bmatrix} \tag{A.10}$$

where $A_{11} \in \mathbb{R}^{n_1\times m_1}$, $A_{12} \in \mathbb{R}^{n_1\times m_2}$, $A_{21} \in \mathbb{R}^{n_2\times m_1}$ and $A_{22} \in \mathbb{R}^{n_2\times m_2}$ with $n_1 + n_2 = n$ and $m_1 + m_2 = m$. In this way the matrix A may be thought of as a two-dimensional square matrix whose elements are themselves matrices. Let

$$B = \begin{bmatrix} B_{11} & B_{12} \\ B_{21} & B_{22} \end{bmatrix} \tag{A.11}$$

where $B_{11} \in \mathbb{R}^{n_1 \times m_1}$, $B_{12} \in \mathbb{R}^{n_1 \times m_2}$, $B_{21} \in \mathbb{R}^{n_2 \times m_1}$ and $B_{22} \in \mathbb{R}^{n_2 \times m_2}$, then the corresponding matrix sum may be written as

$$A + B = \begin{bmatrix} A_{11} + B_{11} & A_{12} + B_{12} \\ A_{21} + B_{21} & A_{22} + B_{22} \end{bmatrix} \tag{A.12}$$

Now suppose the matrix B is partitioned as in equation (A.11) with $B_{11} \in \mathbb{R}^{m_1 \times r_1}$, $B_{12} \in \mathbb{R}^{m_1 \times r_2}$, $B_{21} \in \mathbb{R}^{m_2 \times r_1}$ and $B_{22} \in \mathbb{R}^{m_2 \times r_2}$. The matrix product AB can then be written as

$$AB = \begin{bmatrix} A_{11}B_{11} + A_{12}B_{21} & A_{11}B_{12} + A_{12}B_{22} \\ A_{21}B_{11} + A_{22}B_{21} & A_{21}B_{12} + A_{22}B_{22} \end{bmatrix} \tag{A.13}$$

Let the matrix A represent a linear transformation from the vector space V into the vector space W. Two important vector subspaces of V and W can be associated with the transformation A. The *null space* of A is defined by the set

$$\mathcal{N}(A) = \{v \in V : Av = 0\} \tag{A.14}$$

The *range space* of A is defined by the set

$$\mathcal{R}(A) = \{w \in W : w = Av \text{ for some } v \in V\} \tag{A.15}$$

A.2.3 Rank and Determinant

If A is an arbitrary $m \times n$ matrix over a field K, then the *row space* of A is the subspace of K^n generated by the rows of A, and the *column space* of A is the subspace of K^m generated by the columns of A. The dimensions of the row space and of the column space are called the *row rank* of A and the *column rank* of A respectively. For a general matrix A the following theorem holds.

Theorem A.2 *The row rank and column rank of the matrix A are equal.*

This common value of row rank and column rank is called the *rank* of A and written rank(A). The rank of a matrix determines the maximum number of independent rows and the maximum number of independent columns.

An important inequality relating to matrix rank is

$$\text{rank}(AB) \leq \min\{\text{rank}(A),\ \text{rank}(B)\} \tag{A.16}$$

To every square matrix A over a field K there can be assigned a specific scalar called the *determinant* of A and denoted $\det(A)$. Consider the matrix of dimension 2 given by

$$A = \begin{bmatrix} a_{11} & a_{12} \\ a_{21} & a_{22} \end{bmatrix} \tag{A.17}$$

The determinant of A is given by

$$\det A = a_{11}a_{22} - a_{12}a_{21} \tag{A.18}$$

The determinant of the matrix A of dimension 3 may be defined by

$$\det \begin{bmatrix} a_{11} & a_{12} & a_{13} \\ a_{21} & a_{22} & a_{23} \\ a_{31} & a_{32} & a_{33} \end{bmatrix} = a_{11} \det \begin{bmatrix} a_{22} & a_{23} \\ a_{32} & a_{33} \end{bmatrix} - a_{12} \det \begin{bmatrix} a_{21} & a_{23} \\ a_{31} & a_{33} \end{bmatrix} + a_{13} \det \begin{bmatrix} a_{21} & a_{22} \\ a_{31} & a_{32} \end{bmatrix}$$

The form of the above equation shows how determinants can be defined in terms of determinants of lower order. To develop this idea, the *minor* M_{rs} of any element a_{rs} of a general matrix A of dimension n is the determinant of the array obtained by omitting the row and column containing a_{rs}. The *cofactor* of a_{rs}, denoted by A_{rs}, is given by

$$A_{rs} = (-1)^{r+s} M_{rs}$$

The determinant of the matrix A can then be conveniently expressed in terms of minors and cofactors as

$$\begin{aligned} \det(A) &= a_{r1}A_{r1} + a_{r2}A_{r2} + \ldots + a_{rn}A_{rn}, & r = 1, 2, \ldots, n \\ &= a_{1s}A_{1s} + a_{2s}A_{2s} + \ldots + a_{ns}A_{ns}, & s = 1, 2, \ldots, n \end{aligned}$$

Note that the determinant of a single-element matrix is equal to the element itself. Some useful properties of determinants will now be listed:

1. If A has a row (column) of zeros, then $\det(A) = 0$.
2. If A has two identical rows (columns), then $\det(A) = 0$.
3. If any row (column) of a matrix A is multiplied by a scalar k to produce a new matrix B then $\det(B) = k \det(A)$.
4. Interchanging any two rows (columns) of A to produce the matrix B, then $\det(B) = -\det(A)$.
5. Adding a scalar multiple of a row (column) of A to another to yield B, then $\det(B) = \det(A)$.
6. If A is invertible then $\det(A)$ is nonzero.
7. The determinant of a product of two matrices A and B is equal to the product of their determinants, $\det(AB) = \det(A)\det(B)$.
8. The matrix A has rank equal to its dimension if it has a nonzero determinant.

Remark

The matrix manipulations described in items 3, 4 and 5 are called *elementary row (column) operations.*

A.2.4 Eigenvalues, Eigenvectors and Singular Values

Definition A.5 *If A is an n-dimensional square matrix over a field K, then $\lambda \in K$ is an eigenvalue of A, if for some nonzero vector $v \in K^n$,*

$$Av = \lambda v$$

In this case v is said to be the eigenvector corresponding to the eigenvalue λ.

The matrix $\lambda I_n - A$, where $I_n \in \mathbb{R}^{n\times n}$ denotes the *identity matrix* which has ones on the diagonal elements and zeros elsewhere, is called the *characteristic matrix* of A. The determinant of the characteristic matrix is called the *characteristic polynomial* of A and

$$\det(\lambda I_n - A) = 0 \tag{A.19}$$

is called the *characteristic equation* of A. This is an nth order polynomial in λ. Note that the values of λ which satisfy the characteristic equation determine the eigenvalues of A.

The following property relating to the determinant of a partitioned matrix is useful. Let A be a real symmetric matrix partitioned as

$$A = \begin{bmatrix} A_{11} & A_{12} \\ 0 & A_{22} \end{bmatrix}$$

where the top left and bottom right sub-blocks are square.

Proposition A.1 *The determinant and the eigenvalues of A satisfy:*

1. $det(A) = det(A_{11})\ det(A_{22})$
2. $\lambda(A) = \lambda(A_{11}) \cup \lambda(A_{22})$

A useful decomposition of the nonsquare matrix A defined over the field of complex numbers $\mathbb{C}$ is given by the so-called *singular value decomposition.* Before this decomposition can be defined it is necessary to consider the notion of a *unitary* matrix.

Definition A.6 *The matrix $U \in \mathbb{C}^{n\times n}$ is unitary if*

$$U^{\mathrm{H}} = U^{-1} \tag{A.20}$$

where U^{H} denotes the complex conjugate transpose of the matrix U. In the special case when the matrix U is real $U^{\mathrm{H}} = U^{\mathrm{T}}$ and a real matrix satisfying $U^{\mathrm{T}} = U^{-1}$ is said to be an orthogonal matrix.

Theorem A.3 *Any matrix $A \in \mathbb{C}^{n\times m}$ may be factored into a singular value decomposition*

$$A = U\Sigma V^{\mathrm{H}} \tag{A.21}$$

where $U \in \mathbb{C}^{n\times n}$ and $V \in \mathbb{C}^{m\times m}$ are unitary matrices and the matrix $\Sigma \in \mathbb{R}^{n\times m}$ contains a diagonal submatrix Σ_1 of nonnegative singular values arranged in descending order so that

$$\Sigma_1 = \mathrm{diag}\{\sigma_1, \sigma_2, \ldots, \sigma_k\}; \qquad k = \min(n, m)$$

where

$$\sigma_1 \geq \sigma_2 \geq \ldots \geq \sigma_k$$

The matrix Σ *is expressed as*

$$\begin{aligned} \Sigma &= \begin{bmatrix} \Sigma_1 \\ 0 \end{bmatrix} & n \geq m \\ \Sigma &= \begin{bmatrix} \Sigma_1 & 0 \end{bmatrix} & n \leq m \end{aligned}$$

In the special case when A *is a real matrix then the factors* U *and* V *may be chosen as orthogonal matrices rather than unitary matrices.*

It is useful to note that the singular values may be obtained by computing the eigenvalues of $A^{\mathrm{H}}A$ or AA^{H} and then computing the square root of the result.

It is straightforward to verify that when $n \leq m$ the singular value decomposition provides a useful tool for determining a set of vectors which span the null space of the matrix A. For instance, suppose $A \in \mathbb{R}^{n\times m}$ where $n \leq m$ and A has full rank, i.e. $\mathrm{rank}(A) = m$. Let A be decomposed as in equation (A.21) and let

$$V = \begin{bmatrix} V_1 & V_2 \end{bmatrix}$$

where $V_2 \in \mathbb{R}^{m\times m-n}$ and notice that because V is real $V^{\mathrm{T}}V = I_m$. Using this fact

$$V^{\mathrm{T}}V_2 = \begin{bmatrix} V_1^{\mathrm{T}}V_2 \\ V_2^{\mathrm{T}}V_2 \end{bmatrix} = \begin{bmatrix} 0 \\ I \end{bmatrix}$$

and consequently

$$AV_2 = U\Sigma V^{\mathrm{T}}V_2 = U \begin{bmatrix} \Sigma_1 & 0 \end{bmatrix} \begin{bmatrix} 0 \\ I \end{bmatrix} = 0$$

Since the $m - n$ columns of V_2 are linearly independent and $AV_2 = 0$ it follows that they span the null space of A.

A.2.5 QR Decomposition

Another frequently used decomposition is the QR reduction, whereby an arbitrary matrix $X \in \mathbb{C}^{n\times m}$ can be expressed as the product of an upper triangular matrix R, which will have the same dimension as X, and a unitary matrix Q. It follows that

$$X = QR \tag{A.22}$$

This will be a useful decomposition of the input distribution matrix; it provides a state transformation for partitioning those states upon which the input acts directly from the remainder of the states and it is particularly useful as no loss of numerical accuracy occurs when the inverse transformation is computed.

A.2.6 Norms, Inner Products and Projections

Definition A.7 *A norm is a function which assigns to every vector x in a vector space a real number $||x||$ such that*

1. $||x|| \geq 0$
2. $||x|| = 0$ *if and only if* $x = 0$
3. $||kx|| = |k|\ ||x||$ *where k is a scalar and $|k|$ is the absolute value of k*
4. $||x + y|| \leq ||x|| + ||y||$

An intuitive explanation is that the norm of a vector is an indication of its length or size. Equivalently it might be perceived as representing the distance of the point x from the origin. Viewed in this way, the concept of length would satisfy properties 1 and 2. Likewise property 3 accords with the notion that scaling a vector scales its length by a similar amount. Property 4 is usually called the *triangle inequality*; it states that the distance of the point $x+y$ from the origin cannot be greater than the distance of x plus the distance of y.

Three vector norms or notions of distance are commonly employed by control mathematicians:

1. The 1-norm
$$||x||_1 = |x_1| + |x_2| + \ldots + |x_n| \tag{A.23}$$

2. The Euclidean norm (or 2-norm)
$$||x||_2 = (|x_1|^2 + |x_2|^2 + \ldots + |x_n|^2)^{1/2} \tag{A.24}$$

3. The ∞ norm
$$||x||_\infty = \max |x|_i, \qquad (i = 1, \ldots, n) \tag{A.25}$$

The Euclidean norm is perhaps the most intuitive and corresponds exactly to the notion of distance, i.e. the length of the straight line between two points. All of the above are merely special cases of the p-norm, which is defined by

$$||x||_p = (|x_1|^p + |x_2|^p + \ldots + |x_n|^p)^{1/p} \tag{A.26}$$

where $p \geq 1$.

The vector p-norm in equation (A.26) induces a corresponding matrix norm which is given by

$$||A||_p = \max_{x \neq 0} \frac{||Ax||_p}{||x||_p} \tag{A.27}$$

Such matrix norms satisfy the properties of a vector norm. In addition, the following condition, relating to the norm of a product of matrices, is satisfied:

$$||AB|| \leq ||A||.||B|| \tag{A.28}$$

Corresponding to the 1, 2 and ∞ vector norms are three *induced matrix norms*:

1. The 1-norm
$$\|A\|_1 = \max_j(\sum_i |a_{ij}|) \tag{A.29}$$

2. The 2-norm is given by the maximum singular value of the matrix A where the singular values are as in Theorem A.3.

3. The ∞ norm
$$\|A\|_\infty = \max_i(\sum_j |a_{ij}|) \tag{A.30}$$

An inner product is essentially a rule which assigns to any pair of vectors in a vector space an associated scalar quantity.

Definition A.8 *Let V be a real or complex vector space defined over the field K, which is either the real field $\mathbb{R}$ or the complex field $\mathbb{C}$. Suppose that to each pair of vectors $u, v \in V$ there is assigned a scalar $\langle u, v\rangle \in K$. The mapping is called an inner product in V if it satisfies the following axioms:*

1. $\langle au_1 + bu_2, v\rangle = a\langle u_1, v\rangle + b\langle u_2, v\rangle$
2. $\langle u, v\rangle = \langle v, u\rangle^*$ *where $*$ denotes the complex conjugate*
3. $\langle u, u\rangle \geq 0$; *and* $\langle u, u\rangle = 0$ *if and only if* $u = 0$.

From these axioms it follows that
$$|\langle x, y\rangle| \leq \|x\| \, \|y\| \tag{A.31}$$
where the norm is defined by
$$\|u\| = \langle u, u\rangle^{1/2} \tag{A.32}$$
(It can be readily shown from the axioms of the inner product that the four norm properties are satisfied.) The inequality in (A.31) is often called the *Cauchy–Schwarz inequality.*

In the case when $V = \mathbb{R}^n$ an inner product is given by
$$\langle u, v\rangle = u^{\mathrm{T}} v \tag{A.33}$$
An important fact is that the norm induced by this inner product is the Euclidean norm.

In general, if two vectors u and v satisfy $\langle u, v\rangle = 0$, they are deemed to be *orthogonal* (or perpendicular).

The *orthogonal complement* of a vector subspace $W \subseteq V$ is the vector subspace
$$\{v \in V \ : \ \langle v, w\rangle = 0 \text{ for all } w \in W\}$$
This subspace will be written as $W^{\perp}$.

Note that the Euclidean norm is preserved under an orthogonal transformation. Suppose x is an arbitrary vector and $y = Tx$ where T is orthogonal, then
$$\|y\|^2 = \langle y, y\rangle = y^{\mathrm{T}} y = x^{\mathrm{T}} T^{\mathrm{T}} T x = x^{\mathrm{T}} x = \langle x, x\rangle = \|x\|^2$$

The concept of a projection of a vector into a given subspace will now be introduced. Assume that the columns of the matrix L are independent and span a given subspace $\mathcal{L}$. It is necessary that a vector v_a resides in this prescribed subspace, i.e.

$$v_a = Lx \tag{A.34}$$

for some appropriately dimensioned x. Assume that a desired form v_d is known for the vector v_a. In general, this desired form may not lie in the required subspace. In this case, it is useful to select v_a so that it is as close as possible to the desired form v_d. To this end, an appropriate vector x in equation (A.34) is a vector chosen to minimise

$$J = ||v_d - v_a||^2 = ||v_d - Lx||^2 \tag{A.35}$$

which is a measure of the distance between the desired vector v_d and the achievable vector v_a. It can be shown that

$$\frac{dJ}{dx} = 2L^{\mathrm{T}}(Lx - v_d) \tag{A.36}$$

and thus choosing

$$x = (L^{\mathrm{T}}L)^{-1}L^{\mathrm{T}}v_d \tag{A.37}$$

makes the derivative zero. This implies that the choice of x in (A.37) minimises the difference between the desired form of the vector and the allowable vector. Note that the full column rank assumption on L implies $L^{\mathrm{T}}L$ is invertible and thus the expression in (A.37) is well defined. It follows from equation (A.34) that the allowable vector is given by

$$v_a = L(L^{\mathrm{T}}L)^{-1}L^{\mathrm{T}}v_d \tag{A.38}$$

A.2.7 Quadratic Forms

A quadratic form is a function Q of n real variables $x_1, x_2, \ldots, x_n$ such that

$$Q(x_1, x_2, \ldots, x_n) = \sum_{i,j=1}^{n} q_{ij}x_ix_j \qquad q_{ij} = q_{ji} \tag{A.39}$$

Without loss of generality, the q_{ij} can be thought of as the entries of a particular matrix Q and the x_i can be considered the components of the vector x. The quadratic form (A.39) can thus be alternatively represented by

$$Q(x_1, x_2, \ldots, x_n) = x^{\mathrm{T}}Qx \tag{A.40}$$

where the matrix Q is a *symmetric matrix*, i.e. $Q^{\mathrm{T}} = Q$.

Proposition A.2 *Let A be a symmetric matrix, then*

1. *The eigenvalues of A are all real.*
2. *There exists an orthogonal matrix Q such that*

$$A = Q\Lambda Q^{\mathrm{T}} \tag{A.41}$$

where Λ is a diagonal matrix formed from the eigenvalues of A and the orthogonal matrix Q is formed from the associated eigenvectors.

Quadratic forms always satisfy the Rayleigh principle, namely

$$\lambda_{min}(Q)||x||^2 \leq x^{\mathrm{T}}Qx \leq \lambda_{max}(Q)||x||^2 \tag{A.42}$$

In particular, if $\lambda_{min}(Q) \geq 0$ then it follows that $x^{\mathrm{T}}Qx \geq 0$ for all x.

Definition A.9 *The quadratic form $x^{\mathrm{T}}Qx$ where Q is a real symmetric matrix is said to be positive semidefinite if*

$$x^{\mathrm{T}}Qx \geq 0 \qquad \textit{for all } x$$

In particular if

$$x^{\mathrm{T}}Qx > 0 \qquad x \neq 0$$

then the quadratic form is said to be positive definite

If $x^{\mathrm{T}}Qx$ is positive definite then the matrix Q is said to be a *positive definite matrix*; for convenience, this will be written $Q > 0$. It can be seen from the modal decomposition expression in (A.41) that a symmetric matrix Q will be positive definite if all its eigenvalues are positive. It also follows that a positive definite matrix Q is invertible since from (A.41)

$$\det(Q) = \det(\Lambda) = \prod_{i=1}^{n} \lambda_i > 0$$

Some useful identities relating to partitioned symmetric positive definite matrices will now be presented. Let P be a real symmetric matrix partitioned so that

$$P = \begin{bmatrix} P_{11} & P_{12} \\ P_{12}^{\mathrm{T}} & P_{22} \end{bmatrix}$$

where the matrix sub-blocks P_{11} and P_{22} are square.

Proposition A.3 *The symmetric matrix P satisfies*

$$\begin{aligned} P > 0 \quad &\Leftrightarrow \quad P_{11} > 0 \quad \textit{and} \quad P_{22} > P_{12}^{\mathrm{T}}P_{11}^{-1}P_{12} \\ &\Leftrightarrow \quad P_{22} > 0 \quad \textit{and} \quad P_{11} > P_{12}P_{22}^{-1}P_{12}^{\mathrm{T}} \end{aligned}$$

As a corollary

$$P > 0 \quad \Rightarrow \quad P_{11} > 0 \quad \text{and} \quad P_{22} > 0$$

Another property that will be exploited is that, given a symmetric matrix P and a nonsingular matrix T of the same dimension, then

$$P > 0 \quad \Leftrightarrow \quad T^{\mathrm{T}}PT > 0$$

A symmetric matrix P is said to be *negative definite* if $-P$ is positive definite.

A.3 NOTES AND REFERENCES

Numerous excellent introductions to linear systems theory are currently in print, including Brogan (1991), Chen (1984), Bélanger (1995) and Kailath (1980).

Appendix B

Assorted mfiles

This appendix provides a listing of some mfiles which would have caused a distraction from the main argument had they been placed in the body of the text.

B.1 A VARIATION ON THE *PLACE* COMMAND

The function below has been used in several places, notably Chapters 4 and 7, as a robust pole placement method. It is fundamentally the same as the MATLAB command *place* except for a preliminary transformation to the system and input distribution matrix. This precaution has been added because invariably the command is used on some subsystem – usually (A_{11}, A_{12}) or a variation thereof – which although guaranteed to be controllable does not necessarily have the property that A_{12} is of full column rank. QR reduction is performed on A_{12}^{T} so that

$$A_{12} = \left[\begin{array}{cc} \tilde{A}_{12} & 0 \end{array} \right] Q^{\mathrm{T}}$$

where Q is orthogonal and $\tilde{A}_{12}$ has full column rank. The pole placement is then performed on the pair $(A_{11}, \tilde{A}_{12})$.

mfile: a modified version of the 'place' command

```
function K=vplace(A,B,p)
[nn,mm]=size(B);

rnk=rank(B);
if rnk==mm,
   K=place(A,B,p);
else
  [Q,Btilde]=qr(B');
  Btilde=Btilde';
  Btilde=Btilde(:,1:rnk);
  Ktilde=place(A,Btilde,p);
  K=Q*[Ktilde;zeros(mm-rnk,nn)];
end

return
```

B.2 EIGENSTRUCTURE ASSIGNMENT: THE COMPLEX CASE

The mfile below is a generalisation of the eigenstructure assignment procedure described in Section 4.3. Here a desired modal structure incorporating both real and complex parts may be specified to design a hyperplane using eigenstructure assignment. It should be noted that complex eigenvalues and associated eigenvectors must appear in conjugate pairs. The approach that will be presented relies on decomposing the problem into one requiring only real arithmetic.

Suppose λ_i is a complex eigenvalue and v_i represents an associated complex eigenvector written as

$$\lambda_i = \lambda_{re} + j\lambda_{im} \tag{B.1}$$

and

$$v_i = v_{re} + jv_{im} \tag{B.2}$$

Because the eigenvalue and eigenvector must satisfy

$$(A + BF)v_i = \lambda_i v_i \tag{B.3}$$

by comparing real and imaginary parts it follows that the equations

$$(A + BF)v_{re} = \lambda_{re}v_{re} - \lambda_{im}v_{im} \tag{B.4}$$

$$(A + BF)v_{im} = \lambda_{re}v_{im} + \lambda_{im}v_{re} \tag{B.5}$$

must hold. These equations may be rewritten as

$$\begin{bmatrix} \lambda_{re}I - A & \lambda_{im}I & B \end{bmatrix} \begin{bmatrix} v_{re} \\ -v_{im} \\ -Fv_{re} \end{bmatrix} = 0 \tag{B.6}$$

and

$$\begin{bmatrix} \lambda_{re}I - A & \lambda_{im}I & B \end{bmatrix} \begin{bmatrix} v_{im} \\ v_{re} \\ -Fv_{im} \end{bmatrix} = 0 \tag{B.7}$$

Define

$$G_c(\lambda_i) = \begin{bmatrix} \lambda_{re}I - A & \lambda_{im}I & B \end{bmatrix} \tag{B.8}$$

and let $K(\lambda_i) \in \mathbb{R}^{(2n+m)\times m}$ be a matrix whose columns span the null space of $G_c(\lambda_i)$. If

$$K(\lambda_i) = \begin{bmatrix} N(\lambda_i) \\ P(\lambda_i) \\ M(\lambda_i) \end{bmatrix} \begin{matrix} \updownarrow n \\ \updownarrow n \\ \updownarrow m \end{matrix} \tag{B.9}$$

it follows that

$$\begin{bmatrix} v_{re} \\ -v_{im} \\ -Fv_{re} \end{bmatrix} \in \text{span}\begin{Bmatrix} N(\lambda_i) \\ P(\lambda_i) \\ M(\lambda_i) \end{Bmatrix} \quad \text{and} \quad \begin{bmatrix} v_{im} \\ v_{re} \\ -Fv_{im} \end{bmatrix} \in \text{span}\begin{Bmatrix} N(\lambda_i) \\ P(\lambda_i) \\ M(\lambda_i) \end{Bmatrix}$$

which implies

$$\begin{bmatrix} v_{re} \\ v_{im} \end{bmatrix} \in \text{span}\begin{Bmatrix} N(\lambda_i) \\ -P(\lambda_i) \end{Bmatrix} \cap \text{span}\begin{Bmatrix} P(\lambda_i) \\ N(\lambda_i) \end{Bmatrix} \tag{B.10}$$

To calculate a basis for this vector space numerically, the key observation is that, given any matrix H, if K_H is a matrix which spans the null space H^{T} then

$$H^{\mathrm{T}}K_H = 0 \quad \Leftrightarrow \quad K_H^{\mathrm{T}}H = 0$$

which implies that the $\mathcal{R}(H) = \mathcal{N}(K_H^{\mathrm{T}})$. With this fact in mind, define

$$\alpha = \begin{bmatrix} N(\lambda_i) \\ -P(\lambda_i) \end{bmatrix} \quad \text{and} \quad \beta = \begin{bmatrix} P(\lambda_i) \\ N(\lambda_i) \end{bmatrix} \tag{B.11}$$

and let K_α and K_β be matrices whose columns span the null spaces of the matrices α^{T} and β^{T} respectively. Next define

$$\gamma = \begin{bmatrix} K_\alpha^{\mathrm{T}} \\ K_\beta^{\mathrm{T}} \end{bmatrix} \tag{B.12}$$

and let K_γ be a matrix whose columns span the null space of K_γ. Notice that $K_\alpha^{\mathrm{T}}K_\gamma = 0$ and $K_\beta^{\mathrm{T}}K_\gamma = 0$ and consequently, from the observation above, the columns of K_γ span the intersection of $\mathcal{R}(\alpha) \cap \mathcal{R}(\beta)$. This construction is easily implemented in MATLAB using the singular value decomposition.

Comparing this situation with that in Section 4.3, the matrix K_γ plays the role of the matrix $N(\lambda_i)$; consequently, the rows of K_γ need to be reordered to isolate an equation of the form

$$v_i^s = N_1\delta_i \tag{B.13}$$

where v_i^s is a vector comprising the real and imaginary components which have been specified and δ_i represents a vector which will determine the eigenvector.

In the mfile fragment below, the variable *nocomp* represents the number of complex conjugate pairs specified in *lambda*. The first loop calculates the required δ_i's which determine complex eigenvalues using the method described above. The second loop is identical to the mfile fragment in Section 4.3 and calculates the eigenvectors associated with the remaining real eigenvalues.

Note that the matrix V contains only real values. For complex eigenvalues, V contains the real and imaginary components as two columns of real numbers. Since the eigenvectors must occur in complex conjugate pairs, the real and imaginary components for only one eigenvector of the pair need be stored; in this way, as in the real case, $V \in \mathbb{R}^{n\times(n-m)}$.

mfile: hyperplane design via eigenstructure assignment

```
% For the system pair (A,B), nocomp defines the number of pairs of desired
% complex conjugate poles. The vector lambda defines the n-m desired
% sliding mode poles;the first 2i-1 (i=1, nocomp) entries contain the
% real parts of the complex conjugate poles and the entries 2i (i=1,nocomp)
% contain the imaginary parts. In the matrices described below, the first
% 2i-1 (i=1, nocomp) columns contain information about the desired real
% parts of the eigenvectors and the first 2i (i=1,nocomp) contain the
% imaginary parts. The columns of the nx(n-m) matrix specpos are used
% to distinguish vector entries which are arbitrary (denoted by 0 in the
% relevant matrix entry) from entries which are to be specified (denoted
```

```
% by 1 in the relevant matrix entry). The columns of the nx(n-m) matrix
% specent specify any desired eigenvector entries.

[nn,mm]=size(B);
for i=1:nocomp,

   % Determine the space in which the eigenvector corresponding
   % to a specific desired complex conjugate pair of poles must lie
   glambda1=[lambda(2*i-1)*eye(nn)-A lambda(2*i)*eye(nn) B];
   [u,v,w]=svd(glambda1);
   nlambda1=w(1:nn,nn+1:2*nn+mm);
   plambda1=w(nn+1:2*nn,nn+1:2*nn+mm);
   alpha=[nlambda1; -plambda1]';
   beta=[plambda1; nlambda1]';
   [u1,v1,w1]=svd(alpha);
   [u2,v2,w2]=svd(beta);
   gamma=[w1(:,nn+mm+1:2*nn)'; w2(:,nn+mm+1:2*nn)'];
   [u3,v3,w3]=svd(gamma);
   kgamma=w3(:,2*nn-2*mm+1:2*nn);

   % Find subvector of specified entries
   despos=find(specpos(:,2*i-1));
   numspecr=length(despos);
   n1=[]; desent=[];
   for j=1:numspecr,
      n1(j,:)=(kgamma(despos(j),:));
      desent(j)=specent(despos(j),2*i-1);
   end
   despos=find(specpos(:,2*i));
   numspecc=length(despos);
   for j=1:numspecc,
      n1(j+numspecr,:)=kgamma(despos(j)+nn,:);
      desent(j+numspecr)=specent(despos(j),2*i);
   end

   % Perform least-squares projection
   delta=n1\desent';
   vector=kgamma*delta;
   V(1:nn,2*i-1)=vector(1:nn);
   V(1:nn,2*i)=vector(nn+1:2*nn);
end

for i=2*nocomp+1:nn-mm,

   % Determine the space in which the eigenvector corresponding
   % to a specific desired real pole must lie
   glambda=[lambda(i)*eye(nn)-A B];
   [u,v,w]=svd(glambda);
   nlambda=w(1:nn,nn+1:nn+mm);

   % Find subvector of specified entries
   despos=find(specpos(:,i));
   numspec=length(despos);
```

```
    desent2=[]; n2=[];
    for j=1:numspec,
       n2(j,:)=nlambda(despos(j),:);
       desent2(j)=specent(despos(j),i);
    end

    % Perform least-squares projection
    delta=n2\desent2';
    V(:,i)=nlambda*delta;
end

% Use a singular value decomposition to determine the
% switching function matrix from the selected eigenvectors
[u,v,w]=svd(V);
S=u(:,nn-mm+1:nn)';
```

As an example, consider the aircraft vertical translation manoeuvre using integral action (Section 4.5.2). The model is given in equations (4.120) to (4.124) which, after augmenting with the integral action states, give system and input distribution matrices

$$A = \begin{bmatrix} 0 & 0 & -1 & 0 & -1 & 0 & 0 \\ 0 & 0 & -1 & 0 & 0 & 0 & 0 \\ 0 & 0 & 0 & 0 & 1.74 & 0.08 & 0.59 \\ 0 & 0 & 0 & -1.99 & -13.41 & -18.95 & -3.60 \\ 0 & 0 & 0 & 1.00 & -1.74 & -0.08 & -0.59 \\ 0 & 0 & 0 & 0 & 0 & -20 & 0 \\ 0 & 0 & 0 & 0 & 0 & 0 & -20 \end{bmatrix} \qquad B = \begin{bmatrix} 0 & 0 \\ 0 & 0 \\ 0 & 0 \\ 0 & 0 \\ 0 & 0 \\ 20 & 0 \\ 0 & 20 \end{bmatrix}$$

In Section 4.5.2 the poles of the ideal sliding mode dynamics were chosen to be

$$\{-5.6 \pm 4.2j, -1.0, -0.4, -0.7\}$$

and the desired eigenstructure was indicated in equation (4.144). In terms of the matrices used in the mfiles, this information translates into

$$\text{lambda} = \begin{bmatrix} -5.6 & 4.2 & -1 & -0.4 & -0.7 \end{bmatrix} \tag{B.14}$$

with the corresponding matrices

$$\text{specpos} = \begin{bmatrix} 0 & 0 & 0 & 0 & 0 \\ 0 & 0 & 0 & 0 & 0 \\ 1 & 1 & 1 & 1 & 1 \\ 1 & 0 & 1 & 1 & 1 \\ 0 & 1 & 1 & 0 & 1 \\ 0 & 0 & 0 & 0 & 0 \\ 0 & 0 & 0 & 0 & 0 \end{bmatrix} \qquad \text{specent} = \begin{bmatrix} * & * & * & * & * \\ * & * & * & * & * \\ 0 & 0 & 1 & 0 & 1 \\ 1 & * & 0 & 1 & 0 \\ * & 1 & 0 & * & 0 \\ * & * & * & * & * \\ * & * & * & * & * \end{bmatrix} \tag{B.15}$$

The attainable eigenvectors become

$$\begin{bmatrix} -0.1918 \pm 0.0347j & 0.3333 & -6.2500 & 0.7215 \\ 0.0000 \pm 0.0000j & 0.6667 & 0.0000 & 1.0750 \\ 0.0000 \pm 0.0000j & 0.6667 & 0 & 0.7525 \\ 1.0000 \pm 9.5000j & -0.3333 & 1.0000 & -0.3535 \\ -0.9286 \pm 1.0000j & -0.3333 & -2.5000 & -0.2475 \\ -1.8252 \pm 2.2364j & 0.2886 & 0.2921 & 0.2362 \\ 2.9860 \pm 2.6459j & -0.1860 & 7.3333 & -0.1950 \end{bmatrix}$$

It can be seen that in the vectors in which only two components have been specified, the desired values have been obtained exactly. When three components have been specified, an approximation has been returned.

B.3 WORLD WIDE WEB SITE

The mfile fragments described in this book are available at:

`http://www.le.ac.uk/eg/ce14/vscbook/mfiles.html`

A toolbox which incorporates these fragments is also available.

References

Åström, K.J. and Wittenmark, B., 1989, *Adaptive Control*, Addison-Wesley, Reading MA.

Bag, S.K., Spurgeon, S.K. and Edwards, C., 1997, Output feedback sliding mode design for linear uncertain systems, *Proceedings of IEE, Part D* **144**, 209–216.

Balestrino, A. and Innocenti, M., 1990, The hyperstability approach to VSCS design, in Zinober, A.S.I. (ed.), *Deterministic Control of Uncertain Systems*, Peter Peregrinus, Stevenage, UK, pp. 170–199.

Balestrino, A., DeMaria, G. and Zinober, A.S.I., 1984, Nonlinear adaptive model-following control, *Automatica* **20**, 559–568.

Bélanger, P.R., 1995, *Control Engineering: A Modern Approach*, Harcourt Brace, London.

Bhatti, A.I., 1998, *Advanced Sliding Mode Controllers for Industrial Applications*, PhD thesis, University of Leicester.

Boyd, S.P. and Barratt, C.H., 1991, *Linear Controller Design, Limits of Performance*, Prentice Hall, Englewood Cliffs NJ.

Breinl, W. and Leitmann, G., 1987, State feedback for uncertain dynamical systems, *Applied Mathematics and Computation* **22**, 65–87.

Brogan, W.L., 1991, *Modern Control Theory*, Prentice Hall, Englewood Cliffs NJ.

Broussard, J.R. and O'Brien, M.J., 1980, Feedforward control to track the output of a forced model, *IEEE Transactions on Automatic Control* **AC-25**, 851–853.

Burton, J.A. and Zinober, A.S.I., 1986, Continuous approximation of variable structure control, *International Journal of Systems Science* **17**, 876–885.

Butchart, R.L. and Shakcloth, B., 1966, Synthesis of model-reference adaptive control systems by Lyapunov's second method, *Proceedings of the 2nd IFAC Symposium on the Theory of Self-adaptive Control Systems*, pp. 145–152.

Butts, K., Sivashankar, N. and Sun, J., 1995, Feedforward and feedback design for engine idle speed control using l_1 optimization, *Proceedings of the 1995 American Control Conference*, Seattle WA., pp. 2587–2590.

Chan, Y.T., 1973, Perfect model-following with a real model, *Proceedings of the Joint Automatic Control Conference*, pp. 287–293.

Chen, C.T., 1984, *Linear System Theory and Design*, Holt, Rinehart & Winston, New York.

Corless, M., 1994, Robust stability analysis and controller design with quadratic Lyapunov functions, in Zinober, A.S.I. (ed.), *Variable Structure and Lyapunov Control*, Springer-Verlag, Berlin, pp. 181–203.

Corless, M., Goodall, D.P., Leitmann, G. and Ryan, E.P., 1985, Model-following controls for a class of uncertain dynamical systems, *Proceedings of the 7th IFAC/IFORS Symposium on Identification and System Parameter Estimation*, pp. 1895–1899.

Davies, R. and Spurgeon, S.K., 1993, Robust tracking of uncertain systems via a quadratic stability condition, *Proceedings of the 12th IFAC World Congress*, pp. 10:43–46.

DeCarlo, R.A., Żak, S.H. and Matthews, G.P., 1988, Variable structure control of nonlinear multivariable systems: a tutorial, *Proceedings of IEEE* **76**, 212–232.

deJager, B., 1992, Comparison of methods to eliminate chattering and avoid steady state errors in sliding mode digital control, *Proceedings of the IEEE VSC and Lyapunov Workshop*, Sheffield, pp. 37–42.

Disdell, K.J., Burnham, K.J. and James, D.J.G., 1994, Developments in furnace control for improved efficiency and reduced emissions, *IEE Digest No. 1994/018, Colloquim on improvements in furnace control: current developments and new technologies.*

Dorling, C.M. and Zinober, A.S.I., 1985, Hyperplane design in model-following variable structure control systems, *Proceedings of the 7th IFAC/IFORS Symposium on Identification and System Parameter Estimation*, pp. 1901–1905.

Dorling, C.M. and Zinober, A.S.I., 1986, Two approaches to hyperplane design in multivariable variable structure control systems, *International Journal of Control* **44**, 65–82.

Draženović, B., 1969, The invariance conditions in variable structure systems, *Automatica* **5**, 287–295.

El-Ghezawi, O.M.E., Billings, S.A. and Zinober, A.S.I., 1983, Variable-structure systems and system zeros, *Proceedings of the IEE, Part D* **130**, 1–5.

El-Khazali, R. and DeCarlo, R.A., 1991, Variable structure output feedback control: switching surface design, *Proceedings of the 29th Allerton Conference on Communications, Control and Computing*, pp. 430–439.

El-Khazali, R. and DeCarlo, R.A., 1992, Variable structure output feedback control, *Proceedings of the American Control Conference*, pp. 871–875.

El-Khazali, R. and DeCarlo, R.A., 1993, Output feedback variable structure control design using dynamic compensation for linear systems, *Proceedings of the American Control Conference*, pp. 954–958.

Erzberger, H., 1968, Analysis and design of model-following control systems by state space techniques, *Proceedings of the Joint American Control Conference*, pp. 572–581.

Filippov, A.F., 1964, Differential equations with discontinuous right hand-sides, *American Mathematical Society Translations* **42**, 199–231.

Goodhart, S.G., 1994, An industrial perspective on temperature control, *Proceedings of the 10th International Conference on Systems Engineering*, Coventry University, pp. 399–404.

Gu, G., 1990, Stabilizability conditions of multivariable uncertain systems via output feedback control, *IEEE Transactions on Automatic Control* **35**, 925–927.

Hammond, P.S., 1992, Advanced kiln and furnace control, *Proceedings of the International Gas Research Conference*, Florida.

Harvey, C.A. and Stein, G., 1978, Quadratic weights for asymptotic regulator properties, *IEEE Transactions on Automatic Control* **AC-23**, 378–387.

Heck, B.S., Yallapragada, S.V. and Fan, M.K.H., 1995, Numerical methods to design the reaching phase of output feedback variable structure control, *Automatica* **31**, 275–279.

Hermans, F.J.J. and Zarrop, M.B., 1996, Sliding mode observers for robust sensor monitoring, *Proceedings of the 13th IFAC World Congress*, pp. 211–216.

Hottel, H.C. and Sarofim, A.F., 1967, *Radiative Transfer*, McGraw-Hill, New York.

Hsu, L., Araújo, A.D. De and Costa, R.R., 1994, Analysis and design of input/output based variable structure adaptive control, *IEEE Transactions on Automatic Control* **39**, 4–21.

Huenecke, K., 1987, *Modern Combat Aircraft Design*, Airlife, Shrewsbury, UK.

Hui, S. and Żak, S.H., 1993, Robust output feedback stabilisation of uncertain dynamic systems with bounded controllers, *International Journal of Robust and Nonlinear Control* **3**, 115–132.

Hung, J.Y, Gao, W. and Hung, J.C., 1993, Variable structure control: a survey, *IEEE Transactions on Industrial Electronics* **40**, 2–22.

Itkis, U., 1976, *Control Systems of Variable Structure*, Wiley, New York.

Kailath, T., 1980, *Linear Systems*, Prentice Hall, Englewood Cliffs NJ.

Kautsky, J. and Nichols, N.K., 1983, Robust eigenstructure assignment in state feedback controls, *Numerical Analysis Report NA/2/83*, Department of Mathematics, University of Reading, UK.

Kautsky, J., Nichols, N.K. and Dooren, P. Van, 1985, Robust pole assignment in linear state feedback, *International Journal of Control* **41**, 1129–1155.

Khalil, H.K., 1992, *Nonlinear Systems*, Prentice Hall, Englewood Cliffs NJ.

Kwakernaak, H. and Sivan, R., 1972, *Linear Optimal Control Systems*, Wiley, New York.

Landau, I., 1974, Survey of model-reference adaptive techniques, *Automatica* **10**, 353–379.

Landau, I. and Courtoil, B., 1974, Design of multivariable adaptive model-following control systems, *Automatica* **10**, 483–494.

Landau, I.D., 1979, *Adaptive Control: The Model Reference Approach*, Marcel Dekker, New York.

Luenberger, D.G., 1971, An introduction to observers, *IEEE Transactions on Automatic Control* **16**, 596–602.

Luo, N.S. and Feng, C.B., 1989, A new method for suppressing chattering in variable structure feedback control systems, *Nonlinear Control Systems Design: Science Papers of the IFAC Symposium*, Pergamon, Oxford, pp. 279–284.

Monopoli, R.V., Gilbert, J.W. and Thayer, W.D., 1968, Model-reference adaptive control based on Lyapunov-like techniques, *Proceedings of the 2nd IFAC Symposium on System Sensitivity and Adaptivity*, pp. F24–36.

Moore, B.C., 1976, On the flexibility offered by state-feedback in multivariable systems beyond closed-loop eigenvalue assignment, *IEEE Transactions on Automatic Control* **AC-21**, 689–692.

Mudge, S.K. and Patton, R.J., 1988, Variable structure control laws for aircraft manoeuvres, *Proceedings of the IEE Conference Control '88*, pp. 564–568.

Nichols, N.K., 1987, Robustness in partial pole placement, *IEEE Transactions on Automatic Control* **AC-23**, 728–732.

O'Brien, M.J. and Broussard, J.R., 1978, Feedforward gains to track the output of a forced model, *Proceedings of the 17th IEEE Conference on Decision and Control.*

Ogata, K., 1997, *Modern Control Engineering*, 3rd edn, Prentice Hall, Englewood Cliffs NJ.

Palmer, M.R., 1989, A practical computer package for the thermal design of high temperature industrial plant, *Proceedings of the International Gas Research Conference*, Tokyo.

Patel, R.V. and Toda, M., 1980, Quantitative measures of robustness for multivariable systems, *Proceedings of the American Control Conference*, pp. TP8–A.

Patton, R.J., Frank, P.M. and Clark, R.N., 1989, *Fault Diagnosis in Dynamic Systems: Theory and Application*, Prentice Hall, New York.

Puskorius, G. and Feldkamp, L., 1993, Automotive engine idle speed control with recurrent neural networks, *Proceedings of the 1993 American Control Conference*, San Francisco CA.

Puskorius, G. and Feldkamp, L., 1994, Neurocontrol of nonlinear dynamical systems with Kalman filter trained recurrent networks, *IEEE Transactions on Neural Networks* **5**, 279–297.

Ridgely, D.B., Silverthorn, J.T. and Banda, S.S., 1982, Design and analysis of a multivariable control system for a CCV-type fighter aircraft, AIAA Technical report 82-1350.

Ryan, E.P., 1988, Adaptive stabilisation of a class of uncertain nonlinear systems: a differential inclusions approach, *Systems and Control Letters* **10**, 95–101.

Ryan, E.P. and Corless, M., 1984, Ultimate boundedness and asymptotic stability of a class of uncertain dynamical systems via continuous and discontinuous control, *IMA Journal of Mathematical Control and Information* **1**, 223–242.

Shim, D., Park, J., Khargonekar, P. and Ribbens, W., 1995, Engine idle speed control, *Proceedings of the 1995 American Control Conference*, Seattle WA., pp. 2582–2586.

Sira-Ramírez, H. and Lischinsky-Arenas, P., 1991, The differential algebraic approach in nonlinear dynamical compensator design for dc-to-dc power converters, *International Journal of Control* **54**, 111–134.

Slotine, J.J.E., 1984, Sliding controller design for nonlinear systems, *International Journal of Control* **40**, 421–434.

Slotine, J.J.E. and Li, W., 1991, *Applied Nonlinear Control*, Prentice Hall, Englewood Cliffs NJ.

Slotine, J.J.E., Hedrick, J.K. and Misawa, E.A., 1987, On sliding observers for nonlinear systems, *Transactions of the ASME: Journal of Dynamic Systems, Measurement and Control* **109**, 245–252.

Smith, K.L. and Anderson, D.C., 1982, Optimal control applications to AFTI/F-16 and advanced flight control system design, AIAA Technical report 82-1572.

Sobel, K.M. and Shapiro, E.Y., 1986, Flight control synthesis via eigenstructure assignment–the discrete version, *Proceedings of the the 28th Israel Conference on Aviation and Astronautics*, pp. 1–6.

Soderstrom, T. and Stoica, P., 1989, *System Identification*, Prentice Hall, Hemel Hempstead, UK.

Spurgeon, S.K., 1991, Choice of discontinuous control components for robust sliding mode performance, *International Journal of Control* **53**, 163–179.

Spurgeon, S.K. and Davies, R., 1993, A nonlinear control strategy for robust sliding mode performance in the presence of unmatched uncertainty, *International Journal of Control* **57**, 1107–1123.

Sreedhar, R., Fernández, B. and Masada, G.Y., 1993, Robust fault detection in nonlinear systems using sliding mode observers, *Proceedings of the IEEE Conference on Control Applications*, pp. 715–721.

Strang, G., 1988, *Linear Algebra and its Applications*, Harcourt Brace Jovanovich, London.

Syrmos, V.L., Abdallah, C.T., Dorato, P. and Grigoriadis, K., 1997, Static output feedback – a survey, *Automatica* **33**, 125–137.

Utkin, V.I., 1977, Variable structure systems with sliding modes, *IEEE Transactions on Automatic Control* **22**, 212–222.

Utkin, V.I., 1981, Principles of identification using sliding regimes, *Soviet Physics Doklady* **26**, 271–272.

Utkin, V.I., 1992, *Sliding Modes in Control Optimization*, Springer-Verlag, Berlin.

Utkin, V.I. and Young, K.-K.D., 1978, Methods for constructing discontinuity planes in multidimensional variable structure systems, *Automation and Remote Control* **39**, 1466 –1470.

Vachtsevanos, G., Farinwata, S. and Kang, H., 1992, A systematic design method for fuzzy logic control with application to automotive idle speed control, *Proceedings of 31st IEEE Conference on Decision and Control*, Tuscon AZ.

Walcott, B.L. and Żak, S.H., 1987, State observation of nonlinear uncertain dynamical systems, *IEEE Transaction on Automatic Control* **32**, 166–170.

Walcott, B.L. and Żak, S.H., 1988, Combined observer–controller synthesis for uncertain dynamical systems with applications, *IEEE Transactions on Systems, Man and Cybernetics* **18**, 88–104.

Walcott, B.L., Corless, M.J. and Żak, S.H., 1987, Comparative study of nonlinear state observation techniques, *International Journal of Control* **45**, 2109–2132.

White, B.A., 1990, Applications of output feedback in variable structure control systems, in Zinober, A.S.I. (ed.), *Deterministic Control of Uncertain Systems*, Peter Peregrinus, Stevenage UK, pp. 144–169.

Wilkinson, J.H., 1967, *The Algebraic Eigenvalue Problem*, Oxford University Press, Oxford.

Williams, S., Hrovat, D., Davey, C., Maclay, D., Crevel, J. and Chen, L., 1989, Idle speed control design using an $\mathcal{H}_\infty$ approach, *Proceedings of the American Control Conference*, Pittsburgh PA.

Wonham, W.M., 1967, On pole-assignment in multi-input, controllable linear systems, *IEEE Transactions on Automatic Control* **AC-12**, 660–665.

Woodham, C.A. and Zinober, A.S.I., 1993, Eigenvalue placement in a specified sector for variable structure control systems, *International Journal of Control* **57**, 1021–1037.

Young, K-K.D., 1977, Asymptotic stability of model reference systems with variable structure control, *IEEE Transactions on Automatic Control* **AC-22**, 279–281.

Young, K.-K.D. and Kwatny, H.G., 1982, Variable structure servomechanism design and applications to overspeed protection control, *Automatica* **18**, 385–400.

Young, K.-K.D., Kokotović, P.V. and Utkin, V.I., 1977, A singular perturbation analysis of high-gain feedback system, *IEEE Transactions on Automatic Control* **22**, 931–937.

Żak, S.H. and Hui, S., 1993, Output feedback in variable structure controllers and state estimators for uncertain/nonlinear dynamic systems, *Proceedings of the IEE, Part D* **140**, 41–50.

Żak, S.H. and Walcott, B.L., 1990, State observation of nonlinear control systems via the method of Lyapunov, in Zinober, A.S.I. (ed.), *Deterministic Control of Uncertain Systems*, Peter Peregrinus, Stevenage UK, pp. 333–350.

Zhou, F. and Fisher, D.G., 1992, Continuous sliding mode control, *International Journal of Control* **55**, 313–327.

Zinober, A.S.I., 1984, Properties of adaptive discontinuous model-following control systems, *Proceedings of the 4th IMA International Conference on Control Theory*, pp. 337–346.

Zinober, A.S.I., 1990, An introduction to variable structure control, in Zinober, A.S.I. (ed.), *Deterministic Control of Uncertain Systems*, Peter Peregrinus, Stevenage, UK, pp. 1–26.

Zinober, A.S.I., 1994, An introduction to sliding mode variable structure control, in Zinober, A.S.I. (ed.), *Variable Structure and Lyapunov Control*, Springer-Verlag, Berlin, pp. 1–22.

Zinober, A.S.I. and Dorling, C.M., 1990, Hyperplane design and CAD of variable structure control systems, in Zinober, A.S.I. (ed.), *Deterministic Control of Uncertain Systems*, Peter Peregrinus, Stevenage UK, pp. 52–79.

Zinober, A.S.I., El-Ghezawi, O.M.E. and Billings, S.A., 1982, Multivariable variable-structure adaptive model-following control systems, *Proceedings of the IEE, Part D* **129**, 6–12.

Index